European Consortium for
Mathematics in Industry

M. Brøns, M. P. Bendsøe
and M. P. Sørensen (Eds.)

Progress in Industrial
Mathematics at ECMI 96

European Consortium for Mathematics in Industry

Edited by
Leif Arkeryd, Göteborg
Heinz Engl, Linz
Antonio Fasano, Firenze
Robert M. M. Mattheij, Eindhoven
Pekka Neittaanmäki, Jyväskylä
Helmut Neunzert, Kaiserslautern

Within Europe a number of academic groups have accepted their responsibility towards European industry and have proposed to found a European Consortium for Mathematics in Industry (ECMI) as an expression of this responsibility.

One of the activities of ECMI is the publication of books, which reflect its general philosophy; the texts of the series will help in promoting the use of mathematics in industry and in educating mathematicians for industry. They will consider different fields of applications, present casestudies, introduce new mathematical concepts in their relation to practical applications. They shall also represent the variety of the European mathematical traditions, for example practical asymptotics and differential equations in Britian, sophisticated numerical analysis from France, powerful computation in Germany, novel discrete mathematics in Holland, elegant real analysis from Italy. They will demonstrate that all these branches of mathematics are applicable to real problems, and industry and universities in any country can clearly benefit from the skills of the complete range of European applied mathematics.

Progress in Industrial Mathematics at ECMI 96

Edited by
Morton Brøns, Martin Philip Bendsøe and Mads Peter Sørensen

Technical University of Denmark
Lyngby/Copenhagen, Denmark

B. G. Teubner Stuttgart 1997

Die Deutsche Bibliothek – CIP-Einheitsaufnahme

ECMI <9, 1996, Lyngby-Tårbaek>:
Progress in industrial mathematics at ECMI 96
[European Consortium for Mathematics in Industry].
Ed. by Morton Brøns ... –
Stuttgart : Teubner, 1997
  ISBN 3-519-02607-4

© Copyright 1997 by B. G. Teubner Stuttgart

All rights reserved

No part of this book may be reproduced by any means, or transmitted, or translated into a machine language without the written permission of the publisher.

Printed and bound in Germany by Präzis-Druck GmbH, Karlsruhe
Cover design by Peter Pfitz, Stuttgart.

Preface

In the last week of June 1996 the 9th conference of the European Consortium for Mathematics Industry, *ECMI 96*, took place at the Technical University of Denmark. The present volume of papers is a selection among the almost 200 contributions to the conference.

As a logo on the announcements of the conference the organising committee chose a picture of the connection between Denmark and Sweden which is currently under construction. We chose this picture primarily because of the elegant and decorative lines of the bridge, but for other reasons as well:

Denmark is a country of islands, and the art of building bridges has a long tradition here. Danish civil engineers have built bridges all over the world and have been rated among the most competent experts in their field. Many have acted as consultants as well as professors at the Technical University of Denmark, and one of them once said:

To build a bridge you need steel and mathematics.

We think that this selection of papers with its broad spectrum of industrial topics proves that the importance of mathematics is continuously growing. Mathematics has penetrated subjects far beyond traditional engineering applications and in more and more places it becomes natural to add mathematics to the list of things you obviously need to succeed.

Important financial support was granted to the conference. The COWI Foundation generously supported the publication of this volume. The Thomas B. Thrige Foundation and the "Tips and Lotto" Foundation helped us invite participants from Eastern Europe. We gratefully acknowledge this indispensable help.

In the spirit of the conference and of the work of ECMI in general this volume is primarily organised around industrial topics. The volume does not give a comprehensive picture of the conference nor of European industrial mathematics in general. However, we hope we have provided you with a useful cross-section of the potentials of mathematics in industry.

Lyngby, Spring 1997

Morten Brøns
Martin Philip Bendsøe
Mads Peter Sørensen

Contents

Environmental modelling   11

A Method of Lines Approach for River Alarm Systems   12
    P. Rentrop, G. Steinebach
The use of bubbles in purification ponds for exhaust gases   20
    I.R. Shreiber
A similarity solution for light oil spreading on groundwater   27
    M.I.J. van Dijke, S.E.A.T.M. van der Zee
Estimation of a low pass filter for solar radiation data   35
    J.L. Jacobsen, H. Madsen, P. Harremoes
Propagation of the low-frequency noise generated by power station water-cooling towers   43
    S.P. Fisenko
A Mathematical Model of a Hollow-Fibre Filter   50
    A.F. Jones, J.A. King-Hele, P.T. Cardew

Railway systems   58

A brief survey of wheel-rail contact: Theory, Algorithms, Applications (Invited presentation)   60
    J.J. Kalker
Wear profiles and the dynamical simulation of wheel-rail systems   77
    M. Arnold, H. Netter
Dynamic response of a periodically supported railway track in case of a moving complex phasor excitation   85
    I. Zobory, V. Zoller
A mathematical treatise of periodic structures under traveling loads with an application to railway tracks   93
    T. Krzyzynski, K. Popp
The Bifurcation Behaviour of Periodic Solutions of Impact Oscillators   101
    J.P. Meijaard

Industrial processes   109

Modelling industrial processes involving infiltration in deformable porous media   110
    D. Ambrosi, L. Preziosi
Modelling the Thermal Decomposition of Chlorinated Hydrocarbons in an Ideal Turbulent Incinerator   118
    M. Kraft, H. Fey, C. Procaccini, J.P. Longwell, A.F. Sarofim, H. Bockhorn
Mathematical model for isobaric non-isothermal crystallization of polypropylene   126
    S. Mazzullo, R. Corrieri, C. De Luigi

Stochastic modelling and morphological features of polymer crystallization processes *V. Capasso, I. Gialdini, A. Micheletti*	135
Mathematical model of Deep-bed Grain Layer Drying by ventilation *A. Aboltins*	143
A model of oil Burnout from Glass Fabric *A. Buikis, A.D. Fitt, N. Ulanova*	150
Computing of the trajectories generated by 2D discrete model approximations of the dynamics of differential linear repetitive processes *K. Galkowski, E. Rogers, D. Owens*	158
A certain mathematical model of the glass fibre material production *J. Cepitis, H. Kalis*	166

Electronics, circuits and filters   175

Numerical Solutions for the Simulation of Monolithic Microwave Integrated Circuits *G. Hebermehl, R. Schlundt, H. Zscheile, W. Heinrich*	176
Symbolic Modelling in Circuit Simulation *W. Klein*	184
An extended hydrodynamic model for silicon semiconductor devices *O. Muscato*	192
Oscillations of a miniature single stator synchronous motor *W. D. Collins*	200
White Noise and Nonlinear Damping in the Two-Dimensional Model of Energy Transfer in Molecular Systems *P.L. Christiansen, M. Johansson, K.Ø. Rasmussen, Y.B. Gadidei, I.I. Yakimenko*	208
Mathematics of Denominator-separable Multidimensional Digital Filters with Application to Image Processing *R. Unbehauen, X. Nie*	217
Comparison and Assessment of Various Wavelet and Wavelet Packet Based Denoising Algorithms for Noisy Data *F. Hess, M. Kraft, M. Richter, H. Bockhorn*	223

Ship industry   231

Hydrodynamic Design of Ship Hull Shapes by Methods of Computational Fluid Dynamics (Invited presentation) *H. Nowacki*	232
Superplastic Protective Structures *L.I. Slepyan, M.V. Ayzenberg*	252
Trial methods for nonlinear Bernoulli problem *K. Kärkkäinen, T. Tiihonen*	260
A Direct (Potential Based) Boundary Element Method for the Lifting Bodies Hydrodynamic Calculation *B. Ganea*	268
Procedure for Free Surface Potential Flow Numerical Simulation around Ship Model Hulls using Finite Element Method (Galerkin Formulation) *H. Tanasescu*	275

Oil industry   283

Mathematics for the oil industry: non-mathematical review and views (Invited presentation)   284
    T. Gimse

Discretization on general grids for general media   292
    I. Aavatsmark, T. Barkve, Ø. Bøe, T. Mannseth

Identification of Mobilities for the Buckley-Leverett Equation by Front Tracking   298
    V. Haugse

Flow of waxy crude oils   306
    A. Farina, L. Preziosi

Modelling of Bending Effects in Oil-Water Microemulsions   314
    Y.Z. Povstenko

Optimization in industry   323

Optimal power dispatch via multistage stochastic programming   324
    M.P. Nowak, W. Römisch

Solving the unit commitment problem in power generation by primal and dual methods   332
    D. Dentcheva, R. Gollmer, A. Möller, W. Römisch, R. Schultz

Optimal Trajectory and Configuration of Commuter Aircraft with Stochastic
and Gradient Based Methods   340
    R. Pant, C.M. Kalker-Kalkman

A general multi-objective optimization program for mixed continuous/integer variables
based on genetic algorithms   348
    C.M. Kalker-Kalkman

The optimization of natural gas liquefaction processes   356
    G. Engl, H. Schmidt

Genetic algorithm methodologies for scheduling electricity generation   364
    C.J. Aldridge, S. McKee, J.R. McDonald

Target Zone Models with Price Inertia: A Numerical Solution Method   372
    I. Scheunpflug, N. Köckler

Dynamics of machinery   381

Model Reduction of Random Vibration Systems   382
    R. Wunderlich, J. Gruner, J. vom Scheidt

On Compensation of Vibration of High-Speed Press Automatic Machine   390
    A. Rodkina, V. Nosov

Identification of External Actions on Dynamic Systems as the Method of Technical Diagnostics   398
    Y.L. Menshikov

Fluids in industry   **407**

Capillary effects in thin films   408
    S.B.G. O'Brien
The Linear Stability of Channel Flow of Fluid with Temperature-Dependent Viscosity   416
    D.P. Wall, S.K. Wilson
Adaptive Methods in Internal and External Flow Computations   424
    J. Felcman, V. Doleisi
Bifurcation, symmetry and parameter continuation in some problems about
capillary-gravity waves   432
    B.V. Loginov, S.V. Karpova, V.A. Trenogin

Wacker price   **441**

Evolution of Sedimentation Profiles in the Transport of Coal Water Slurries through a Pipeline   442
    A. Mancini

Environmental modelling

A Method of Lines Approach for River Alarm Systems

Peter Rentrop

Fachbereich Mathematik, TH Darmstadt, Schloßgartenstr. 7,

D-64289 Darmstadt, Federal Republic of Germany.

Gerd Steinebach

Bundesanstalt für Gewässerkunde, Postfach 309,

D-56003 Koblenz, Federal Republic of Germany,

e-mail: steinebach@koblenz.bfg.bund400.de.

Abstract

River alarm systems are designed for the forecasting of water stages during floods or low flow conditions or the prediction of the transport of pollution plumes. The basic model equations are introduced and a Method of Lines approach for their numerical solution is discussed. The approach includes adaptive space-mesh strategies and a Rosenbrock-Wanner scheme for the time integration. It fits into a PC environment and fulfills the requirements on an implementation within river alarm systems.

1 Introduction

In 1986 at the Sandoz chemical factory near Basel a fire happened. With the extinguishing water large quantities of chemicals were washed into the river Rhine, causing a lethal pollution of the water and severe restrictions of the water supply up to several hundred kilometers downstream of Basel. This severe accident was the starting point for the development of an alarm model, which should predict quickly and reliably the transport and dispersion process of a pollution plume.

A different alarm system is necessary to predict the water stages in case of flood events. On the other side at low flow conditions the forecasting of the water depth is important for the ship traffic. Usually there is a strong linkage between the calculation of flow, water stages and transport.

After a brief introduction of the model equations for flow and transport in rivers, we concentrate on a special taylored Method of Lines approach for their solution. Adaptive space-mesh strategies and a Rosenbrock-Wanner scheme for the time integration are discussed. Simulation results for a Rhine-Moselle model are presented and compared with measurements at gauging stations.

2 Basic equations

In large river basins it is possible to predict the water stages by flow routing models without taking into account precipitation data. The water flow in the river can be described by the Saint-Venant equations or further simplifications. Here we restrict the considerations to one space dimension in order to handle models of several hundreds of kilometers river-reach on PC environments.

The hyperbolic Saint-Venant equations (1), (2) are based on the conservation of mass and momentum and additional assumptions [13, 15].

$$A_t = -Q_x + q, \tag{1}$$

$$v_t = -vv_x - g(h_x + S_f) - \frac{q}{A}(v - v'). \tag{2}$$

The subscripts t and x denote the partial derivatives with respect to time t, $t \geq 0$ and space x, $x \in [x_0, x_1]$. The unknowns are the cross-section area $A(x,t)$ and the flow velocity $v(x,t)$. $q(x,t)$ denote the lateral in- or outflows per unit length with the velocity component v' in the direction of the main flow. $h(x,t)$ represents the water elevation above sea level. For a given geometry of the river-bed, e. g. by measurement data, h can be expressed as a function of x and A: $h(x,t) = H(x, A(x,t))$. The friction slope S_f is derived from the empirical resistence law of Manning-Strickler:

$$S_f = \frac{1}{K_s^2} \frac{v|v|}{R^{\frac{4}{3}}}. \tag{3}$$

The hydraulic radius R is defined as the ratio of cross-section area A to the wetted perimeter U. The constant K_s is assumed to depend mainly on the roughness of the river-bed. Usually K_s is used as a parameter for model calibration and therefore it is dependent on the flow $Q = vA$ for each river segment located between two gauging stations.

Moreover it is assumed that the associated eigenvalues of (1), (2) possess opposite sign, i.e. the flow is subcritical. As a consequence of direction of the characteristics one has to impose one left and one right boundary condition on (1), (2). The left boundary condition usually represents measured water elevations $h(x_0,t)$ or flows $Q(x_0,t)$. At the right boundary a known stage-flow relationship $0 = f(h(x_1,t), Q(x_1,t))$ at a gauging station has to be fullfilled.

The model equations for transport of soluble substances in rivers are of advection-diffusion type:

$$(Ac)_t = -(Qc)_x + (DAc_x)_x - K_1 Ac, \tag{4}$$

where $c(x,t)$ is the concentration of the substance and K_1 is a linear decay rate. The diffusion term $(DAc_x)_x$ describes the transport caused by molecular and turbulent diffusion and more importantly the dispersion, which is caused by different flow velocities in the cross-section area through the influences of the river bottom and banks. Therefore $D(x)$ is called dispersion coefficient. It may depend on flow Q additionally.

(4) can be simplified by substituting (1):

$$c_t = -vc_x + \frac{1}{A}(DAc_x)_x - (\frac{q}{A} + K_1)c . \quad (5)$$

Because of the parabolic nature of (5) two boundary conditions are necessary. The left condition should represent observed concentrations $c(x_0, t)$. The right boundary condition can be omitted, if (5) is advection dominant and the discretization of the spatial derivatives is done by left sided finite differences near the right boundary [11].

Appropriate initial conditions $v(x,0)$, $A(x,0)$ and $c(x,0)$ complete the problem formulation for flow and transport in rivers. The junction of rivers can be handled by the coupling of two or more models through the boundary conditions. Extended models are possible and may contain lateral retention capacities or dead-zones, sedimentation and resuspension or biochemical processes [6].

3 Method of Lines approach

The Method of Lines (MOL) is a general approach for the numerical solution of time dependent PDEs. The semi-discretization in space can be based on finite differences, finite elements or other schemes. The following investigations are restricted to the semi-discretization by finite differences.

In the most simple case the space derivative of a solution component $u_x(X_i, t)$ is approximated on an equidistant grid $a = X_0 < \ldots < X_N = b$ of grid size $\delta_x = X_{i+1} - X_i$ by

$$u_x(X_i, t) = \frac{u(X_i, t) - u(X_{i-1}, t)}{\delta_x} + O(\delta_x) .$$

Finite difference formulae of higher order for the approximation of space-derivatives on non-equidistant grids can be obtained by polynominal interpolation. One has to choose $r+1$ grid points and function values $\{(X_j, u(X_j, t)), \ldots, (X_{j+r}, u(X_{j+r}, t))\}$ with $j \leq i \leq j+r$, to obtain the interpolation polynominal $p_r(x)$ of degree r. The space derivative of u is then approximated by

$$u_x(X_i, t) = p'_r(X_i) + O(\delta_x^r) ,$$

where δ_x stands for the maximal grid size. The computation of $p'_r(x)$ can be performed efficiently by Newton's interpolation formula via divided differences.

Depending on the $r+1$ interpolation points one obtains upwind, central or any other stencils. In practical computations r must be restricted to $r \leq 4$. A more sophisticated approach is a semi-discretization by essentially non-oscillatory (ENO) schemes [16]. The approximation of cell mean values in space leads to numerical solutions without spurious oscillations and can resolve sharp gradients or discontinuities, like shocks.

The discretization process in space generates large systems of stiff ODEs or differential-algebraic equations (DAEs). For their solution many robust and efficient implicit or semi-implicit methods of higher order are available [3]. These integration schemes combine good stability properties like A- or L-stability with an automatic step-size control. Unfortunately many methods suffer from order-reductions if they are applied in the MOL context [7, 9]. In [12] a modified coefficient set was derived for the well known Rosenbrock-Wanner method RODAS [3]. The new method preserves the classical order 4 of convergence for certain classes of semidiscretized PDEs. Combined with a careful evaluation of sparse Jacobians and the solution of sparse linear systems, this ROW-method could be employed efficiently to several river models.

Other useful strategies for a MOL package are **adaptive meshes**. In the classical MOL approach a space mesh is chosen in the beginning and the resulting ODEs or DAEs are integrated. If during the integration step a moving front occurs, a very fine mesh for the whole space interval $[x_0, x_1]$ is necessary. In this case adaptive meshes would be preferable. The mesh is adapted during the integration in such a way, that a fine resolution is obtained near the fronts and a coarse one in regions of smooth solution components.

One has to distinguish between static and dynamic remeshing [2, 4]. In static remeshing a new grid is fixed after one or m integration steps depending on the actual solution behaviour. The solution has to be interpolated from the old mesh onto the new one and the integration procedure can be continued.
To define the new discretization points strategies like equidistribution of mesh-functions and locally bounded meshes can be useful [14, 5, 1]. The disadvantage of static remeshing is, that the time-integration has to handle systems of varying order.
In the following example a comparison between fixed and adaptive grid computations is shown. The advection dominated problem

$$c_t = -vc_x + Dc_{xx} \qquad t \geq 0 , \ x \in [a, b]$$

with Peclet-number

$$P_e = \frac{v}{D}(b - a) = 10^3$$

and typical constants $a = 0$, $b = 30000$, $v = 1$, $D = 30$ is treated. The left boundary condition

$$c_0(t) = 10 \exp(-0.001 \frac{(t - 7500)^2}{t})$$

represents a wave coming from left into the computational domain $[a, b]$. The problem is integrated until $t = 60.000$. At this time the wave has left the interval $[a, b]$.

		N=40	N=80	N=160	N=320	N=640
fixed grid	ϵ	2.714	1.222	0.3741	0.0951	0.0214
	CPU	4.130	9.060	18.05	35.10	72.50
	NSTEP	70	79	79	77	76
adaptive grid	ϵ	1.560	0.3649	0.1093	0.0406	0.0263
	CPU	5.730	10.66	19.59	38.20	77.39
	NSTEP	87	88	87	87	87

Table 1: Comparison of fixed and adaptive grid computations

The equidistribution strategy of [5], which is also implemented in the SPRINT-software [1], has been used. In Table 1 the numerical solution computed on a fixed grid of size N is compared with the adaptive grid solution of maximally allowed size N. ϵ is the maximum discrete absolute solution error, NSTEP the number of used integration steps and CPU the consumed time in seconds. The computations were performed on a seven year old CADMUS 9700 under FORTRAN77. The results show, that this type of problems can benefit from adaptive grid strategies. Fixing the CPU-time the maximum error is halfened, or describing an equal error, the number of grid points is smaller for the adaptive strategy. For larger Peclet-numbers the improvements are even increasing.

In dynamic remeshing the space-discretization points are considered to be time dependent and they move with the solution. This allows step-sizes as great as possible or an equidistribution of a mesh-function [8]. The disadvantage is the introduction of new unkwons (the meshpoints) and the altered structure of the semidiscretized equations [11]. Moreover to avoid collisions and to smooth the mesh motion, dynamic remeshing is often used in connection with static remeshing.

Since every static remeshing step introduces a new system of equations with possibly different dimension, the application of one-step methods is preferable.

4 Simulation results

Some typical results of river models are presented and compared with measurements. The numerical solution techniques of the previous section are applied to two models:

The first model is used for the prediction of water stages at river Rhine at low and medium flow conditions. The model is based on Saint-Venant equations and one additional equation for the description of water exchange between the main-stream and lateral retention capacities. The model covers the Rhine between gauging station Maxau at km 362 and gauging station Emmerich at km 852 and the Moselle downstream of Trier, located 195 km from the confluence with river Rhine. These river-reaches are discretized by 2700 grid points. In Figure 1 some

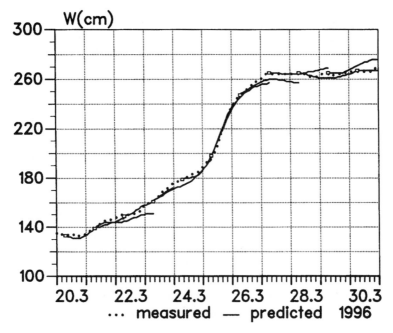

Figure 1: Gauging station Düsseldorf: Observed and predicted (48 hours) water stages (March 1996).

typical prediction results are compared with measured water stages at gauging station Düsseldorf, which is located at Rhine-km 744. The computing time of one prediction over 48 hours and the simulation of five days before the start of the prediction is about 240 sec on a HP 9000/770, the number of integration steps is about NSTEP=110. The model is used operationally for a daily water level prediction. Model inputs are the measured stages of 28 gauging stations at river Rhine and its tributaries.

During 1989 and 1992 several tracer experiments at Rhine and Moselle were performed to get data for the calibration and verification of an alarm model of river Rhine [10]. This official alarm model is based on an analytical approximation of an advection-diffusion equation combined with an equation describing the exchange of concentration with dead-zones. A disadvantage of this model (running on PCs) is, that no unsteady flow conditions can be taken into account and that the flow velocities $v(x,t)$ are assumed to be constant in river-reaches up to 25 kilometers. An improvement of the model is possible. The flow model of the Rhine and Moselle can be coupled with the advection-diffusion equation (4). Furthermore, the simulation of the transport of soluble substances can be improved by taking into account dead-zones [6]. As an example Figure 2 shows computed and observed concentrations of a tracer at three monitoring stations at the Moselle from the year 1992. The tracer were released at km 59.4 into the Moselle. It should be mentioned, that the flow in the Moselle is highly regulated by the operation of power plants.

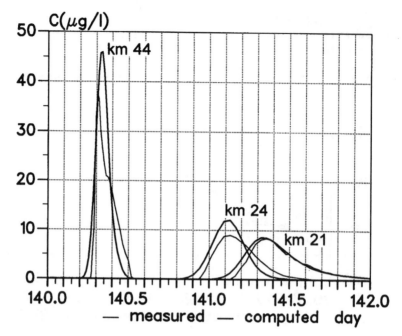

Figure 2: Observed and computed tracer concentrations at river Moselle (1992).

Conclusion

The Method of Lines equipped with a robust and efficient time-integrator and adaptive mesh strategies is a suitable numerical solution tool for river models. It allows the construction of models for the simulation and prediction of flow and transport covering several hundred kilometers of the river-course. The models fit into a PC environment and fulfills the requirements on an implementation within river alarm systems.

Acknowledgement

The authors like to thank Dr. R. Busskamp of the Bundesanstalt für Gewässerkunde Koblenz for helpful comments and discussions.

References

[1] Berzins, M., Furzeland, R.M.: A user's manual for sprint - A versatile software package for solving systems of algebraic ordinary and partial differential equations: Part 1 (TNER.85.058), 2 (TNER.86.050) and 3 (TNER.88.034), Thornton Research Centre, Shell Maatschappij (1985, 1986, 1989).

[2] Flaherty, J.E., et al: Adaptive methods for partial differential equations. SIAM Publications (1989).

[3] Hairer, E., Wanner, G.: Solving ordinary differential equations II, Springer Series in Computational Mathematics, Berlin, Heidelberg (1991).

[4] Hyman, J.M.: Moving mesh methods for partial differential equations, in Mathematics Applied to Science, eds. J. Goldstein, S. Rosencrans, G.Sod, Academic Press, 129-154 (1988).

[5] Kautsky, J., Nichols, N. K.: Equidistributing meshes with constraints. SIAM J. Sci. Stat. Comput. 1, 499-511 (1980).

[6] Rentrop, P., Steinebach, G.: Model and numerical techniques for the alert system of river Rhine. To appear in Surveys Math. Industry.

[7] Ostermann, A., Roche , M.: Rosenbrock methods for partial differential equations and fractional orders of convergence. SIAM J. Numer. Anal. 30, 1084-1098 (1993).

[8] Petzold, L.R.: Observations on an adaptive moving grid method for one-dimensional systems of partial differential equations. Applied Numer. Math. 3, 347-360 (1987).

[9] Sanz-Serna, J.M., Verwer, J.G., Hundsdorfer, W.H.: Convergence and order reduction of Runge-Kutta schemes applied to evolutionary problems in partial differential equations. Numer. Math. 50, 405-418 (1986).

[10] Spreafico, M., van Mazijk, A.: Alarmmodell Rhein. Ein Modell für die operationelle Vorhersage des Transportes von Schadstoffen im Rhein. KHR-Bericht Nr. I-12, Lelystad (1993).

[11] Steinebach, G.: Die Linienmethode und ROW-Verfahren zur Abfluß- und Prozeßsimulation in Fließgewässern am Beispiel von Rhein und Mosel. TH Darmstadt, Thesis (1995).

[12] Steinebach, G.: Order-reduction of ROW-methods for DAEs and method of lines applications. To appear in Numer. Math.

[13] Stoker, J. J.: Water Waves, the mathematical theory with applications. Interscience Publishers Inc., New York (1957).

[14] Verwer, J.G. et al: A moving grid method for one-dimensional PDEs based on the method of lines. In Flaherty, J.E., et al: Adaptive Methods for Partial Differential Equations, SIAM Publications, 160-175 (1989).

[15] Vreugdenhil, C.B.: Computational hydraulics, an introduction. Springer, Berlin (1989).

[16] Walsteijn, F. H.: Essentially non-oscillatory (ENO) schemes. In Vreugdenhil, C.B., Koren, B.: Numerical methods for advection-diffusion problems. Vieweg Notes on Numerical Fluid Mechanics, 45, 27 - 54, Braunschweig (1993).

The use of bubbles in purification ponds for exhaust gases

Isaac R. Shreiber

Institute for Industrial Mathematics, Beer Sheva, 84213, Israel,
isaak@math.bgu.ac.il

Abstract

The paper deals with a model for desulphurization of waste water after its saturation by exhaust toxic gases from the combustion chamber of an electric power station. This model develops the cell model approach for bubbles ensemble mass-transfer and provides for a quantitative estimation of the H_2SO_4 concentration field formation in a purification pond saturated with air bubbles. The cell model permits the mass flux only to be calculated and then, using some exact solutions for the bubble diffusion problem, the volume concentration of H_2SO_4 in the purification pond to be evaluated. Also there are two processes accompanying the technology that can be modeled, namely helix stream formation in the ponds and CO_2 cavitation.

1   Motivation of research

Desulphurization of waste water at a power plant can be accomplished in several ways [1], [2], but the most attractive to Israel or other countries wishing to use sea water is as follows. The exhaust gas in the absorber reacts with water:

$$SO_2 + H_2CO_3 \rightarrow H_2SO_3 + CO_2 \qquad (1)$$

resulting in the highly toxic sulphuric acid. In order to reduce toxicity it is desirable to oxidize the SO_3^{--} into SO_4^{--}, namely

$$2H_2SO_3 + O_2 \rightarrow 2H_2SO_4 \qquad (2)$$

after which it is possible to dispose of the waste water, now containing mainly SO_4^{--}, by discharging it into the sea. The acid H_2SO_4 is strong enough to react with some common salts, producing environmentally more acceptable materials. For instance,

$$H_2SO_4 + CaCO_3 \rightarrow CaSO_4 + H_2O + CO_2 \qquad (3)$$

or similarly with Mg.

Reaction (1) takes place in the absorber, reaction (2) in a special purification pond, and (3) in a purification pond as well as in the sea. The method requires relatively large ponds and gas pipelines for creating air bubbles that float in the waste water and react with the sulphuric acid. This process involves mass-transfer of the oxidant at the boundaries of the floating bubbles. Oxygen reaches the boundaries by diffusion through the air inside the bubbles, which gradually acquire higher nitrogen content. The problem is to develop a model for evaluation in the purification pond based on modeling mass transfer processes.

Goal. The goal of this project was to elaborate a mathematical tool for the evaluation of concentration field formation in bubbly pond purification exhaust gas and estimation of the real optimal sizes of the purification ponds.

2 General assumptions and research strategy

First of all it is necessary to consider two different schemes for the description of mass transfer which can take place in the purification ponds. Only one of them can provide correct mathematical estimations. There are two possible physical pictures of mass transfer. One is as follows: the diffusion boundary layer near the bubble is rather small and does not interact with the boundary diffusion layers of the neighboring bubbles; the H_2SO_4 concentration is defined only by a solution obtained for a single fixed bubble, and the solution contains no information about the air flux in the ponds. Another disadvantage of the solution is that for the mass flux $j_{H_2SO_4}$ it depends strongly on the coefficient of H_2SO_4 diffusion in water, which in its turn depends on the temperature of the water and can be found in each case with the help of experimental methods. The other picture deals with the conditions when each diffusion layer interacts with the boundary diffusion layers of the neighboring bubbles. In this case the mass flux does not depend on the coefficient of diffusion and enables us to link all technical parameters of the equipment to the acid concentration in the ponds. The duty of the equipment is to involve all the water in the pond in the purification process and provide for its better performance. This paper focuses therefore on the second view on mass transfer. As it follows from our estimation of the characteristic times and comparison between the chemical reaction time and the diffusion time, the former can be neglected, so that reaction (2) is assumed to take place immediately. It follows that the H_2SO_4 concentration is defined only by the O_2 flux from the air bubbles. The chemical balance and mass transfer models will be used for calculating the H_2SO_4 concentration and for a detailed description of O_2 microdiffusion in the bubble. The "cell model approach" will be developed for estimation of the mass-transfer flux in the bubbles ensemble. The general idea of the approach is to define the solution concentration caused by the bubbles ensemble as the value of the concentration on the cell boundary. So, we expect to study mass transfer processes for a single fixed cell and the concentration value in the cell boundary calculation is identical with the total concentration of H_2SO_4 dissolved in the pond.

The strategy of the research is to find the mass flux on the bubble providing for the existence of the cell model, namely the flux on the cell boundary must be equal to zero. After that the flux found will be used in some solutions to define the concentration of H_2SO_4 on the cell boundary. The chain of our calculation can be presented as follows.

We estimate the O_2 diffusion flux from the bubbles ensemble and calculate the H_2SO_4 flux produced due to reaction (1) on the bubble boundary. The problem is to define the H_2SO_4 concentration on the bubble boundary and then to define the H_2SO_4 concentration on the cell boundary. Really, it is necessary to calculate the oxygen mass flux. The central moment of the flux calculation and its relation with the concentration on the bubble boundary is to take into account the condition for the cell model existence, namely the mass flux through the cell boundary is equal to zero. So, if the concentration for the bubble boundary is defined, the field of the concentration can be calculated using the general solution of the diffusion equation for an unlimited space including the cell boundary,

$$C_2(r,t) = F[C_2(r=R), D_2], \qquad (4)$$

where C_2 is the H_2SO_4 concentration and D_2 the oxygen diffusion coefficient in water. It is $C_2(r=R)$ on the bubble boundary which provides the mass flux on the cell boundary $j=0$.

3 Cell model for bubbles ensemble

Let a bubble with the radius R be surrounded by a cell with the radius $\Re$. Assume that the cell is independent of other cells. It means that the mass flux on the cell boundary is equal to zero. The model permits transition from a single bubble to the continual parameter of the bubbling medium, namely the gas-void $\varphi_g = R^3/\Re^3$.

Mass flux of O_2 transfer. The problem posed mathematically reads as the internal diffusion problem for a sphere.

$$\frac{\partial C_1}{\partial t} = D_1(\frac{\partial^2 C_1}{\partial r^2} + \frac{2}{r}\frac{\partial C_1}{\partial r}); \quad (\frac{\partial C_{O_2}}{\partial r})_{r=0} = 0; \quad (C_1)_{r=R} = 0; \quad j_{O_2} \to j_{H_2SO_4} \qquad (5)$$

Mass flux of H_2SO_4 transfer. The problem posed mathematically reads as

$$R < r < \Re; \quad \frac{\partial C_2}{\partial t} = D_2 \frac{1}{r}\frac{\partial^2 (rC_2)}{\partial r^2}; \quad (j)_{r=R} = -D_2\frac{\partial C_2}{\partial r}; \quad (\frac{\partial C_2}{\partial r})_{r=\Re} = 0; \qquad (6)$$

The aim of solving system (6) is calculation of H_2SO_4 concentration in the water space between the bubble radius and the cell radius. The acid concentration in the boundary of the space cell is equal to the average acid concentration in the bubbly liquid component of the ponds. This is the main point of the approach.

Let us study the internal diffusion problem for the bubble in the cell: the problem can be solved by the Laplace transform method, and we obtain the expression for the mass flux providing for the "adiabatic" condition for the cell. We will focus on the low frequency case, $t \to \infty, w \to 0$, where the mass flux from the bubble ensemble reads:

$$j_{H_2SO_4} = -D_2(\frac{\partial C_2}{\partial r})_{r=R} = \frac{R(1-\varphi-\varphi^{2/3})}{3\varphi}(\frac{\partial C_2}{\partial t})_{r=R} \qquad (7)$$

where $j_{H_2SO_4}$ is defined by the solution of problem (5). $j_{H_2SO_4} = Xj_{O_2}$, where X is defined by the reaction (2). $X \approx 0.2$.

Using the O_2 mass flux on the bubble boundary, let us define the acid H_2SO_4 flux:

$$j_{O_2} = Xj_{H_2SO_4} = XD_2(\frac{\partial C_2}{\partial r})_{r=R}; \quad D_1\frac{\partial C_1}{\partial r} = X\frac{R(1-\varphi-\varphi^{2/3})}{3\varphi}(\frac{\partial C_2}{\partial t})_{r=R} \qquad (8)$$

The problem can be solved after solving the internal diffusion problem (5). Let us return to Eqn. (5). System (5) is the simplest model for O_2 diffusion because it does not take into account the rising bubble's internal hydrodynamics, namely the Hill vortex formation, the bubble boundary deformation. The aim of our communication is to present an approach for the bubble ensemble mass flux calculation. Thus, the simplest system (5) can be solved by any conventional method and the oxygen mass flux from a single bubble reads:

$$j = -D_1(\frac{\partial C_1}{\partial r})_{r=R} = -\frac{2D_1C_{1(t=0)}}{R}\sum_{n=1}^{\infty}\exp[-\frac{n^2\pi^2 D_1}{R^2}t] \qquad (9)$$

Let us introduce the characteristic time $t^* = \frac{R^2}{D_1}$ ($D_1 = 0.178 * 10^{-4}\frac{m^2}{sec}$) and rewrite the expression for the mass flux

$$j = \frac{2RC_1}{t^+}\sum_{n=1}^{\infty}\exp[-\frac{n^2\pi^2}{t^+}t] = \frac{2RC_1}{t^+}\sum_{n=1}^{\infty}\exp[-n^2\pi^2\xi]; \quad \xi = \frac{t}{t^+} \qquad (10)$$

Note that $\sum_{n=1}^{\infty}\exp[-n^2\pi^2\xi] = 0.5[\Theta_3(0,\xi)-1]$ and Eqn. (10) can be expressed through the Θ_3-function: $j = -\frac{RC_1}{t^+}[\Theta_3(0,\xi)-1]$.

In reality the characteristic time of diffusion t^* is equal to $t^+ = R^2/D_1\pi^2$ and differs from t^* by 10 times.

It is useful for us to take into account the asymptotics of (Θ_3-1)-function. They read:

$$\text{for } \xi \ll 1 \quad [\Theta_3(0,\xi)-1] \approx 1/\sqrt{\pi\xi}$$
$$\text{for } \xi \gg 1 \quad [\Theta_3(0,\xi)-1] \approx 2\exp(-\xi) \qquad (11)$$

The expression for mass flux (10) can be presented in the form of two terms with the same accuracy:

$$-j = \frac{D_1C_1}{R}[\exp(-\xi)+1/\sqrt{\pi\xi}] \text{ for } 0 < \xi < \xi^* \qquad (12)$$

where $\xi^* = T/t^+$ and T is the bubble's rising time in the ponds. Let us take only the first part of (12), and (8) can be rewritten in the slow case:

$$-\frac{D_1 C_1}{R}\exp(-\xi) = X\frac{R(1-\varphi-\varphi^{2/3})}{3\varphi}(\frac{\partial C_2}{\partial \xi})_{r=R},$$

$$(C_2(\xi))_{r=R} = \frac{D_1 C_1}{R^2 X}\frac{3\varphi}{1-\varphi-\varphi^{2/3}}\exp(-\xi) \qquad (13)$$

Let us substitute (13) into the conventional solution for the diffusion equation using the flux on the boundary which provides the adiabatic conditions on the cell boundary and find the H_2SO_4 concentration value on the cell boundary:

$$C_2(\Re, t) =$$

$$\frac{D_1 C_1}{R^2 X \sqrt{\pi D_2}}\frac{(1-\varphi)^{1/3}3\varphi}{1-\varphi-\varphi^{2/3}}\int_{-\infty}^{t}\frac{\exp[-\pi D_2/4R^2(t-\tau)]}{(t-\tau)^{3/2}}\exp[-\frac{R^2(1-\varphi^{1/3})^2}{4D_2(t-\tau)\varphi^{2/3}}]d\tau \qquad (14)$$

The solution concentration depends on the gas-void, the bubbles size and time. There are two characteristic times, namely the bubbles rising time t_1 and diffusing time $t_2 = R_0^2/\pi D_1$. If the bubbles rising time $t_1 \gg t_2$, we must integrate up to $t = t_2$.

4 CO_2 Cavitation

Due to reactions (1), (3) CO_2 cavitation takes place, leading to the formation of a CO_2 boundary layer near the bubble walls. Small bubbles of CO_2 coalesce with air bubbles and form within them an inert gas boundary layer, thus hindering O_2 diffusion from the air bubbles. This requires an improvement in the scheme presented above, which must be supplemented by an approach taking into account the inert gas layer near the bubble walls. The phenomenon of CO_2 cavitation can be described in the framework of a real liquid model, taking into account the mass flux according to reactions (1) and (3). A real liquid consists of a mass of small or invisible bubbles. All the CO_2 production is concentrated within the small bubbles and is linked to the bubbles' growth. The coefficient of O_2 diffusion through N_2 does not differ from the diffusion through CO_2, and we do not take into account the CO_2 layer in the air bubble, formed by the CO_2 cavitation. The governing equation for the bubbles energy reads:

$$\frac{dP_2}{dt} + \frac{3\gamma P_2}{R}\frac{dR}{dt} = \frac{3(\gamma-1)c_{P_2}T_2}{R}(-D_3\frac{\partial C_3}{\partial r})_{r=R} + \frac{3(\gamma-1)}{r}q_R = 0, \qquad (15)$$

where D_3 is the diffusion coefficient of CO_2 and C_3 is its concentration in the water surrounding the bubble. If q_R changes to the heat release of chemical reaction (3) Q, the complete model will be obtained for CO_2 cavitation: the heat release of chemical reaction (3) Q produces mass flux into the bubble and forms the pressure field $P_2 = P(R)$ in the Rayleigh equation for the bubble dynamics, causing the bubble radius to grow.

5 Helix Stream Formation

The method for desulphurization requires an intensive bubbly liquid stream. It is possible to get it by using special walls separating the ponds, so that the bubbling process takes place only in one part of the pond. Moving upwards, the bubbly liquid loses the air bubbles and then moves downwards to the bottom of another part of the pond. A model of the helix formation can be developed by using the drift wave conception for the floating bubbles [3].

Here we will present an approximate model of the helix stream formation in the pond which is based on the idea of the existence of "charge waves" and the assumption that a filtration flow among a bubbles takes place. Let us assume that a cloud of bubbles moves rather slowly in the nonsteady regime to the pond surface. We employ the system of "one-velocity" equation for flotation and will try to find the solution for so-called "drift waves". The system reads:

$$\frac{\partial \varepsilon \rho_w}{\partial t} + \frac{\partial \varepsilon \rho_w U}{\partial x} = 0 \quad (16)$$

$$\frac{dU}{dt} = -\frac{1}{\rho_w}\frac{\partial P_w}{\partial x} + g + \frac{\varepsilon \nu U}{K(\varepsilon)} \quad (17)$$

where K is permeability of the bubble cloud, $\varepsilon = 1 - \varphi$, U is the bubble floating velocity, ρ_w is water density, ν is water viscosity, g is gravity acceleration.

The idea of "charge wave" (also so-called "drift wave") enables us to find $U = U(\varepsilon)$ using the Darcy approximation of Eqn.17. Then we can substitute to obtained relation $U = K(\varepsilon)g/\varepsilon \nu$ into Eqn.16. Eqn.17 is thereby reduced to an equation for "charge waves" which reads:

$$\frac{\partial \varepsilon}{\partial t} + [\frac{K(\varepsilon)g}{\varepsilon \nu} + \frac{g}{\varepsilon \nu}\frac{dK(\varepsilon)}{d\varepsilon} - \frac{K(\varepsilon)g}{\varepsilon^2 \nu}]\frac{\partial \varepsilon}{\partial x} = 0 \quad (18)$$

A reversion of the drift wave can be interpreted as conditions for the helix stream formation. It occurs when the bracket changes sign. For numerical estimation of ε, it is necessary to define the function $K(\varepsilon)$. We will employ the Cozeny–Karman formula for $K(\varepsilon)$ with variating bubble radius. It reads:

$$K(\varepsilon) = (\frac{3}{4\pi n})^{2/3}\frac{(1-\varepsilon)^{2/3}\varepsilon^3}{180(1-\varepsilon)^2} \quad (19)$$

where n is the number of bubbles in relative units. Substituting (19) into Eqn.18 we obtain the condition for signum change in the bracket. The coefficient of the derivative $\partial \varepsilon/\partial x$ changes the sign when $\varepsilon \approx 0.69$. It corresponds to the value of the gasvoid $\varphi = 0.31$. Thus the condition $\varphi > 0.31$ provides the helix stream formation in the pond.

6 Conclusion

The aim of our technical note is to present an approach for evaluating purification ponds efficiently and to discuss models for mass transfer. We focused on only one approximation, the so-called low frequency approximation. It assumes that the mass diffusion wave reaches the cell boundary and forms the concentration field.

References

[1] Nolan, P. S. Flue-Gas Desulfurization in Thermal Power Plants. J. Thermal Engineering, 1994 **41**, No. 6, pp. 434-438.

[2] Shmigol' L. N. & Nekrasov B. V. Pilot Sulfur-Trapping Plants, J. Thermal Engineering, 1994 **41**, No. 6, pp. 427-433.

[3] Gelman E., Pridor A., Shreiber I. Phenomenon of swelling and emergency situation formation. In book: Two-Phase Flow Modeling and Experimentation, Edizioni ETS Pisa, 1995, pp. 1177-1180.

A similarity solution for light oil spreading on groundwater

Marinus I.J. van Dijke and Sjoerd E.A.T.M. van der Zee
Department of Soil Science and Plant Nutrition
Wageningen Agricultural University
PO Box 8005
6700 EC Wageningen
The Netherlands
e-mail: rink.vandijke@bodhyg.benp.wau.nl

January 30, 1997

Abstract

Spreading of a LNAPL lens (oil) in the vicinity of the groundwater table in a 2-D planar or axisymmetric homogeneous domain is described using a multiphase flow model including capillary forces and oil entrapment by water. Assuming vertical flow equilibrium the equations are vertically integrated. Hence, the free oil volume per unit lateral area and the vertically averaged relative permeability are explicitly related to the vertical position of the interface between zones with either two or three phases. The trapped oil volume is approximated by a linear hysteretic relation that is based on the maximum in time of the free oil volume. The resulting nonlinear diffusion equation admits a similarity solution for the free oil volume per unit lateral area as a function of time and the lateral space coordinate. Additionally, expressions for the extension and amount of trapped oil are obtained. The applicability of the analytical solution is demonstrated by comparison to a simulation, which is based on the nonreduced flow model. Special attention is paid to criteria to determine the remaining volume parameter and the time scale of the similarity solution. After determination of these parameters the analytical solution is a good approximation of the spreading at all larger times.

1 Introduction

Spills of nonaqeous phase liquids, that are less dense than water (henceforth called oil for brevity) in aquifers, may accumulate at the water table and the resulting lenses spread laterally. At locations where an oil lens thins, oil may become fixed as discrete drops surrounded by water. This trapped oil causes considerable problems for remediation of the contaminated aquifers. Prediction of the lateral extension of the oil lenses usually requires computations with complicated multiphase flow models, but for relatively thin lenses these can be approximated well by models that involve vertically integrated variables.

2   Model

We model spreading of an oil lens at the water table in a 2-D planar or axisymmetric homogeneous domain as shown in Figure 1. Oil is either mobile (free) or

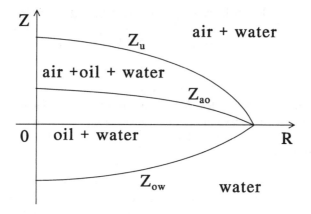

Figure 1: Schematic of oil lens.

trapped by water, whereas the free oil is assumed to be present in the two-phase region, bounded by the levels $Z_{ow}(R,T)$ and $Z_{ao}(R,T)$, and in the three-phase region, bounded by $Z_{ao}(R,T)$ and $Z_u(R,T)$. We use the mass balance equations

$$\phi\frac{\partial S_j}{\partial T} + \nabla \cdot \vec{U}_j = 0, \qquad (1)$$

for both water and oil ($j = w, o$), see e.g. [2, 7], whereas air (a) is considered as infinitely mobile. Phase velocities $\vec{U}_j$ are specified by Darcy's law

$$\vec{U}_j = -\frac{K k_{rj}}{\mu_j} \nabla (P_j + \rho_j g Z), \quad j = w, o, \qquad (2)$$

where the relative permeabilities k_{rj} and pressures P_j are nonlinear functions of the saturations S_j. In equations (1) and (2) we have introduced the constants porosity ϕ, absolute permeability K, viscosity μ_j, density ρ_j and gravitational accelaration g. The saturations satisfy the constitutive relation $S_w + S_o + S_a = 1$, where oil saturation is the sum of the free and the trapped fraction [2, 6]: $S_o = S_{of} + S_{ot}$. We consider spreading in case that the lateral extension of the oil lens is much larger than the vertical extension and we assume that in the vertical direction the capillary forces (corresponding to the first term on the righthandside of equation (2)) and the gravitional forces balance [8], such that we can neglect the vertical velocities. This vertical equilibrium assumption yields hydrostatic vertical pressure distributions (i.e. pressures vary linearly with height). Consequently, once we know the oil pressure at a certain height, all phase pressures and saturations with the same lateral coordinate are known. We get for the oil pressure

$$P_o(R, Z, T) = \rho_o g(Z_{ao}(R, T) - Z). \tag{3}$$

Furthermore, because of vertical equilibrium we need to consider the mass balance equation for oil only [5] with the lateral oil velocity given by (2). The resulting equation can be vertically integrated. This yields

$$\frac{\partial W_o}{\partial T} = \kappa \frac{1}{R^{N-1}} \frac{\partial}{\partial R}(R^{N-1} \bar{k}(W_f) \frac{\partial Z_{ao}}{\partial R}) \quad N = 1, 2, \tag{4}$$

with $W_o = \phi \int S_o dZ$, $W_f = \phi \int S_{of} dZ$, the total and free oil volume per unit lateral area respectively, and $\bar{k} = \int k_{ro} dZ$, the vertically integrated oil relative permeability. Based on the assumption that the oil lens is relatively thin, we approximate the complicated relations between saturations and pressures, see e.g. [3, 6], such that we can write explicitly

$$Z_{ao} = \lambda_1 W_f^{\frac{1}{n+1}}, \quad \bar{k} = \lambda_2 W_f^{\frac{5n-2}{2(n+1)}}, \tag{5}$$

with $\lambda_1, \lambda_2 > 0$ and $n > 1$, which yields the constant $\kappa = \dfrac{K \rho_o g \lambda_1 \lambda_2}{\mu_o}$ in equation (4). The trapped oil volume per unit lateral area $\phi W_t = \phi \int S_{ot} dZ$ is approximated by the hysteretic relation

$$W_t = \begin{cases} c_t \cdot (W_m - W_f) & \text{if } \dfrac{\partial W_f}{\partial T} < 0 \\ 0 & \text{if } \dfrac{\partial W_f}{\partial T} \geq 0 \end{cases} \tag{6}$$

with $c_t \in [0, 1)$ and $W_m(R, T) = \max_{T' \leq T} W_f(R, T')$ as shown in Figure 2. We assume that $\partial W_f / \partial T$ has at most one sign change.

For a lens with designated volume V_1 we introduce dimensionless variables

$$r = \frac{R}{L_c}, \quad w_f = \frac{W_f}{L_c}, \quad t = \frac{T}{T_c}, \tag{7}$$

where the characteristic length L_c and the characteristic time T_c are defined by

$$L_c = (\frac{V_1}{(2\pi)^{N-1}})^{\frac{1}{N+1}}, \quad T_c = \frac{p(n+1) L_c^{\frac{n+6}{2(n+1)}}}{\kappa}, \tag{8}$$

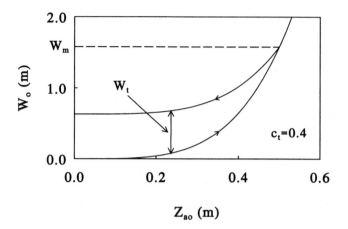

Figure 2: Oil volume per unit lateral area W_o as a function of the level Z_{ao}. For increasing Z_{ao} W_o follows a different path than for decreasing Z_{ao}. The lower path specifies the non-hysteretic relation $W_f(Z_{ao})$, whereas the difference between the two curves yields W_t.

respectively, where $p = 1/(1-c_t) \geq 1$. Hence, we consider for $r > 0$ and $t > t_1 > 0$ the nonlinear diffusion equation

$$F(\frac{\partial w_f}{\partial t}) = \frac{1}{r^{N-1}} \frac{\partial}{\partial r} (r^{N-1} w_f^m \frac{\partial w_f}{\partial r}), \qquad (9)$$

where

$$F(s) = \begin{cases} s & \text{for } s < 0 \\ ps & \text{for } s \geq 0 \end{cases} \qquad (10)$$

and $m = \frac{3n-2}{2(n+1)}$, $\frac{1}{4} < m < \frac{3}{2}$. We assume that at $t = t_1$ only free oil is present and we impose the compactly supported (free) oil distribution $w_f(r, t_1) = w_i(r)$, which satisfies the volume condition

$$\int_0^\infty r^{N-1} w_i(r) \, dr = 1. \qquad (11)$$

Because the 'diffusion' coefficient w_f^m degenarates for $w_f = 0$, a free boundary $r = r_l(t) > 0$ exists, separating the regions where $w_f > 0$ and $w_f = 0$. By symmetry, we impose at $r = 0$ the no-flow condition $\partial w_f / \partial r = 0$.

3 The similarity solution

Equation (9) is the modified porous medium equation, that admits a similarity solution (identified by the subscript s) of the form

$$w_{fs}(r, \bar{t}) = \bar{t}^{-\mu} h(r\nu^{\frac{1}{2}} \bar{t}^{-\nu}) \quad \text{for } r \geq 0, \qquad (12)$$

with $\mu, \nu > 0$ [1], where we have introduced $\bar{t} = t - t_0$ and $t_0 < t_1$, which represents the time at which the solution becomes singular. h has compact support with free boundary $r_A = A\nu^{-\frac{1}{2}}\bar{t}^\nu$, $0 < A < \infty$. The function F in equation (9) gives rise to the definition of $r_B = B\nu^{-\frac{1}{2}}\bar{t}^\nu$, $B \in (0, A)$, beyond which $\partial w_{fs}/\partial \bar{t} > 0$, the boundary that separates the regions with and without trapping. Substitution of (12) into equation (9), shows that the positive part of the similarity profile $h(\eta)$, with variable $\eta = r\nu^{\frac{1}{2}} t^{-\nu}$, satisfies the nonlinear ordinary differential equation

$$\eta^{1-N}(\eta^{N-1}h^m h')' = F(-\eta h' - kh) \quad \text{for } 0 < \eta < A \tag{13}$$

(primes ' denote differentiation with respect to η) and that μ and ν satisfy $2\nu + m\mu - 1 = 0$. Therefore, we have defined the ratio $k = \mu/\nu$, which reflects the influence of trapping on the similarity profile. We obtain the boundary conditions $h'(0) = 0$, $h(A) = 0$, whereas we can derive the extra condition $h^{m-1}h'(A) = -pA$. To simultaneously compute the similarity profile and the correct value of k, we scale h according to

$$\tilde{h}(\xi) = A^{-\frac{2}{m}} h(A\xi), \quad \xi = \frac{\eta}{A}, \tag{14}$$

where A follows from a mass balance condition.

For $p = 1$, k equals N and equation (13) has the explicit Barenblatt-Pattle point source solution [1], i.e. in terms of $\tilde{h}$

$$\tilde{h}(\xi) = \left(\frac{m}{2}(1 - \xi^2)\right)^{\frac{1}{m}}. \tag{15}$$

To compute the profile and the value of k for $p > 1$ we introduce $y_1 = \tilde{h}^m/m$ and $y_2 = \tilde{h}^{m-1}\tilde{h}'$, which yields the system

$$\begin{cases} y_1' = y_2 \\ y_2' = -\dfrac{N-1}{\xi} y_2 - \dfrac{y_2^2}{my_1} + F\left(-\xi \dfrac{y_2}{my_1} - k\right) \end{cases} \tag{16}$$

for $\xi \in (0, 1)$, with boundary conditions $y_2(0) = 0$, $y_1(1) = 0$ and $y_2(1) = -p$. System (16) is solved by shooting backward from $\xi = 1$ using a fourth order Runge-Kutta scheme. As $y_2(0)$ is a monotone function of k, we can use a simple iteration to vary k until the solution satisfies $y_2(0) = 0$. In Figure 3 we show k as a function of p for several m-values.

The similarity solution (12) is generally not a solution of equation (9), because the initial condition $w_i(r)$ has not necessarily the similarity profile. However, if we can obtain appropriate values of A and t_0, the similarity solution provides a good approximation of the solution of equation (9) with its boundary and initial conditions, after a sufficiently large time. We define the 'moment' $\int r^{k-1} w_{fs} \, dr$, which yields for the similarity solution in terms of $\tilde{h}(\xi)$ the time-independent expression

$$m_a = \nu^{-\frac{1}{2}k} A^{k+\frac{2}{m}} \int_0^1 \xi^{k-1}\tilde{h} \, d\xi. \tag{17}$$

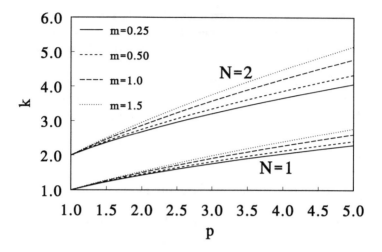

Figure 3: Parameter k as a function of p for several values of m and the lateral dimension N.

By comparison to the corresponding 'moment' of the general solution, which becomes approximately time-independent as well, we obtain the appropriate value of A independent of t_0. In case of entrapment, $p > 1$, the volume of free oil in the entire domain $\int r^{N-1} w_{fs}\, dr$ is time-dependent and is given for the similarity solution by

$$v_a(\bar{t}) = \bar{t}^{N\nu-\mu} \nu^{-\frac{1}{2}N} A^{N+\frac{2}{m}} \int_0^1 \xi^{N-1} \tilde{h}(\xi)\, d\xi. \tag{18}$$

Comparison to the free oil volume of the general solution at a given time yields the value of t_0.

4 A numerical example

As an example of the usefulness of the similarity solution, we made a simulation with a numerical multiphase flow model [7] based on (1) and (2), using the following values of the parameters: $\phi = 0.40$, $K = 1.77 \cdot 10^{-11}\text{m}^2$, $\mu_w = 1.00 \cdot 10^{-3}$ Pa s, $\mu_o = 5.00 \cdot 10^{-4}$ Pa s, $\rho_w = 1.00 \cdot 10^3$ kg m^{-3}, $\rho_o = 7.00 \cdot 10^3$ kg m^{-3}, $n = 3.0$ and $c_t = 0.49$. Furthermore, we obtained $\lambda_1 = 0.446$ m$^{0.75}$ and $\lambda_2 = 2.79$ m$^{-0.625}$. In a 2-D domain (i.e. $N = 1$) of 4.5 m high and 40 m wide with the water table at 1.5 m from the bottom, we simulated oil injection through a strip at the left boundary, such that at time $T_1 = 200$ hrs the oil volume $V_1 = 1.00$ m^3 was present in the flow domain. For $T > T_1$ we simulated spreading by imposing no-flow conditions on every boundary, except the right one, where we imposed hydrostatic water pressures.

We computed the value of $k = 1.47$ and the scaled similarity profile $\tilde{h}$ for $\xi \in [0,1]$. Comparison of m_a and the corresponding 'moment' m_n (the subscript n identifies the numerical solution), where the latter was approximately constant at $T = T_2 = 3000$ hrs ($\bar{t}_2 = 1600$), yielded $A = 1.33$. Comparison of v_a and v_n led to $t_o = -542$. To demonstrate that the obtained similarity solution was an approximation of the numerical solution for all larger times, we computed for $T = 11000$ hrs ($\bar{t} = 6409$) w_{fs} and w_{ts}. Computation of w_{ts} according to the trapping model (6) requires the maximum oil volume per unit lateral area w_{ms}. Considering the 'maximum' time $\bar{t}_m = (\frac{r\nu^{\frac{1}{2}}}{B})^{\frac{1}{\nu}}$ at which for a given lateral position the free boundary $r_B(\bar{t})$ passes, we get $w_{ms}(r,\bar{t}) = w_{fs}(r,\bar{t}_m)$ for $\bar{t} \geq \bar{t}_m$. However, for small r both w_{tn} and w_{mn} are strongly influenced by the initial condition, such that w_{ms} does not lead to accurate approximations of these functions. Therefore, using the numerical w_{tn} we computed at $\bar{t} = \bar{t}_2$ a modified $\tilde{w}_{ms}(r) = w_{fs}(r,\bar{t}_2) + w_{tn}(r,\bar{t}_2)/c_t$. Hence, we obtained

$$w_{ms}(r,\bar{t}) = \begin{cases} \tilde{w}_{ms}(r) & \text{if } 0 \leq r < r_B(\bar{t}_2) \\ w_{fs}(r,\bar{t}_m) & \text{if } r_B(\bar{t}_2) \leq r \leq r_B(\bar{t}), \end{cases} \quad (19)$$

for $\bar{t} \geq \bar{t}_2$. From a practical point of view this adaption of the maximum oil volume is reasonable if the adaption-time ($\bar{t}_2$) is small compared to the time-scale of the entire spreading process. In Figure 4 we show the profiles of both the numerical

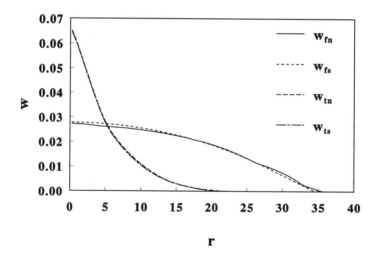

Figure 4: Numerical (w_{fn}, w_{tn}) and analytical (w_{fs}, w_{ts}) free and trapped oil volumes per unit lateral area at $\bar{t} = 6409$.

and the similarity solution for $\bar{t} = 6409$.
The results indicate that the similarity solution provides a good approximation of the free oil spreading. Furthermore, the amount and location of trapped oil is accurately derived from this similarity solution.

5 Conclusions

The complicated 2-D multiphase flow process of spreading of a LNAPL lens was well described by a 1-D equation for oil only, as a similarity solution provided a good approximation of the long term behaviour of an oil spill with prescribed volume. Especially, the influence of oil entrapment by water on spreading was determined. To evaluate the similarity profile a simple numerical procedure was developed and to fit the remaining unknown parameters from the early stages of a realistic contamination case suitable criteria were presented.

Acknowledgements
This project was sponsored by the Netherlands Organization for Scientific Research (NWO project NLS 61-251). The research was partly funded by European Communities Program ENVIRONMENT, EV5V-CT94-0536 (S. van der Zee).

References

[1] Hulshof, J. & Vazquez, J.L.: Self-similar solution of the second kind for the modified porous medium equation, Eur. J. Appl. Math. 5 (1994), pp. 391-403.

[2] Kaluarachchi, J.J. & Parker, J.C.: Multiphase flow with a simplified model for oil entrapment, Transport in Porous Media 7 (1992), pp. 1-14.

[3] Lenhard, R.J. & Parker, J.C.: A model for hysteretic constitutive relations governing multiphase flow, 2. permeability-saturation relations, Water Resour. Res. 23 (1987), pp. 2197-2206.

[4] Lenhard, R.J. & Parker, J.C.: Estimation of free hydrocarbon volume from fluid levels in monitoring wells, Ground Water 28 (1990), pp. 57-67.

[5] Miller, C.A. & Van Duijn, C.J.: Similarity solutions for gravity-dominated spreading of a lens of organic contaminant, in: Environmental studies: mathematical, computational and statistical analysis, ed. M.F. Wheeler, IMA Volumes in Mathematics and its Applications, Vol. 79, Springer-Verlag, New York, 1995.

[6] Parker, J.C. & Lenhard, R.J.: A model for hysteretic constitutive relations governing multiphase flow, 1. saturation-pressure relations, Water Resour. Res. 23 (1987), pp. 2187-2196.

[7] Van Dijke, M.I.J., Van der Zee, S.E.A.T.M. & Van Duijn, C.J.: Multi-phase flow modeling of air sparging, Adv. Water Resour. 18 (1995), pp. 319-333.

[8] Yortsos, Y.C.: A theoretical analysis of vertical flow equilibrium, Transport in Porous Media 18 (1995), pp. 107-129.

Estimation of a low pass filter for solar radiation data.

Judith L. Jacobsen
IMM, Bldg. 321, DTU, DK – 2800 Lyngby and
PH-Consult, Ordruphøjvej 4, DK – 2920 Charlottenlund.
E-mail: jlj@imm.dtu.dk

Henrik Madsen
Institute of Mathematical Modelling, IMM.
E-mail: hm@imm.dtu.dk

Poul Harremoës
IMT, Bldg. 115, DTU, DK – 2800 Lyngby and PH-Consult.
E-mail: ph@imt.dtu.dk and phcon@intet.uni-c.dk

Abstract

In the paper a method is suggested for a simultaneous estimation of parameters in some differential equations, describing the variation of oxygen in a small creek, and the parameter in a low pass filtering of the solar radiation input. The low pass filtering is obtained by using a simple first order filter.

The oxygen dynamics are modelled by using linear and non-linear stochastic differential equations. The modelling involves a description of the influence from variations in water temperature and solar radiation. The latter affects the oxygen level via the photo synthesis process.

It is shown that the simultaneous low pass filtering of the solar radiation data provides more precise estimates. The low pass filter compensates for the fact that the solar radiation is a high frequency signal and a point measurement, which probably is measured some distance away from the main body of the river. The estimated filter enables an on-line monitoring of water quality. Thus, immediate assessment of receiving waters, as well as of data quality, is possible.

Key Words: Data filtration; Water quality; On-line monitoring; Parameter estimates.

1 Introduction

To describe the oxygen variation in an aquatic river system, like a river, the photosynthesis must be taken into account. This process is highly dependent on the solar radiation (for a thorough description of the oxygen dynamics in rivers (see [8, 10, 5, 2]). However, the solar radiation measurements are most frequently obtained as point measurements and often show high frequency variation.

In this paper a method is suggested for estimating the parameter of low pass filter along with the estimation of the parameters of a model for the oxygen dynamics. The simultaneous estimation is carried out by setting up a state space model for a set of stochastic differential equations describing both the oxygen dynamics and the low pass filter. For the oxygen dynamics, two models, a linear and a non-linear model, are investigated. Estimating the filter directly in a state space model enables the filter to be used on-line, which may be of use in monitoring river water quality.

2 The model and the filter

For a description of the formulation of the differential equation for the oxygen dynamics see [5, 2]. The data used in this investigation is from a small creek in Denmark. Oxygen concentration, solar radiation, temperature, and the water level are sampled at five minute intervals for a period of one week. The collection of data is thoroughly described in [3]. The river, and similar data from the river, is described in [9, 10] and [1].

2.1 The model for the oxygen dynamics

The differential equation for the oxygen dynamics is:

$$\frac{dC}{dt} = K(C_m - C) + P(I) - R \tag{1}$$

where C is the oxygen concentration, C_m is the saturated oxygen concentration, K the reaeration coefficient, and R, the respiration. The term $P(I)$ describes how the solar radiation, I, affects the oxygen concentration. The quantities K, R and the function $P(I)$ are temperature dependent as described in [3].

Very often a simple linear relation is used for $P(I)$, i.e. $P(I) = E_o I$. However, in [3] several possibilities for the function $P(I)$ were investigated and it was shown that monotone and concave functions gave the best results. Among those, the Michaelis-Menten type model yield good results:

$$P(I) = P_m \frac{1}{(P_m/E_o I) + 1} \tag{2}$$

which clearly is a non-linear model. The function describes the photosynthesis as a function of solar radiation to a saturation level.

It is well known that spectrum analysis can be used to describe how the variations of a signal is distributed on frequencies. An analysis has shown that for the considered sampling time the variations of the solar radiation contains a rather large amount of high frequency variation, and it is shown that using a low pass filtered solar radiation as input leads to a considerable improvement of models describing the dynamics of the oxygen concentration [3]. The conclusion is that only the low frequency part of the variation correlates well with the variation of oxygen (via the photosynthetic process). The low pass filtering is possibly acting as a simple description of the persistence of the photosynthetic process of the plants.

The fact, that the measurements of the solar radiation are very often point measurements and measured some distance away from the main body in the river, might partly explain why the use of a low pass filtered solar radiation caused a considerable improvement in the description of the dynamics of the oxygen concentration.

The method used in [3] was simply to use low pass filtered solar radiation in the function $P(I)$ in eq. (2). Since a traditional low pass filter requires also future solar radiation measurements, the method can only be used in off-line estimates; but for monitoring purposes an on-line estimation is required. In the following we suggest a simple low pass filter which can be used for on-line estimation of the parameters, and hence for monitoring the states of the river.

2.2 The low pass filter

In order to enable a simultaneous estimation of the parameters, associated with the differential eq. (1) and the filter, the filter is formulated in continuous time by the following differential equation:

$$\frac{dI_f}{dt} = aI_f + bI \qquad (3)$$

where I_f is the filtered solar radiation.

Some requirements for the coefficients (a,b) in eq. (3) are readily found by considering the transfer function from the measured solar radiation, I, to the filtered solar radiation, I_f. The demand of a stationary gain equal to one, specifies $b = -a$. Furthermore, low pass filtering, or stability of the system (3), is obtainable only for $a < 0$.

The observations are given in discrete time and the sampling time is $\tau = 5$ min. Using the requirement $b = -a$, the discrete time version of eq. (3) is obtained by integration through the sample interval $[t, t+\tau]$:

$$I_f(t+\tau) = \phi(\tau) I_f(t) + (1 - \phi(\tau))I(t) \qquad (4)$$

where $\phi(\tau) = e^{a\tau}$. Since $a < 0$ (and $\tau > 0$) we have that $0 < \phi(\tau) < 1$.

No filtering is obtained for $a \to -\infty$, and in this case $I_f(t+\tau) = I(t)$. Hence, in order to avoid a time lag in the differential equation for the oxygen equation, $I_f(t+\tau)$ is used as input variable. This means that the "on-line" calculations actually are $\tau = 5$ min. behind the actual measurements.

2.3 The combined model and filter

The parameter, a, of the filter are estimated by including a state variable, that describes the filtered solar radiation. Finally, stochastic terms are added to the deterministic differential equations. For the linear model, $P(I_f) = E_o I_f$ the stochastic differential equations then can be written as the state space model:

$$\begin{bmatrix} dC \\ dI_f \end{bmatrix} = \begin{bmatrix} -K(T) & E_0(T) \\ 0 & a \end{bmatrix} \begin{bmatrix} C \\ I_f \end{bmatrix} dt + \begin{bmatrix} -R(T) & K(T) & 0 \\ 0 & 0 & -a \end{bmatrix} \begin{bmatrix} 1 \\ C_m \\ I \end{bmatrix} dt + \begin{bmatrix} dw \\ 0 \end{bmatrix} \quad (5)$$

where $a < 0$ and dw is a stochastic term describing the deviation between the true oxygen concentration and the oxygen concentration as described by the model (1).

The observed oxygen concentration is:

$$Y(t) = \begin{bmatrix} 1 & 0 \end{bmatrix} \begin{bmatrix} C(t) \\ I_f(t) \end{bmatrix} + e(t) \quad (6)$$

where $e(t)$ is the measurement error. The state space model may be written in matrix notation:

$$dX = AX\,dt + BU\,dt + dw(t) \quad (7)$$
$$Y(t) = CX(t) + e(t) \quad (8)$$

where X is the state vector containing the state of oxygen and the filter. The input vector U contains the measured solar radiation and the other inputs. The matrix A characterises the dynamical behaviour of the system and B is a matrix which specify how the input signals enter the system. In order to be able to find maximum likelihood estimates of the parameters based on Gaussian distributions, the process $w(t)$ is restricted to be a Wiener-process. The measurement error $e(t)$ is assumed to be Gaussian distributed white noise with zero mean. Furthermore, it is assumed that $w(t)$ and $e(t)$ are mutually independent.

A better (closer to an ideal low pass) filter is expected if we consider a second or higher order filter. This would involve three or more coupled differential equations as this is comprised of two or more first order filters in series.

In the next section it is briefly outlined how the parameters of the model (state space model (5)), including the parameter in the low pass filter, are estimated by a maximum likelihood method.

3 Maximum likelihood estimates

All observations are equidistantly spaced at five minute intervals. However, in order to simplify the notation, we shall assume that the time index t belongs to the set $\{0, 1, 2, ..., N\}$, where N is the number of observations. Introducing

$$\mathcal{Y}(t) = [Y(t), Y(t-1), \ldots, Y(1), Y(0)]' \quad (9)$$

i.e. $\mathcal{Y}(t)$ is a vector containing all the observations up to and including time t.

All the unknown parameters, denoted by the vector $\boldsymbol{\theta}$, are embedded in the continuous time state space model (eq.s (7) and (8)). The observations are, however, given in discrete time. Hence, the state space model has to be sampled in order to calculate the likelihood function. Under the assumption that the input variables are constant through the sampling time, the discrete time state space model can be written [6, 7, 2]:

$$\begin{aligned} \boldsymbol{X}(t+1) &= \boldsymbol{\Phi}\,\boldsymbol{X}(t) + \boldsymbol{\Gamma}\,\boldsymbol{U}(t) + \boldsymbol{v}(t) & (10) \\ \boldsymbol{Y}(t) &= \boldsymbol{C}\,\boldsymbol{X}(t) + \boldsymbol{e}(t) & (11) \end{aligned}$$

where $\boldsymbol{\Phi} = e^{\boldsymbol{A}\tau}, \boldsymbol{\Gamma} = \int_0^\tau e^{\boldsymbol{A}s}\boldsymbol{B}ds$ and $\boldsymbol{v}(t)$ is a white noise sequence.

The likelihood function is the joint probability density of all the observations assuming that the parameters are known, i.e.

$$\begin{aligned} L'(\boldsymbol{\theta}; \mathcal{Y}(N)) &= p(\mathcal{Y}(N)|\boldsymbol{\theta}) \\ &= p(\boldsymbol{Y}(N)|\mathcal{Y}(N-1),\boldsymbol{\theta})p(\mathcal{Y}(N-1)|\boldsymbol{\theta}) \\ &= \left(\prod_{t=1}^N p(\boldsymbol{Y}(t)|\mathcal{Y}(t-1),\boldsymbol{\theta})\right)p(\boldsymbol{Y}(0)|\boldsymbol{\theta}) & (12) \end{aligned}$$

where successive applications of the rule $P(A \cap B) = P(A|B)P(B)$ is used to express the likelihood function as a product of conditional densities.

Since both the model error and the measurement error are normally distributed the conditional density is also normal. The normal distribution is completely characterized by the mean and the variance. Hence, in order to parameterize the conditional distribution, we introduce the conditional mean and the conditional variance as

$$\hat{\boldsymbol{Y}}(t|t-1) = E[\boldsymbol{Y}(t)|\mathcal{Y}(t-1),\boldsymbol{\theta}] \quad \text{and} \quad \boldsymbol{R}(t|t-1) = V[\boldsymbol{Y}(t)|\mathcal{Y}(t-1),\boldsymbol{\theta}] \quad (13)$$

respectively. It may be noticed that these corresponds to the one-step prediction and the associated variance, respectively. Furthermore, it is convenient to introduce the one-step prediction error (or innovation)

$$\epsilon(t) = \boldsymbol{Y}(t) - \hat{\boldsymbol{Y}}(t|t-1) \qquad (14)$$

Using (12) – (14) the conditional likelihood function (conditioned on $\boldsymbol{Y}(0)$) becomes

$$L(\boldsymbol{\theta}; \mathcal{Y}(N)) = \prod_{t=1}^N \left((2\pi)^{-m/2} \det \boldsymbol{R}(t|t-1)^{-1/2} \exp(-\tfrac{1}{2}\epsilon(t)'\boldsymbol{R}(t|t-1)^{-1}\epsilon(t))\right) \qquad (15)$$

where m is the dimension of the $\boldsymbol{Y}$ vector (in the present case m= 1). Traditionally the logarithm of the conditional likelihood function is considered

$$\log L(\boldsymbol{\theta}; \mathcal{Y}(N)) = -\tfrac{1}{2}\sum_{t=1}^N \left(\log \det \boldsymbol{R}(t|t-1) + \epsilon(t)'\boldsymbol{R}(t|t-1)^{-1}\epsilon(t)\right) + \text{const} \qquad (16)$$

In the linear case, the conditional mean and variance eq. (13) can be calculated recursively by using a Kalman filter. In the non-linear case, an extended Kalman filter is used for calculating the conditional mean and variance. The numerical details of these algorithms can be found in [6].

The maximum likelihood estimates are found by maximization of the log likelihood function (16), and the uncertainties of the parameter estimates are found using the observed curvature of the log likelihood function, evaluated at the final estimates [6, 7].

4 Results and discussion

Tables 1 and 2 show the obtained maximum likelihood estimates for the parameters, and their standard variations, for the linear and the non-linear case, respectively. The un- and pre-filtered estimates were obtained previously (see [3]) using a filter that was an approximation to an ideal low pass filter.

Linear model:

Par.	Unit	Un-filtered	Pre-filtered	On-line filter
E_o	$\frac{mgO_2/l\,hr}{Ly/min}$	$0.179\ 10^{-2}$	$0.196\ 10^{-2}$	$0.178\ 10^{-2}$
		$(0.363\ 10^{-4})$	$(0.305\ 10^{-4})$	$(0.356\ 10^{-4})$
R	$mgO_2/l\,hr$	1.000	1.078	1.050
		(0.022)	(0.018)	(0.024)
K	hr^{-1}	0.248	0.269	0.259
		$(0.670\ 10^{-2})$	$(0.495\ 10^{-2})$	$(0.682\ 10^{-2})$
a	hr^{-1}	–	–	−16.208
				(1.828)
σ^2	–	$0.700\ 10^{-3}$	$0.306\ 10^{-3}$	$0.656\ 10^{-3}$

Table 1: Estimated parameters for the linear model. Standard deviations are shown in parentheses and σ^2 is the variance of the prediction error.

First of all, it is noted that all parameter estimates for the on-line filter lie between the values obtained by using the raw solar radiation data and values obtained by pre-filtering the solar radiation data. This means that the very simple on-line filter clearly gives better estimates than those obtained by using the raw (un-filtered) solar radiation data. In both cases the variance of the prediction error is only slightly smaller than those found when not filtering at all.

As previously pointed out in the litterature [3], some deviations are seen between the estimated parameters using the linear and the non-linear functions to describe the influence of the solar radiation. Considering the variance of the one-step prediction error, it is clearly seen that the non-linear model gives a much better description.

As pointed out in [3] the model for the oxygen dynamics is a very simple model, that only takes the very basic physical and biological processes into account.

Non-linear model:

Par.	Unit	Un-filtered	Pre-filtered	On-line filter
E_o	$\frac{mgO_2/l\,hr}{Ly/min}$	$0.467\ 10^{-2}$	$0.667\ 10^{-2}$	$0.545\ 10^{-2}$
		$(0.167\ 10^{-3})$	$(0.165\ 10^{-3})$	$(0.232\ 10^{-3})$
P_m	$mgO_2/l\,hr$	3.503	2.727	3.173
		(0.128)	(0.042)	(0.096)
R	$mgO_2/l\,hr$	1.377	1.509	1.451
		(0.025)	(0.016)	(0.027)
K	hr^{-1}	0.307	0.328	0.319
		$(0.615\ 10^{-2})$	$(0.384\ 10^{-2})$	$(0.611\ 10^{-2})$
a	hr^{-1}	–	–	-39.088
				(4.096)
σ^2	–	$0.514\ 10^{-3}$	$0.155\ 10^{-3}$	$0.501\ 10^{-3}$

Table 2: Estimated parameters for the non-linear model. Standard deviations are shown in parentheses and σ^2 is the variance of the prediction error.

Including other physical and biological processes may improve the total model [4].

5 Conclusion

A method for simultaneously estimating the parameters of a simple model for the oxygen dynamics and the parameter of a first order low pass filter of the solar radiation input is suggested. The approach uses a stochastic state space model formulation of the combined model and filter. A maximum likelihood method for estimating the parameters of the combined model is used. In the linear case, an ordinary Kalman filter, and in the non-linear case an extended Kalman filter, was used to evaluate the likelihood function.

The estimated filter was not as effective as pre-filtering the solar radiation data. However, pre-filtering can only be performed off-line, whereas the method proposed in this paper provides the possibility for on-line monitoring.

The possibility for on-line monitoring is a great advantage at a time where industry is being asked to monitor their waste water and the effect it has on the water quality in the receiving waters. Also, it is very useful to monitor data collection for these purposes. Enormous investments are placed in monitoring programs and the measuring equipment is more and more sophisticated. It is prudent that these investments are looked after and protected.

On-line measurements are often measured with small sampling intervals, which then provides very high frequency data. A part of the high frequency variation might be a result of stochastic fluctuations, not necessarily representative of the dynamics in the system. Thus, filtering the data directly may result in a better monitoring program, as demonstrated in this paper.

References

[1] B.J. Cosby, G.M. Hornberger, and M.G. Kelly. Identification of photosynthesis-light models for aquatic systems. II. Application to a macrophyte dominated stream. *Ecological Modelling*, 23:25–51, 1984. Elsevier Science Publishers B.V., Amsterdam.

[2] Judith L. Jacobsen and Henrik Madsen. Grey box modelling of oxygen levels in a small stream. *EnvironMetrics*, 7:109–21, 1996.

[3] Judith L. Jacobsen, Henrik Madsen, and Poul Harremoës. Modelling the dynamical effect of solar radiation on the oxygen content of a small creek. *In press: EnvironMetrics*, 1996. (Presented at the 6th International Conference on Environmetrics in Malaysia, Dec. 1995.).

[4] Judith L. Jacobsen, Henrik Madsen, and Poul Harremoës. Modelling the transient impact of rain events on the oxygen content of a small creek. *Water, Science and Technology*, 33(2):177–187, 1996. (In: Uncertainty, Risk and Transient Pollution Events. Selected proceedings of the IAWQ Interdisciplinary Symposium).

[5] Judith L. Jacobsen and L.M. Voss. Time series analysis and modelling of corrupted data. Master's thesis, Institute of Mathematical Statistics and Operations Research, Technical University of Denmark, Lyngby, Denmark, 1994. No. 12/94.

[6] Henrik Madsen and Henrik Melgaard. The mathematical and numerical methods used in CTLSM - a program for ML-estimation in stochastic, continuous time dynamical models. Technical Report No. 7/1991, Institute of Mathematical Statistics and Operations Research, Technical University of Denmark, Lyngby, Denmark, 1991.

[7] Henrik Melgaard and Henrik Madsen. CTLSM continuous time linear stochastic modelling. In J.J. Bloem, editor, *In: Workshop on Application of System Identification in Energy Savings in Buildings*, pages 41–60. Institute for Systems Engineering and Informatics, Joint Research Centre, 1993.

[8] J. Simonsen. *Oxygen fluctuation in streams*. PhD thesis, Department of Technical Hygiene, Technical University of Denmark, 1974.

[9] Erik Thomsen and Niels Thyssen. Monitoring of important state variables in rivers. *Vand*, 29(2):46–53, 1979. (In Danish: Måling og registrering af vigtige tilstandsvariabler i vandløb).

[10] Niels Thyssen. Aspects of the oxygen dynamics of a macrophyte dominated lowland streams. In J.J. Symoens, S.S. Hooper, and P. Compére, editors, *Studies on Aquatic Vascular Plants*, pages 202–213. 1982.

Propagation of the low-frequency noise generated by power station water-cooling towers

SERGEI P. FISENKO

A.V.Luikov Heat&Mass Transfer Institute,
15 P.Brovka St., 220072, Minsk, Belarus, CIS, E:mail:fsp@hmti.ac.by

Abstract

The propagation of low-frequency noise generated by air turbulent motion in water-cooling towers is investigated by the use of geometrical acoustics of moving media. It is shown that a cooling tower plum acts as the sonic channel for some acoustic rays. In some cases the acoustic rays are bent by the interaction of the plum and wind. Qualitative and numerical results of mathematical simulations are presented.

1 Introduction

The ecological problems of power plants are of growing importance. Among these problem the acoustic contamination of the environment by low-frequency noise has been recognized in the last few years. Cooling towers at thermal and nuclear power stations are sources of such noise. If the height of a cooling tower over spray system is equal to h, the expression for the eigenfrequencies, v_n, of the noise generated in a tower can be accurately described by [1]

$$v_n = c(2n+1)/4h. \tag{1}$$

where n =0,1,2,... and c is the sound velocity for air-vapor mixture inside a cooling tower . Using expression (1), it is possible to show, that for a tower with h =_60_m and air-vapor mixture temperature of 20^0 C the eigenfrequencies are: 1.19 Hz, 3.53 Hz, 5.9 Hz and etc. Thus, the a cooling tower act as an acoustic resonator for infrasonic and low-frequency acoustic waves. Acoustic noise with a frequency around 1 Hz is generated by a 0.1m turbulent vortex in air with a velocity fluctuation of about 0.1m/s. It is obvious, that such vortexes will frequently occur within the air-vapor flow in a cooling tower. [1], [2].

The propagation of acoustic noise, generated in cooling tower, is described in this work with the help of the geometrical acoustics of moving media. The approach of the geometrical acoustics is formally inapplicable to calculation of the propagation of the first eigenmodes of a standard cooling tower. But it can give qualitatively correct results as is known from the application of geometrical optics and acoustics [3], [4], [5].

2 The Hamiltonian formulation of equations of geometrical acoustics of moving media.

Let us to establish a coordinate system with the positive direction of the abscissa coinciding with the wind direction, the axis z directed vertically upward along the axis of the cooling tower. Then a wave equation for scalar value φ, which describes the sound propagation, can be written as

$$(1-\beta^2)\partial_{xx}\varphi + \partial_{yy}\varphi + \partial_{zz}\varphi - \frac{1}{c^2}\partial_{tt}\varphi - \frac{2\beta}{c}\partial_{tx}\varphi = 0 \qquad (2)$$

where β(z) = w(z)/c(z) with w(z) and c(z) the speeds of wind and sound at height z, respectively. Using the approach of geometrical acoustics [3] for a phase, θ, of an acoustic wave extending in the direction of wind, from (2) we receive the eiconal equation for θ

$$(1-\beta^2)\theta_x^2 + \theta_y^2 + \theta_z^2 = n^2 - 2\beta\theta_x \qquad (3)$$

where n(z)= c_0/c(z) is the acoustic index of the refraction, with c_0 the speed of a sound near to a surface of the ground. For an acoustic wave, extending opposite the direction of the wind it is necessary to change the sign of the last term on the right side equation (3).

Writing the Hamiltonian of the system, H, using the eiconal equation (3) [4]

$$H = \frac{1}{2}\{\sum_1^3 p_i^2 - (n - \beta p_x)^2\} \qquad (4)$$

where the momentum, p_i, is defined as $p_i = \theta_i$. The position of an acoustic ray is characterized by a point in the six-dimensional phase space. The space variables are x, y and z with corresponding momentums, p_x, p_y and p_z. Choosing coordinate x as the independent variable, we receive the equations {,} determining the trajectories of acoustic rays by means of the usual technique of the Hamiltonian dynamics [4]. The model of flatly layered media is used. The assumption of a flatly layered media in which sound propagates means that the acoustic index refraction n(x) and parameter β(x) depend only on the height.

Calculations result in the system of equations

$$\frac{dy}{dx} = \frac{p_y}{(1-\beta^2)p_x + n\beta}; \quad \frac{dz}{dx} = \frac{p_z}{(1-\beta^2)p_x + n\beta}; \qquad (5)$$

$p_x = const; \ p_y = const;$

$p_z = \pm[(n(z) - \beta p_x)^2 - p_x^2 - p_y^2]^{1/2}.$

The initial conditions for the system (5) are:
$(x_0, y_0) \in D, \ y = y_0, \ z = h.$

$p_x = n_1 \sin \psi \cos \varphi; \qquad p_y = n_1 \sin \psi \sin \varphi;$ \hfill (6)

where D – top cross-section of the cooling tower, n_1 is the acoustic refraction index near top cross- section a cooling tower, φ and ψ are angular coordinates of the defined at the spherical system of coordinates. The system of the differential and algebraic equations (5) with the initial conditions (6) is investigated analytically and numerically, below.

3 Results of mathematical modeling.

Next a qualitative investigation of a system (5) will be conducted. The dependence of sound speed in air on temperature T(z) is given by

$c[T(z)] \sim \sqrt{T(z)},$ \hfill (7)

therefore

$n(T(z)) = [T_0 / T(z)]^{1/2}$ \hfill (8)

where T_0 is the air temperature near the ground surface. From the equation (8) it follows, that the distribution of air temperature influences on propagation of a sound near the cooling tower, because, as a rule, β<<1. The temperature field in the boundary layer of atmosphere is, as a rule, such that air temperature drops [6]. At average latitudes temperature of air drops about on 0.6-0.8 of degree on one kilometer of height. This effect causes the main influence on the distribution of acoustic rays in atmosphere.

In the problem to be considered there is one exception to the air temperature distribution: near of the cooling tower plum . At the exit of a cooling tower, the mixture of the air-vapor plum with cold air, because of the turbulent injection, an expansion of the plum borders occurs. Because of high water vapor concentration in a cooling tower plum a partial condensation of water vapor results from mixing with cold air. As a result of the latent heat of condensation, the cold air is warmed, so the plum temperature remains nearly constant [7], [8], despite of increasing of the geometrical sizes of the plum.

Within the framework of the flat – layered media model the average temperature, $T_1(z)$, over an area, S, is determined by

$$T_1(z) = \frac{2}{\pi \lambda^2} \int_S T(x,y,z) ds \qquad (9)$$

The integration is conducted over a half-circle S, whose radius is equal to the wavelength of sound, λ, and the center lies on the trajectory of the acoustic rays [9]. It is easy to show, from simple geometrical reasoning, that the average temperature, $T_1(z)$, increases with increasing h, or remains unchanged, provided that the area S completely lies within the cooling tower plum. The average temperature remains constant if the characteristic size of a section of the plum in a plane parallel to the surface of the ground is greater than or equal to the acoustic wavelength. The increase of average temperature occurs due to increasing of the plum size at its constant temperature [8], and takes place when the initial size of the plum is smaller of a acoustic wavelength.. If the acoustic wavelength is equal to 30m, the temperature leaving air-vapor mixture 26^0C, and the initial radius of a plum is equal to 18m and $T_0 = 15^0C$, than $T_1 = 19^0C$.

The increase of average temperature $T_1(z)$ with height results in refraction of acoustic rays [3,4]. Using expressions (5-9), it can be shown that there is a condition for the air temperature at which a turning point in the acoustic ray results at a height, z^*. At the turning point $p_z = 0$, from this condition we receive the following expression

$$T(z^*) = T_1 / \{\sin^2(\psi)[1 + \beta(z)\cos\varphi]\}^2 \qquad (10)$$

From equation (10) follows that the height of the point of turn of an acoustic rays depends strongly on the initial angular conditions. The dependence on wind velocity is weaker. It is interesting to note, that there is the critical angle, ψ_c, such that at $\psi > \psi_c$ acoustic rays cannot leave the cooling tower. For such angle the acoustic momentum $p_z = 0$. The critical angle, ψ_c, is determined from

$$\psi_c = \arcsin[1 / (1 + \beta \cos\varphi)]. \qquad (11)$$

This effect arises only when taking into account the influence of wind (i. e. the parameter β is not equal to zero) It is natural, that the acoustic rays which are most subject to trajectory changes are those for which the initial angle ψ is close to the critical angle. Thus, if the increase in temperature of air near the cooling tower at height z^* is greater than or equal to the temperature deviation expressed by equation (10), a bending of the acoustic rays will take place. This phenomenon is the principle result of this paper.

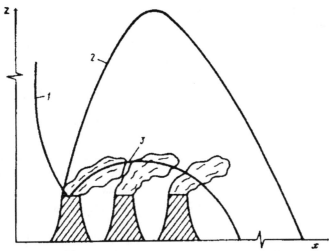

Figure 1. Classification of the acoustic rays radiated from the cooling tower. Wind direction along axis x. Trajectory 1-against wind, trajectory 2 -big initial angle, trajectory 3- acoustic rays are capture by plum.

Introducing characteristic scales x^* and y^*, we shall treat the results of numerical simulation of acoustic rays trajectories by means of theses parameters. During time, t, the acoustic wave front exiting the cooling tower at less than the critical angle ψ_c will pass the path s along an axis x , which is equal to

$s = ct \sin\psi_c \, s\varphi + wt$ (12)

For the problem of bending of an acoustic ray the characteristic time scale, t, is equal to
$t_* = h/c \, \cos\psi_c$

Substituting (13) in (12), and taking into account that $\beta \ll 1$, receive

$$x^* = h \, tg\psi_c \, \cos\varphi;$$
$$y^* = h \, tg\psi_c \, \sin\varphi \quad (14)$$

The angular parameters for acoustic rays leaving a cooling tower and geometrical scales x^* and y^* were used in the numerical simulations. The numerical simulation of the system equations (5) with the initial conditions (6) shows, that for standard atmospheric conditions the rays, leaving the cooling tower against wind velocity at any angle turn to a vertical direction and propagate towards the upper layers of atmosphere (trajectory 1 on fig. 1), where the sound energy dissipates. The acoustic rays, leaving from a cooling tower on wind direction, can be divided into two classes. The first class consists of acoustic rays leaving in a nearly vertical direction and not falling in a cooling tower plum. The numerical simulation shows the behavior of these rays is rather complicated and depends on ratio of gradients of temperature and wind velocity in the lower layers of the atmosphere (at heights about up to thirty kilometers). In limit of an isothermal atmosphere or the availability of layers with an inverse temperature distribution these rays turn and fall on the ground a distance of a few tens of kilometers from the sound source (trajectory 2 at the Fig.1).

The greatest interest from the point of view of acoustic "contamination" of an environment, is presented by acoustic rays, which fall on top a cooling tower in the air-vapor plum. For such acoustic rays the angle ψ should satisfy to a ratio [8], $ctg\psi > 0.5 (w_1/v_c)^2$, where w_1 is the wind speed at the top of the cooling tower and v_c is the vertical speed of the plum. As shown by the numerical simulation, the rays, which travel a long enough distance in the cooling tower plum, turn downward and hit the ground a short distances (about one kilometer) from a cooling tower (trajectory 3 at Fig.1). The attenuation of high frequency acoustic waves in plums due to absorption of sound by water drops can be considered on the basis of the approach stated in [10]

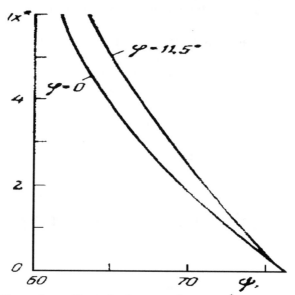

Figure 2. Dependence dimensionless coordinate x_e/x^* versus initial angular parameters of the acoustic rays.

In Fig. 2. dependence of the dimensionless coordinate x_e/x^* on the initial angular parameters of an acoustic ray are presented, where x_e is the distance between the cooling tower and the point where the rays intersect the ground, ($x^* = 319m$). In these simulation the height of the a cooling tower was taken to be 80m, the wind velocity was 10 m/s at the top of the cooling tower, the air temperature was to 15^0C, and the initial plum temperature is equal to 25^0C. The angle ψ_c is equal to $68,75^0$.

4 Conclusions

The problem of propagation of acoustic noise, generated by a ascending air-vapor flow in a cooling tower has been investigated. This problem has great environmental significance given the influence of low frequency noise on human organism.

Hamiltonian dynamics were used to calculate the trajectories of the acoustic rays. It's shown that strong winds cause the refraction of the acoustic rays resulting in a turning of there motion. As a result of the turning the acoustic rays fall on the surface of the ground a small distance from a cooling tower. The direction of the falling acoustic rays is close to vertical.

5 References

[1] Landau,L.D., Lifshits,E.M.: Hydrodynamics, Nauka, Moskau, 1977. (in Russian)
[2] Garmize,L.X. et al: Laboratory modeling of the enhancement of heat and mass transfer process in chimney-type evaporative cooling tower, J. of Engineering Physics and Thermophysics, 1994, **66**, pp.126-132 .
[3] Blokhintsev,D.I.: Acoustics of Moving Media, Nauka, Moskau,1980 (in Russian)
[4] Ostashev,V.E.: Sound Propagation in Moving Media, Nauka, Moskau,1992 (in Russian)
[5] Abdulaev,S.S.: Ray dynamics of the propagaton of sound in nonuniform moving media, Chaos, 1993, **3**, pp.101-106.
[6] Khirgian,A.Kh.: Physics of Atmosphere, Gidrometeoizdat, Leningrad, 1978. (in Russian
[7] Scorer,R.S.: Enviromental Aerodynamics, John Wiey&Sons, New York, 1978.
[8] Ivanov,V.V., Fisenko,S.P.: Simulation of the process of aerosol washout from vent pipe plum of a nuclear power station on interaction with the steam–air plum of a water–cooling tower, J. of Engineering Physics and Thermophysics, 1993, **65**, pp.1064-1072 .
[9] Feynmann,R.P.,Hibbs,A.R.: Quantum Mechanics and Path Integrals, McGray-Hill, New York, 1965.
[10] Khubaidullin,D.A., Ivandaev,A.I.: Mass transfer in acoustics of polydispersed mists., Siberian Physico- Technical Zurnal, 1993, N3, pp 16-21.

A Mathematical Model of a Hollow-Fibre Filter

A.F. Jones, J.A. King-Hele and P.T. Cardew*
Mathematics Department, Manchester University,
Oxford Road, Manchester, UK. Email: King-Hele@ma.man.ac.uk

Abstract

A typical hollow-fibre water filter consists of a long hollow circular cylindrical casing inside which is packed a large number of hollow fibres each with its axis parallel to the casing axis. Water is pumped across the casing in a radially symmetric fashion, enters the region outside the fibres, and then flows into the interior of the hollow fibres leaving impurities at the fibre surface. In an ideal filter the fluid pressure outside the fibres is uniform (equal to the inlet pressure) and the fluid pressure inside the fibres is also uniform (equal to the low exit pressure). The water flow into the fibres is then uniform and impurities are deposited uniformly throughout the filter. This ideal situation cannot be achieved exactly in practice since pressure gradients are required to drive the flow against inertia and viscous forces. Using a Mathematical Model design criteria are established under which the filter can be regarded as ideal. The Model is also used to estimate the pressure drop at a small inlet, a potential source of inefficiency.

1 Introduction

Hollow-fibre filters are widely used in separation devices for example in artificial kidneys [1], gas separators [2] and water filters [3], [4]. In this paper we study a particular water filter of interest to North-West Water plc., and derive criteria under which it is efficient.

The filter consists of a long hollow circular cylindrical casing, typically about 1m long, inside which is packed a large number of hollow porous cylindrical fibres each with its axis parallel to that of the casing. The structure of the filter is illustrated in Figure 1. The fibres can be made of a variety of materials depending on the application. Water is forced radially into the system through an inlet on the surface of the outer casing near one end. This inlet jet is diverted azimuthally

---

*North West Water plc, Gorsey Lane, Widnes, U.K.

before entering the filter so that water enters in an axi-symmetric fashion. The water flows through the region outside the fibres and then into the interior of the hollow fibres before flowing axially to leave the system at the same end as the water was injected. In this particular device the other end of the filter is completely sealed so that water can only leave the system through one end via the hollow fibres. Impurities are deposited on or within the fibres. The filter can be cleaned by reversing the direction of water flow.

In an ideal filter the fluid pressure equals the given inlet pressure p_i everywhere outside the fibres and equals p_e, the exit pressure, everywhere inside the fibres. In this situation the flow into the interior of each hollow fibre, driven by the pressure difference $p_i - p_e$, is the same everywhere in the system. Impurities are then deposited uniformly within the filter.

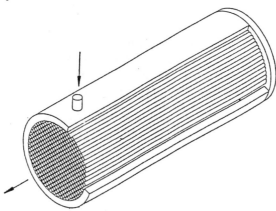

Figure 1: A cut-away section showing the fibres within the casing. The arrows indicate the direction of fluid flow.

In practice this ideal arrangement cannot be achieved. It is evident that pressure gradients are required both inside and outside the fibres to drive the flow against viscous or inertia forces. Viscous forces are particularly important within the small diameter fibres. There may also be a substantial pressure drop near the entry region if the diameter of the inlet is small compared with the diameter of the cylindrical casing. In this latter case, although it may be possible to arrange for the pressure to be almost constant outside the fibres in most of the device, its value is less than the imposed inlet pressure p_i. A consequence is that for a given pressure difference $p_i - p_e$, less fluid flows through the system than in an ideal system, leading to a loss of efficiency.

This paper focuses on a simplified mathematical model of the interaction between the fluid flow outside the fibres (the shell side) and the flow inside the fibres (the lumen side). We deduce some important dimensionless parameters, and study in particular the pressure drop near a small inlet.

2 A Mathematical Model

The fluid flow within a filter is extremely complex so our aim here is to produce a simplified model which includes the important physical processes. We consider a cylindrical filter of length l_0 and radius r_0. Fluid flows into filter in an axisymmetric fashion across the inlet surface $r = r_0$, $0 \leq z \leq d_0$ where r and z are the radial and axial coordinates respectively. The inlet pressure is p_i. The fluid flows into the fibres and then flows axially leaving the filter at $z = 0$ where the exit pressure is p_e. There is no fluid flow across the filter surface except at the inlet surface. The filter contains a large number of identical axially orientated cylindrical fibres each of length l_0, external radius a_0 and internal radius c_0. The geometrical structure is illustrated in Figure 2.

As there are a large number of fibres per unit cross-sectional area of filter we assume that the flow field can be represented by a continuum model in which $u_1(r,z)$ and $u_2(r,z)$ are the mean superficial fluid velocities for flow outside and inside the fibres respectively. The mean superficial velocity is defined as the fluid flux per unit area averaged over an area which is large compared with a fibre diameter. We do not therefore attempt to model the flow near a particular fibre.

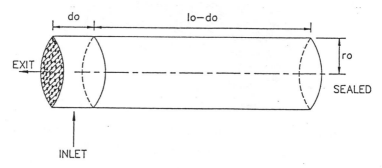

Figure 2: The geometrical structure of the simplified model.

Consider first the axial flow within a fibre so that $\mathbf{u}_2 = w(r,z)\hat{z}$. We can show that as the fibres are slender the flow is of Poiseuille type so that there is a close balance between the axial pressure gradient and the viscous forces. If follows that if $p_2(r,z)$ is the mean pressure field for flow within a fibre

$$w = -k_2 \frac{\partial p_2}{\partial z}, \qquad (1)$$

where $k_2 = (1-\eta)c_0^4/8\mu a_0^2$. Here η is the void fraction and μ is the fluid viscosity. We note that the factor $(1-\eta)/a_0^2$ allows for the fact that w is defined as the axial flux within the fibres per unit area of filter.

The flow outside the fibres is more complex. There appears to be little experimental evidence to establish how the mean axial and radial fields interact.

We shall in effect consider the axial and radial flow separately and then simply combine them assuming they are independent.

In the axial direction we have channel flow parallel to the fibres. The boundary of the channel consists of arcs of circles (the fibres) where no-slip conditions apply, and the region between the fibres where a no-stress condition is appropriate. Happel and Brenner [5] state that for $0.4 \leq \eta \leq 0.8$ a Poiseuille type formula is reasonably accurate, so that if $v(r,z)$ is the axial velocity and $p_1(r,z)$ is the mean pressure field for flow outside the fibres

$$v = -k_1 \frac{\partial p_1}{\partial z}, \tag{2}$$

where $k_1 = (a_0^2/8\mu\eta^2(1-\eta))(-2\ln(1-\eta) - 2\eta - \eta^2)$, a positive constant.

Consider now radial flow normal to an array of cylindrical fibres. The pressure gradient force will be proportional to the fluid speed $u(r,z)$ if viscous forces dominate and proportional to $u|u|$ if inertia forces dominate. It was first proposed by Reynolds [6] that a formula of wide validity has the form

$$-\frac{\partial p_1}{\partial r} = \frac{u}{\alpha} + \frac{u|u|}{\beta}. \tag{3}$$

Few experiments have been performed to test this formula, but it has been justified experimentally for flow past glass spheres (Ergun [7]). For an array of cylinders, Kuwabara [8] has predicted that $\alpha = k_1/2$ and this is the value we have used. The value of β is related to the drag coefficient c_D for flow past an array of cylinders by the formula $\beta = \pi a_0 \eta^2 / c_D(1-\eta)\rho$ where ρ is the fluid density. The ratio of the quadratic to the linear term in (3) is of order the fibre Reynolds number $R_a = 2u a_0 \mu / \rho$. In our particular filter the linear term in (3) is dominant in most of the filter but we would expect the quadratic term to dominate near the inlet where u is larger. We estimate that if $\eta = 0.5$ and $c_D = 1.0$ the linear and quadratic terms in (3) are comparable if $R_a = 50$.

The equations of mass continuity for fluid flow outside and inside the fibres are respectively

$$\frac{1}{r}\frac{\partial}{\partial r}(ru) + \frac{\partial v}{\partial z} = -\lambda(p_1 - p_2), \tag{4}$$

$$\frac{\partial w}{\partial z} = \lambda(p_1 - p_2). \tag{5}$$

In these equations $\lambda(p_1 - p_2)$ is the volume flux into the fibres per unit volume of filter. This appears as a sink term in (4) and a source term in (5). The constant λ is determined by solving the problem of radial flow across the wall of a porous fibre. We find that $\lambda = 2(1-\eta)K/a_0^2 \ln(a_0/c_0)$ where K is a material constant which has been determined experimentually by North West Water.

In the linear case where $u\alpha/\beta << 1$, equations (1) - (5) can be combined to yield a coupled set of linear equations for p_1 and p_2, namely

$$\frac{1}{2r}\frac{\partial}{\partial r}\left(r\frac{\partial p_1}{\partial r}\right) + \frac{\partial^2 p_1}{\partial z^2} = \frac{\lambda}{k_1}(p_1 - p_2), \tag{6}$$

$$\frac{\partial^2 p_2}{\partial z^2} = -\frac{\lambda}{k_2}(p_1 - p_2). \tag{7}$$

The inlet and exit pressures are given so that

$$p_1 = p_i, \quad r = r_0, \quad 0 \leq z \leq d_0,$$
$$p_2 = p_e, \quad z = 0. \tag{8}$$

The remaining boundary conditions are that there is no flow normal to the filter casing elsewhere. These equations have been solved numerically, but details will be given elsewhere.

In an ideal filter $p_1 \equiv p_i$ and $p_2 \equiv p_e$ and this will be solution if we formally let k_1 and k_2 tend to infinity. It is evident therefore that the filter will be almost ideal if the source/sink terms in (6) and (7) give rise to small perturbations from the ideal solution. For our filter $r_0 \ll l_0$, so this requirement will be satisfied if

$$\epsilon_1 = \frac{l_0^2 \lambda}{k_1} \ll 1, \quad \epsilon_2 = \frac{l_0^2 \lambda}{k_2} \ll 1. \tag{9}$$

For the purpose of illustration we use the following parameter values:- $a_0 = 4 \times 10^{-4}$m, $c_0 = 2.5 \times 10^{-4}$m, $d_0 = 2.5 \times 10^{-2}$m, $l_0 = 1$m, $r_0 = 7.5 \times 10^{-2}$m, $\eta = 0.5$, $K = 2 \times 10^{-13}$ms^3/kg, $\mu = 1.0 \times 10^{-3}$kg/ms, $\rho = 10^3$kg/m^3, $c_D = 1.0$, $p_i = 2$atm, $p_e = 1$atm. Then we find that $k_1 = 2.2 \times 10^{-5}$m^{3}s/kg, $k_2 = 1.5 \times 10^{-6}$m^{3}s/kg, $\lambda = 2.7 \times 10^{-6}$ms/kg, $\epsilon_1 = 0.12$, and $\epsilon_2 = 1.8$.

The small value of ϵ_1 shows that the pressure field will be almost uniform outside the fibres. However since ϵ_2 is not small the pressure inside the fibres will not be close to the exit pressure. Substantial pressure gradients are required to drive the fluid axially out of the filter against viscous forces. We note that since k_2 is proportional to c_0^4, a modest increase in c_0 will substantially reduce the value of ϵ_2. Numerical calculations show that the volume flow through the filter is about half the value obtained for an ideal filter. Away from the inlet the characteristic radial velocity $\lambda r_0 (p_i - p_e) \simeq 2 \times 10^{-2}$m/s so that the fibre Reynolds number $R_a = 2a_0 u \rho / \mu \simeq 16$. Thus the linear term in (3) is about three times the quadratic term. On the other hand we can show that the characteristic inlet radial velocity is $(\lambda r_0 (p_i - p_e))(l_0/2d_0) \simeq 0.2$m/s. The fibre Reynolds number $R_a \simeq 160$ now so the quadratic term in (3) is dominant.

In the next section we estimate the pressure drop in the inlet region on the assumption that the quadratic term in (3) is dominant.

3 The Pressure Drop at the Inlet

In the small region near the inlet the quadratic term in (3) dominates, the source/sink terms in (6) and (7) are small, and moreover we can work with local cartesian coordinates $(\tilde{x}, \tilde{y})$ defined by $\tilde{y} = r_0 - r$, $\tilde{x} = z$.

As we are concerned here with inlet flow we shall take $u = -\tilde{u}(\tilde{x},\tilde{y})$ and $v = \tilde{v}(\tilde{x},\tilde{y})$, so that the governing equations are

$$\tilde{u}^2 = -\beta\frac{\partial p_1}{\partial \tilde{y}}, \quad \tilde{v} = -k_1\frac{\partial p_1}{\partial \tilde{x}}, \quad \frac{\partial \tilde{v}}{\partial \tilde{x}} + \frac{\partial \tilde{u}}{\partial \tilde{y}} = 0. \tag{10}$$

These equations are to be solved in the quadrant $\tilde{x} \geq 0$, $\tilde{y} \geq 0$. The boundary conditions are that the normal velocity is zero on the filter casing $\tilde{x} = 0$, $\tilde{y} > 0$ and on $\tilde{y} = 0$, $\tilde{x} > d_0$. At the inlet $\tilde{y} = 0$, $0 \leq \tilde{x} \leq d_0$ the pressure is p_i. At large distances from the corner the solution must match with the outer linear flow and we expect that the pressure will tend to a constant value $p_i - (\Delta p)_i$, where $(\Delta p)_i$ is the quantity of interest.

Suppose that u_i is the mean inlet velocity, defined by

$$u_i = \frac{1}{d_0}\int_0^{d_0} \tilde{u}(\tilde{x},0)d\tilde{x}. \tag{11}$$

Then if we define new dimensionless variables by

$$X = \tilde{x}/d_0, \quad Y = (k_1 u_i/\beta d_0^2)^{1/2}\tilde{y} \tag{12}$$

$$U = \tilde{u}/u_i, \quad V = (\beta/k_1 u_i^2)^{1/2}\tilde{v}, \quad P_1 = (p_1 - p_i)\left(\frac{\beta k_1}{u_i^3 d_0}\right)^{1/2},$$

we find that

$$U^2 = -\frac{\partial P_1}{\partial Y}, \quad V = -\frac{\partial P_1}{\partial X}, \quad \frac{\partial V}{\partial X} + \frac{\partial U}{\partial Y} = 0 \tag{13}$$

where now $P_1 = 0$ on the inlet $Y = 0$, $0 < X < 1$, $U = 0$ on $Y = 0$, $X > 1$, $V = 0$ on $X = 0$, $Y \geq 0$ and

$$\int_0^1 U(X,0)dX = 1. \tag{14}$$

The key fact to note is that equation (13) and the boundary conditions involve no dimensionless parameters. This means in particular that P_1 must be an order one quantity and that therefore the order of magnitude of the pressure drop at the inlet is determined by the scaling in (12).

If we eliminate V from (13), we find that

$$2U\frac{\partial^2 P_1}{\partial X^2} + \frac{\partial^2 P_1}{\partial Y^2} = 0 \tag{15}$$

which is a non-linear elliptic equation. It is possible to find a similarity source solution for this problem. Thus we require that the solution tends to this similarity source solution as $X^2 + Y^2 \to \infty$, satisfies the appropriate no flux condition on the filter wall, and satisfies the given inlet pressure condition $P_1 = 0$

on $Y = 0$ for $0 \leq X \leq 1$. Equation (15) was solved by an iterative method using finite differences over a square region of side 10 units. It was found that for $\sqrt{X^2 + Y^2} > 2$, the velocity field was indistinguishable from the corresponding similarity solution. However the pressure was a constant value, 1.6, below the similarity pressure field. This is illustrated in Fig.3 which shows the graphs of $-P_1(X,Y)$ and $-P_1(X,Y) + P_s(X,Y)$ as a function of $R = X\sqrt{2}$, along the line $Y = X$. Here $P_s(X,Y)$ is the similarity pressure field which is singular at the origin. Note that $P_1(X,Y)$ is a negative quantity.

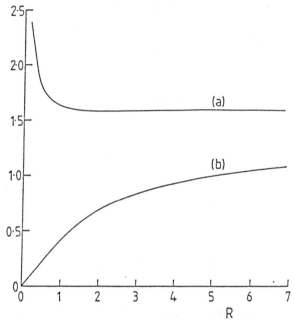

Figure 3: Graphs of $(a) - P_1(X,Y) + P_s(X,Y)$ and $(b) - P_1(X,Y)$ as functions of R along the line $Y = X$ where $R = X\sqrt{2}$. Here $P_s(X,Y)$ is the similarity solution.

In dimensional terms the pressure drop at the inlet is therefore

$$(\Delta p)_i = 1.6 \left(\frac{u_i^3 d_0}{\beta k_1} \right)^{1/2}. \tag{16}$$

Now the precise value of u_i is unknown since it is clearly determined by matching the corner flow with the linear flow in the rest of the filter. Details of this matching procedure will be given elsewhere, but for the present purpose we note thast u_i is always less than the value for an ideal filter and this value is known. This gives

$$u_i \leq \frac{\lambda l_0 r_0}{2 d_0}(p_i - p_e). \tag{17}$$

This can be used to estimate the maximum pressure drop at the inlet. In the present case we find from (16) and (17) that $(\Delta p)_i \leq 4\times 10^{-2}$atm which is negligible compared with $p_i - p_e = 1$atm. We conclude that in this case the inlet pressure drop is negligible.

4 Conclusions

We have formulated a simplified model of fluid flow in a filter and have established criteria under which it may be regarded as an ideal filter. These criteria are not satisfied for this particular filter but clearly indicate the way in which the design of the system can be optimised. The key dimensionless groups derived here will have relevance to other hollow-fibre filter devices.

We have also derived a formula which allows us to estimate the maximum pressure drop in the small entry region. For the parameters used this pressure drop is negligible.

References

[1] Sigdell, J.E. "A mathematical theory for the capillary artificial kidney", Verlag, Stuttgart, 1974

[2] Pich J. "Gas Filtration Theory", Chapter 1, "Filtration, Principles and Practices". Ed. Orr. C. Marcel Dekker, 1990.

[3] Rautenbach, R., Albrecht, R. "Membrane Processes", Willey, 1989.

[4] Porter, M.C. "Ultrafiltration", Chapter 3. "Handbook of Industrial Membrane Tech". Ed. Porter, M.C. 1990.

[5] Happel, J. & Brenner, H., *Low Reynolds Number Hydrodynamics*, Prentice Hall, (1965).

[6] Kuwabara, S., *J. Phys. Soc. Japan*, **14**, 527 (1959).

[7] Reynolds, O., *Papers on Mechanical and Physical Subjects*, CUP (1900).

[8] Ergun, S., *Chem. Eng. Progr.*, **48.2**, 89, (1952).

Railway systems

A brief survey of wheel-rail contact Theory, Algorithms, Applications

J.J.Kalker,
Delft University of Technology, Mekelweg 4,
2628 CD Delft, the Netherlands

Abstract

The technology of rail vehicles consists of three parts: vehicle technology, track technology, and the contact between wheel and rail. This paper, on wheel-rail contact, also consists of three parts: 1. The theory of contact (not necessarily wheel-rail) 2. Algorithms 3. Applications

1 Introduction

The science of contact has several branches.

First there is contact mechanics, which studies bodies in contact, and is connected to elasticity, viscoplasticity, and similar material constitutive laws. Also it can be joined to fracture mechanics.

Secondly, there is contact physics, which concerns itself with friction, wear, normal adhesion, and, in particular, cracks. A typical problem is the nature and the value of the coefficient of friction.

In this paper, we will consider contact mechanics, *viz.* the kinematics and dynamics of the wheel-rail system, which we will consider to be made of an elastic material, namely, steel. This is a well-rounded theory, with many algorithms and applications.

In the part "Applications" the high speed train is discussed. Notably the concept of critical velocity is introduced as the velocity at which the bogies of the train become unstable. Other applications in which successes have been achieved are the wear of wheel and rail, although the theory is far from complete.

All these applications take the form of contact evolutions, i.e. sequences of contact problems. The contact problems themselves are efficiently handled by the algorithms, which were developed with the wheel-rail application in mind. We also give two more applications, which are not on wheel-rail contact, but which are interesting for the Industrial Mathematician. They are on Ball and Roller Bearings, and on Copiers.

PART 1: THE THEORY OF CONTACT

2 Description of two bodies in contact [1]

The theory of this section is based on Shabana [1]. Consider an elastic body. Its unstressed state is the reference state. A Cartesian coordinate system is introduced which is rigidly connected to the body in the unstressed state. The particles of the body are denoted by their coordinates in this coordinate system, *viz.* **y**. The body is deformed; the displacement of particle **y** is denoted by **w**. **w** is a function of **y** and the time t; the particle in the deformed state has the coordinates

$$\mathbf{y} + \mathbf{w}, \quad \mathbf{w} = \mathbf{w}(\mathbf{y}, t) \tag{1}$$

We assume that **w** is small as well as its derivatives with respect to **y** and t. The fact that **w** is small and that its gradients are likewise small is denoted by

$$|\mathbf{w}| \ll |\mathbf{y}| \; ; \; |\partial \mathbf{w}/\partial \mathbf{y}| \ll 1 \,. \tag{2}$$

A small time derivative means that the body is assumed to move quasi-statically, or, in other terms, that inertial effects may be neglected. In sum, the constitutive equations are elastostatic.

A second body with similar properties is introduced. A similar analysis is performed. The particles of the bodies are denoted by $\mathbf{y}_i$, the displacements by $\mathbf{w}_i(\mathbf{y}_i, t)$, the position in the deformed state $\mathbf{y}_i + \mathbf{w}_i$.

We want to compare the quantities of the two bodies. To that end we must refer them to a single coordinate system. So we introduce a third, global coordinate system, in which the particles in the deformed state are given by $\mathbf{x}_i + \mathbf{u}_i$, that is, in the global coordinate system we also distinguish between the particles of body 1 and body 2.

The global coordinate system is connected to the two locals by the rotation matrices $A_i[t]$ and distance between the origins $\mathbf{R}_i[t]$, which are functions of time alone. The rotation matrices are orthogonal matrices. We have the following connection between the $\mathbf{y}_i + \mathbf{w}_i$ and the $\mathbf{x}_i + \mathbf{u}_i$

$$\mathbf{x}_i + \mathbf{u}_i = A_i[t](\mathbf{y}_i + \mathbf{w}_i) + \mathbf{R}_i[t] = \{A_i \mathbf{y}_i + \mathbf{R}_i\} + \mathbf{A}_i \mathbf{w}_i(\mathbf{y}_i, t) \tag{3}$$

This is an identity in **x** and **y**, **u** and **w**; so, if we omit the indices,

$$\mathbf{x} = A\mathbf{y} + \mathbf{R} \Rightarrow \mathbf{y} = A^T(\mathbf{x} - \mathbf{R}) \tag{4}$$

$$\mathbf{u} = A\mathbf{w}[\mathbf{y}, t] \tag{5}$$

(2.5) is the definition of **u**.
It is clear from (2.4) that a variable belonging to body i can be written either as a function of $\mathbf{x}_i$ and t or as as function of $\mathbf{y}_i$ and t.

Let $\mathbf{m}_i$ be the outer normal on body i at $\mathbf{y}_i$; since $\mathbf{w}_i$ is small with small gradients, $\mathbf{m}_i$ is also the outer normal on body i at $\mathbf{y}_i + \mathbf{w}_i$. In the global coordinates, $\mathbf{m}_i$ becomes $\mathbf{n}_i$, with

$$\mathbf{n}_i = A_i \mathbf{m}_i \tag{6}$$

The points $\mathbf{y}_1 + \mathbf{w}_1$ and $\mathbf{y}_2 + \mathbf{w}_2$ are in contact if

$$\mathbf{x}_1 + \mathbf{u}_1 = \mathbf{x}_2 + \mathbf{u}_2 \quad \text{and} \quad \mathbf{n}_1 = -\mathbf{n}_2 \tag{7}$$

Example Let $\mathbf{x}_2$ lie in a part of the surface of body 2 which is continuously differentiable, and suppose that in that part the surface is given by the analytic equation $F(\mathbf{x}_2) = 0$. Then $\mathbf{n}_2$ is proportional to $\partial F(\mathbf{x}_2)/\partial \mathbf{x}_2$

Example A rail surface consists of the union of circular cylinders with parallel axes. These cylinders intersect, and at the intersection the first derivatives are continuous. The equation of one circular cylinder is

$$(y-a)^2 + (z-b)^2 - R^2 = 0, \quad \mathbf{n}_2 = q(0, 2\{y-a\}, 2\{z-b\})$$

where q is a scalar quantity that renders $\mathbf{n}_2$ the unit outer normal on body 2.

3 The distance [2]

We now suppose that the bodies are not in contact but that their distance is of the order of magnitude of $\mathbf{u}_i$, and that the surface is smooth, all in the neighborhood of $\mathbf{x}_i$, while the relationship $\mathbf{n}_1 = -\mathbf{n}_2$ is approximately valid. In a small neighborhood of $\mathbf{x}_i$ the bodies may be visualised as two flat and parallel slabs. Let the distance between these slabs (1)-(2) be h in the unstressed state, and e in the deformed state. Then we have:

$$h = \mathbf{n}_2^T(\mathbf{x}_1 - \mathbf{x}_2), \quad e = \mathbf{n}_2^T\{(\mathbf{x}_1 + \mathbf{u}_1) - (\mathbf{x}_2 + \mathbf{u}_2)\} \tag{1}$$

We simplify the expression for e. As we will see later on, the displacement in a contact problem only occurs in the form of $\{\mathbf{u}_1 - \mathbf{u}_2\}$. This expression is called the displacement difference, and we denote it by $\mathbf{u}$. Indeed, we encounter the displacement difference in the undeformed distance e. We also encounter the deformed distance h in e. We obtain:

$$e = h + \mathbf{n}_2^T \mathbf{u} = h - \mathbf{n}_1^T \mathbf{u} \tag{2}$$

e: deformed, h: undeformed distance, (1)-(2)
We analyze the deformed distance e.

- $e > 0$: There is a gap between the bodies at $\mathbf{x}_1$; $\mathbf{x}_2$; $\mathbf{x} \stackrel{def}{=} (\mathbf{x}_1 + \mathbf{x}_2)/2$

- $e = 0$: The bodies are in contact at $\mathbf{x}$;

- $e < 0$: It seems that the bodies penetrate at $\mathbf{x}$. Impossible.

Summarizing, only $e(\mathbf{x}) = h(\mathbf{x}) + \mathbf{n}_2^T \mathbf{u}(\mathbf{x}) = h(\mathbf{x}) - \mathbf{n}_1^T \mathbf{u}(\mathbf{x}) \geq 0$ is possible.

4 The Slip [2]

The velocity $\mathbf{v}_i$ of body i at the point $\mathbf{x}_i + \mathbf{u}_i$ with respect to $\mathbf{y}_i$ is derived as follows. $\mathbf{x}_i$ is written as a function of $\mathbf{y}_i$ and t, that is, of the particle under consideration and the time: $\mathbf{x}_i = \mathbf{x}_i(\mathbf{y}_i, t)$. $\mathbf{u}_i$ can be regarded as a function of $\mathbf{x}_i$, the position in the contact area in the global system, and the time t: $\mathbf{u}_i = \mathbf{u}_i(\mathbf{x}_i, t)$. The time derivative $(\;)_i$ is defined as the particle fixed time derivative $(\partial/\partial t)_{y_i}$. Hence the velocity $\mathbf{v}_i$ is

$$\begin{aligned}
\mathbf{v}_i &= \{\mathbf{x}_i(\mathbf{y}_i, t) + \mathbf{u}_i(\mathbf{x}_i, t)\}_i^{\cdot} = \\
&= \dot{\mathbf{x}}_i + \{\partial \mathbf{u}_i(\mathbf{x}_i, t)/\partial \mathbf{x}_i\}\{\partial \mathbf{x}_i(\mathbf{y}_i, t)/\partial t\} + \{\partial \mathbf{u}_i(\mathbf{x}_i, t)/\partial t\} = \\
&= \dot{\mathbf{x}}_i + \mathbf{u}_i(\mathbf{x}_i, t)' \dot{\mathbf{x}}_i(\mathbf{y}_i)_i + \partial \mathbf{u}_i(\mathbf{x}_i, t)/\partial t; \quad \mathbf{u}_i(\mathbf{x}_i, t)' \stackrel{def}{=} \partial \mathbf{u}_i(\mathbf{x}_i, t)/\partial \mathbf{x}_i \,.
\end{aligned} \quad (1)$$

The slip s is defined as the local velocity of body 1 with respect to body 2, that is, $\mathbf{s} = \{\mathbf{v}_1(\mathbf{x}_1) - \mathbf{v}_2(\mathbf{x}_2)\}$. Hence

$$\begin{aligned}
\mathbf{s} &= \dot{\mathbf{x}}_1 - \dot{\mathbf{x}}_2 + \mathbf{u}_1(\mathbf{x}_1, t)' \dot{\mathbf{x}}_1 - \mathbf{u}_2(\mathbf{x}_2, t)' \dot{\mathbf{x}}_2 + \partial(\mathbf{u}_1 - \mathbf{u}_1)/\partial t = \\
&= (\dot{\mathbf{x}}_1 - \dot{\mathbf{x}}_2) + (1/2)(\mathbf{u}_1' - \mathbf{u}_2')(\dot{\mathbf{x}}_1 + \dot{\mathbf{x}}_2) + (1/2)(\mathbf{u}_1' + \mathbf{u}_2')(\dot{\mathbf{x}}_1 - \dot{\mathbf{x}}_2) \\
&\quad + \partial(\mathbf{u}_1 - \mathbf{u}_2)/\partial t
\end{aligned} \quad (2)$$

Since $\mathbf{u}_1' + \mathbf{u}_2' \ll 1$, the term $(1/2)(\mathbf{u}_1' + \mathbf{u}_2')(\dot{\mathbf{x}}_1 - \dot{\mathbf{x}}_2)$ (the third term of (4.2)) may be neglected with respect to $(\dot{\mathbf{x}}_1 - \dot{\mathbf{x}}_2)$, the first term. We call

$$\begin{aligned}
\mathbf{c} &= \dot{\mathbf{x}}_1 - \dot{\mathbf{x}}_2 = \text{the creep} & (3) \\
\mathbf{v} &= -(1/2)(\dot{\mathbf{x}}_1 + \dot{\mathbf{x}}_2) = \text{the rolling velocity} & (4) \\
\mathbf{u} &= \mathbf{u}_1 - \mathbf{u}_2 = \text{the displacement difference} & (5) \\
\mathbf{s} &= \dot{\mathbf{x}}_1 - \dot{\mathbf{x}}_2 + \frac{1}{2}(\mathbf{u}_1' - \mathbf{u}_2')(\dot{\mathbf{x}}_1 + \dot{\mathbf{x}}_2) + \partial(\mathbf{u}_1 - \mathbf{u}_2)/\partial t \\
&= \mathbf{c} - (\mathbf{u}_1' - \mathbf{u}_2')\mathbf{v} + \partial(\mathbf{u}_1' - \mathbf{u}_2')/\partial t & (6)
\end{aligned}$$

The minus sign in the definition of the rolling velocity calls for comment. In wheel-rail contact the global coordinate system has its origin in the centre of the contact area, its x-axis in the direction of rolling, and its z-axis pointing normally upwards into the wheel (body 1). The y-axis lies in the plane of contact, and completes the right-handed coordinate system. The global coordinate system is therefore contact fixed. The material of wheel and rail (bodies 1 and 2) flows through this coordinate system, see Fig. 4.1; it is noted that this happens in the direction opposite to the rolling direction.

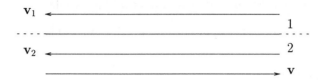

Fig. 4.1: The material of wheel and rail (bodies 1 and 2) flowing opposite to the rolling direction

Further,
$$\mathbf{x}_1 = \mathbf{x}_2 + \mathbf{u}_2 - \mathbf{u}_1; \quad \mathbf{x} \stackrel{def}{=} (1/2)(\mathbf{x}_1 + \mathbf{x}_2) \approx \mathbf{x}_1 \approx \mathbf{x}_2 \tag{7}$$
Owing to the smallness of $\mathbf{u}_i$, $\mathbf{u}'_i(\mathbf{x}_i, t) \approx \mathbf{u}'_i(\mathbf{x}, t) \Rightarrow \mathbf{u}'_1 - \mathbf{u}'_2 = \mathbf{u}'$
Hence
$$\mathbf{s} = \mathbf{c} - \mathbf{u}'\mathbf{v} + \partial \mathbf{u}/\partial t. \tag{8}$$

- The contact area B_c can be divided in the region where $\mathbf{s} = \mathbf{0}$: the area of adhesion or stick region B_a, and in the complementary region where $s \neq 0$: the slip region, or area of slip B_s.

- When $\mathbf{v} = \mathbf{0}$, one speaks of a shift.

- When $\mathbf{v} \neq \mathbf{0}$ one speaks of rollling.

- When $\mathbf{v} \neq \mathbf{0}$, and $\mathbf{c} = \mathbf{0}$, one speaks of free rolling.

- When $\mathbf{v} \neq \mathbf{0}$, and $|\mathbf{c}| \ll |\mathbf{v}|$, then the 2nd and 3rd term of the right-hand side of (4.8) may compensate the 1st term $\mathbf{c}$, so that $\mathbf{s}$ may vanish, and a stick region B_a with finite area may come into being. One speaks of rolling with small creep. Otherwise one speaks of rolling with large creep.
  Characteristic of rolling with small creep is the presence of an adhesion area. Characteristic of rolling with large creep is the absence of an area of adhesion B_a.

- When $\mathbf{u}$ is independent of t for a non-zero period, this implies that the motion is steady state during that period. Notably the third term of the slip equation, $\partial \mathbf{u}/\partial t = \mathbf{0}$. This is a very important special case.

Example: A simplified wheel-rail system
Consider a wheel in the form of a cylinder with radius r, and a rail in the form of a half-space, see fig. 4.2

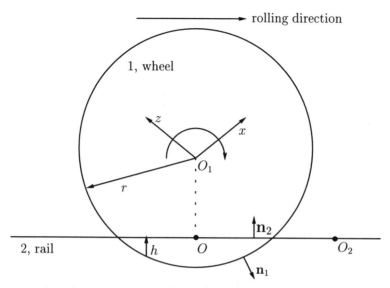

Fig. 4.2 A simplified wheel-rail system

The origin O_1 of the wheel (body 1) lies on the axis of the cylinder. The origin of the rail O_2 lies on its surface. The origin O of the global coordinates is contact fixed, it lies on the surface of the rail, perpendicularly below O_1. The wheel rotates about its y-axis, parallel to the rail, counterclockwise, so that the wheel rolls towards the right. The y-axis of the rail lies on the projection of the y-axis of the wheel. The x-axis lies in the rolling direction, $\perp$ the y-axis. The z-axis points normally upwards.

We have
$$\mathbf{x}_1 = A_1 \mathbf{y}_1 + \mathbf{R}_1 \Leftrightarrow \mathbf{y}_1 = A_1^T(\mathbf{x}_1 - \mathbf{R}_1)$$

The particle-fixed velocity at $\mathbf{x}_1$ is
$$\begin{aligned} \dot{\mathbf{x}}_1 &= \dot{A}_1 \mathbf{y}_1 + \dot{\mathbf{R}}_1 \\ &= \dot{A}_1 A_1^T (\mathbf{x}_1 - \mathbf{R}_1) + \dot{\mathbf{R}}_1 \end{aligned}$$

This means: substitute $\mathbf{x}_1$, $\mathbf{R}_1$ in the right-hand side; you find $\dot{\mathbf{x}}_1$ without calculating $\mathbf{y}_1$.

We implement the example.

Body 1.

$$\mathbf{R}_1 = (0,0,r)^T \Rightarrow \dot{\mathbf{R}}_1 = \mathbf{0}$$

$$A_1 = \begin{pmatrix} \cos wt & 0 & +\sin wt \\ 0 & 1 & 0 \\ -\sin wt & 0 & \cos wt \end{pmatrix}; \quad \dot{A}_1 = w \begin{pmatrix} -\sin wt & 0 & +\cos wt \\ 0 & 0 & 0 \\ -\cos wt & 0 & -\sin wt \end{pmatrix}$$

$$\dot{A}_1 \, A_1^T \quad = \quad \begin{pmatrix} 0 & 0 & 1 \\ 0 & 0 & 0 \\ -1 & 0 & 0 \end{pmatrix};$$

$$\dot{A}_1 \, A_1^T(\mathbf{x}_1 - \mathbf{R}_1) \quad = \quad \begin{pmatrix} wx_{1z} \\ 0 \\ -wx_{1x} \end{pmatrix} - \begin{pmatrix} wr \\ 0 \\ 0 \end{pmatrix} = \begin{pmatrix} \dot{x}_{1x} \\ \dot{x}_{1y} \\ \dot{x}_{1z} \end{pmatrix}$$

Body 1.

$$\mathbf{x}_2 \; = \; \mathbf{y}_z + \begin{pmatrix} -v_2 t \\ 0 \\ 0 \end{pmatrix} \; \Rightarrow \; \begin{pmatrix} \dot{x}_{2x} \\ \dot{x}_{2y} \\ \dot{x}_{2z} \end{pmatrix} = \begin{pmatrix} -v_2 \\ 0 \\ 0 \end{pmatrix}$$

When $\mathbf{x}_1 = 0$:
$\mathbf{c} = \dot{\mathbf{x}}_1 - \dot{\mathbf{x}}_2 = \begin{pmatrix} v_2 - wr \\ 0 \\ 0 \end{pmatrix}$; $\mathbf{v} = -\frac{1}{2}(\dot{\mathbf{x}}_1 - \dot{\mathbf{x}}_2) = \frac{1}{2}\begin{pmatrix} v_2 + wr \\ 0 \\ 0 \end{pmatrix}$
creep   rolling velocity

5 The surface traction and the boundary conditions of contact [2]

Let the surface traction on body 1 be defined as $\mathbf{X}_1$. Then the normally directed traction at $\mathbf{x}$, positive if compressive, is given by $\mathbf{X}_N = \mathbf{n}_2^T \mathbf{X}_1$. The tangential component of the traction is

$$\mathbf{X}_N(\mathbf{x}) \;\; = \mathbf{n}_2^T(\mathbf{x})\mathbf{X}_1(\mathbf{x}) \qquad \text{Normal component} \qquad (1)$$
$$\mathbf{X}_T(\mathbf{x}) \; = \mathbf{X}_1(\mathbf{x}) - \mathbf{n}_2^T(\mathbf{x})\mathbf{X}_N(\mathbf{x}) \qquad \text{Tangential component} \qquad (2)$$

The normal boundary conditions read:

$e(\mathbf{x}) \geq 0:$ either a gap or contact; (a)
$X_N(\mathbf{x}) \geq 0:$ either no normal traction,
or a compressive normal traction; (b)
$e(\mathbf{x})X_N(\mathbf{x}) = 0:$ in contact, X_N may be positive, $e(\mathbf{x}) = 0$
outside contact, the normal traction vanishes, $e(\mathbf{x}) \geq 0$ (c)
(3)

The tangential boundary conditions are as follows.

Usually, in wheel-rail contact, the friction is dry. Friction is called dry when a traction bound $g(\mathbf{x})$ can be distinguished:

- $|\mathbf{X}_T(\mathbf{x})| < g(\mathbf{x})$ implies that the slip $\mathbf{s}(\mathbf{x}) = 0$;

- $\mathbf{s}(\mathbf{x}) \neq 0$ can happen only when $|\mathbf{X}_T(\mathbf{x})| = g(\mathbf{x})$, and then $\mathbf{s}$ and $\mathbf{X}_T$ are oppositely directed.

- Note that $\mathbf{X}_T$ is exerted on body 1, and $\mathbf{s} = (\text{velocity})_1 - (\text{velocity})_2$

- $|\mathbf{X}_T(\mathbf{x})| > g(\mathbf{x})$ is impossible.

We can formulate this in the following manner:

$$\mathbf{s}(\mathbf{x}) = -q(\mathbf{x})X_T(\mathbf{x}); \quad q \geq 0; \quad g^2 - \mathbf{X}_T^T\mathbf{X}_T \geq 0; \quad q\{g^2 - \mathbf{X}_T^T\mathbf{X}_T\} = 0 \quad (4)$$

which can be easily verified from what went before. One of the most popular expressions for $g(\mathbf{x})$ is

$$g(\mathbf{x}) = fX_N(\mathbf{x}), \quad (5)$$

with f the coefficient of friction which may be variable.

6 The half-space approximation and its consequences [2]

It will be clear that the normal traction influences the tangential traction, if only through the traction bound g. But then it is of crucial importance whether the tangential traction influences the normal traction or not. If it does, it is necessary to calculate the normal traction and the tangential traction simultaneously. If the tangential traction does not influence the normal traction, one can calculate the normal traction first, and then the tangential traction. This is much simpler and faster than the simultaneous calculation of normal and tangential traction. Also, the very important Fastsim algorithm does not work when the separation of the calculation of normal and tangential traction cannot take place.

We will show that in so-called tread contact of wheel-rail theory the normal traction is independent of the tangential traction. Consider Fig. 6.1. In it are shown the profiles of a wheel and a rail. The contact region has an area of about 1 cm² or, the contact region has a radius of .5 cm. The flat part of the tread has a length of about 8 cm. The depth is about 6 cm. This means the following:

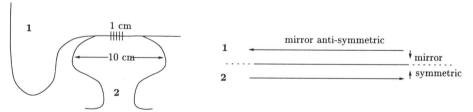

Fig.6.1. The half-space approximation

Fig.6.2. Mirror-antisymmetric and mirror-symmetric loading

1) We determine the boundary conditions for the real wheel and rail;

2) We calculate the elastic field for two half-spaces whose separating plane coincides with the tread surface, see Fig. 6.2. We give no proof of this important assumption.

This is called the half-space approximation. It will be used throughout this paper. Now we consider the half-spaces, see Fig.6.2.

Wheel and rail are both made of steel, and so are the half-spaces. Consider a mirror-antisymmetric loading of the half-spaces. This loading is typical of a tangential loading. Owing to the fact that both half-spaces are made of the same material, each half-space is loaded in accordance to the laws of elastostatics. The normal traction, on the other hand, is mirror symmetric, so no normal traction is generated by any tangential traction. This means that there are two traction systems that obey the laws of elastostatics and form equilibrium systems: the mirror-symmetric system about the separating plane of the two half-spaces, which is governed by the normal traction X_N, and the mirror-antisymmetric system about the same plane, which is governed by the tangential traction $\mathbf{X}_T$.

The calculation proceeds as follows: Set the tangential traction $\mathbf{X}_T = \mathbf{0}$; that is a frictionless loading; calculate the corresponding normal traction X_N (frictionless contact problem); calculate the final tangential traction $\mathbf{X}_T$, that does not influence X_N; READY.

6.1 The Hertz problem [2], [3]

The oldest elastic contact problem ever to be solved exactly is the Hertz problem. It dates from 1882. It concerns the frictionless (normal) problem of an elastic punch with a quadratic base that is pressed into an quadratic elastic surface. Since in wheel-rail contact the normal problem can be regarded as frictionless, the Hertz problem is of especial interest to wheel-rail theory. Indeed, very often one solves the wheel-rail normal problem by a Hertz problem. We give some key results of the Hertz theory.

The contact area is an ellipse with principal axes (a,b); the normal pressure $X(\mathbf{x}, t)$ vanishes outside the contact ellipse; the x-axis lies in the rolling direction, and inside contact the normal pressure is given by

$$X_N(\mathbf{x}, t) = p_0 \{1 - (x/a)^2 - (y/b)^2\}^{\frac{1}{2}} \tag{1}$$

The total normal, compressive force is denoted by F_N. For more details, we refer to [2],[3],[4].

7 The true boundary conditions [2]

There are two schools of classical thought in wheel-rail theory, and, indeed in contact mechanics. The first is centered about the Finite Element Method (FEM). The second school is centered about the half-space approximation and consequent simplifications. It is this latter school of thought that is represented in this paper.

Both schools have advantages and disadvantages. The half-space school yields relatively fast results that are exact within their limitation, which is the half-space approximation. The FEM yields results that cost much time, but its geometry is more flexible. In order to give some idea of the FEM school we present the boundary conditions of wheel-rail contact that are satisfied by the FEM, and which we have termed the true boundary conditions, and compare them to those of the half-space theory.

7.1 The wheel, i.e. body 1, is mounted on a rigid axle, with local coordinates $\mathbf{y}_1$. The displacement $\mathbf{w}_1$ vanishes on the rigid axle.

7.2 On the remainder of the wheel, with the exception of the potential contact area, see 7.4, the tractions vanish: $\mathbf{X}_1 = 0$.

7.3 At the foot of the rail, i.e. body 2, $\mathbf{w}_2 = 0$, and on the free surface of the rail, with the exception of the potential contact area, the traction vanishes: $\mathbf{X}_2 = 0$.

7.4 In the potential contact area B_t, which is a region encompassing the contact area, we have, see (5.3), (5.4):

$$e \geq 0, \quad X_N \geq 0, \quad eX_N = 0 \tag{1}$$
$$g^2 - (\mathbf{X}^T \mathbf{X}^T) \geq 0, \quad q \geq 0, \quad q\{g^2 - (\mathbf{X}_T^T \mathbf{X}_T)\} = 0; \tag{2}$$
$$\mathbf{s} = -q\mathbf{X}_T \tag{3}$$

Note that the half-space boundary conditions are given by Point 7.4 alone. e and $\mathbf{s}$ are 1st degree functions of all X_N resp. all $\mathbf{X}_T$. The algorithm Contact solves (7.1), (7.2) and (7.3) directly.

REMARK. There is a very good reason for the introduction of the potential contact area.

1) For the FEM: Outside the potential contact area, boundary conditions of a non-contact nature are active. Here, the potential contact area serves as a delimiter of the contact boundary conditions. Note that it follows from 7.4 that contact conditions prevail not only in the (a priori unknown) contact area, but also in the (a priori known) potential contact area.

2) For the half-space: Outside the contact area B_c, and, a fortiori, outside the potential contact area B_t, the surface traction vanishes. Now, as we will see later on, the displacement, the slip and the deformed distance can be expressed in the surface traction by 1st degree functions. Tractions outside the contact area and the potential contact area are therefore irrelevant. So we have:

a) B_t can be arbitrary, but it must include B_c

b) With the same number of elements, a smaller B_t yields a finer discretization, and therefore a more accurate result. Therefore B_t should be as small as possible while still encompassing B_c.

PART 2: ALGORITHMS

The preceding algorithms are very slow. To give a practical example, the con93 (=Contact) algorithm costs 1/2 second per case for an acceptable accuracy ($N = 100$; accuracy: 1 %) on an HP 9000-735. A FEM algorithm, by the way, is at least 100 times as slow. Later on in this paper we will express the speed of various algorithms into each other.

At any rate, at the present date (1996) the HP 9000-735 is a fast computer, but 1/2 sec per case is too slow for the calculation of practical problems.

In 1980, British Rail came with the idea to calculate a table book with the aid of the Duvorol code. They did not disseminate their table book, so that the remainder of the railway world had to find a substitute. Notably the Simplified Theory [2][4] and the Fastsim algorithm [2][5] were popular. These are methods that are essentially approximate in the sense that the error stems not only from the discretization, but also from the model. Indeed, Fastsim has a maximum error of 15 % (rarely attained), but it is 2000 times as fast as con93.

Of late, key customers started to complain about Fastsim. It was too slow, said one, too inaccurate, another, it lacked the torsional moment, a third. On the other hand a large memory space used was no object. Note that Fastsim occupied 40 kB, and con93 350 kB. So we set out to make a large Table Book [17] in which one could interpolate linearly. It turned out to be a table of 115,000 entries.

We expected several failures of con93 during the calculation of these 115,000 cases. Such a failure could not be tolerated, however. So we kept an account of the calculated cases; when a failure came, con93 ran into an alarm, but we knew exactly at which case this occurred. Then we changed to another method of calculation, for just this case, which, hopefully, would yield the correct result. When it did, we took up the original calculation right behind the former error.

We actually encountered 4 errors during the 115,000 cases, i.e. one in every 40,000. For all of them, the procedure outlined above worked well. We have now a table book with service programs to match; it is called Usetab and occupies 4.5 MB. A case takes 1/8 x fastsim, 1/15,000 x con93.

There is another type of algorithm that deserves mention. It is based on the Vermeulen-Johnson-Kalker [6][2] algorithm, which is confined to longitudinal and lateral creepage. To understand these terms, it should be comprehended that in almost all of these wheel-rail algorithms, the creep $\mathbf{c}$ is represented as a plane motion with three degrees of freedom, viz. longitudinal creep c_x in the rolling direction, lateral creep c_y in the lateral direction, and spin creep m_z which is a rotation about the z-axis. "creepage" or "creep ratio" is the term employed for $\gamma = \mathbf{c}/|\mathbf{v}|$ ($\mathbf{v}$: rolling velocity).

The algorithm of Vermeulen-Johnson-Kalker cannot handle spin, but it handles the spinless case to perfection.

The Vermeulen-Johnson type algorithms do not employ a discretization, but have the character of curve fitting. There have been a number of attempts to generalize Vermeulen-Johnson-Kalker to the complete case with spin. We mention Shen-Hedrick-Elkins, extensively discussed in [2]. Actually, this type of algorithm which are intended for unrestricted creepage and spin is very fast, but also very inaccurate.

We summarize the properties of Con93, Fastsim, and Usetab.

	ERROR	TIME/CALL	MEMORY SPACE
CON93	1%	0.35 sec	350 kB
FASTSIM	sometimes 15%	0.180 m sec	40 kB
USETAB	1.5%	0.024 m sec	4.5 MB

It would appear from this Table that FASTSIM does not measure up. This is indeed true, but sometimes it is used in USETAB.

For the sake of completeness we mention the linear theory of rolling contact, which occurs when the area of adhesion concides with the interior of the contact area. Under these circumstances, the local tangential force **F** is given by

$$\mathbf{F} = GD\gamma ,$$

with

$\mathbf{F}$ $\quad = (F_x, F_y, M_z)$, total force and moment
G $\quad =$ modulus of rigidity
D $\quad =$ matrix of creepage coefficients
γ $\quad = c/|\mathbf{v}|$, the creepage ($=$ creep/rolling velocity) $= (\nu_x, \nu_y, \varphi)$

The linear theory has the form

$$\begin{pmatrix} F_x \\ F_y \\ M_z \end{pmatrix} = \begin{pmatrix} C_{11} & 0 & 0 \\ 0 & C_{22} & C_{23} \\ 0 & -C_{23} & C_{33} \end{pmatrix} \begin{pmatrix} \nu_x \\ \nu_y \\ \varphi \end{pmatrix} \quad (C_{11}, C_{22}, C_{23}, C_{33}) > (0,0,0,0)$$

A generalization is found if, starting from a certain finite creep, one perturbs the force while keeping the area of adhesion fixed. One determines

"Sensitivities" $\partial F(\gamma, B_a)/\partial \gamma$ [2]

Another generalization is the theory of Knothe, Kalker & Gross-Thebing which are sensitivities for complex creepages [2].

PART 3: APPLICATIONS

8 The high-speed train

Vehicle dynamics and wheel-rail theory were developed simultaneously with computer technology. In vehicle dynamics, the equations of motion of a vehicle are set up by hand, and later by computer, and solved, mostly numerically by computer. As the wheel-rail force is the most important single force that acts on a rail vehicle, wheel-rail theory is fundamental in vehicle dynamics. Wheel-rail theory was explored simultaneously, but by another person, so that the vehicle dynamicist was not distracted by the study of contact mechanics.

We will consider the application of vehicle dynamics on the theory of the high-speed train.

By now, (1996), the high speed train is commonplace. In 1965, this was not so: only one high-speed train was in existence, viz. the Japanese Shinkansen. The theoretical work for the Shinkansen was performed by Mr. Matsudaira. At the time, no practical wheel-rail theory was in existence; there only existed the notion, due to Carter (1926), [16] that for small creep the creep-force law was linear. Matsudaira's work was linear in character, and this fitted well with this concept. Matsudaira made a linear law, and later Kalker and de Pater (1962), [15] did the same, which latter law had a sound theoretical background, and was much used by later investigators.

The work of Matsudaira centred on the so-called critical velocity, that is, the velocity at which the motion of the bogie becomes unstable. The investigators de Pater (TU Delft) and Wickens (British Rail) later investigated whether the critical velocity could become infinite by a special construction of the bogie. This question was finally resolved in the affirmative, which virtually ended the research into the critical velocity.

All this work was performed under the supposition that the track was straight. De Pater and Wickens also investigated the stability of the bogie on curved track.

By then (1985), the stability of a railway bogie was well understood, whatever its velocity and whatever the track and attention was directed toward the actual motion of the bogie. Technically interesting trains based on these techniques were: The Japanese Shinkansen, the British APT, the British HST, the French TGV, the German ICE, etc.

The construction of a fast train had become a commonplace matter.

9 Wheel-rail wear

Let us start with a definition.
Wear is the undesired destruction of an object by a technological process.
A rail may be destroyed by so-called spalling, which is the origination of a long crack parallel to the running direction. Other formations of cracks in the rail are headchecks and squats. These formations are pretty destructive, which is one of the hallmarks of wear. The common property of these forms of wear is that they

start suddenly. There is very little theory about these sudden forms of wear.

A slow form of wear of wheels and rails is the abrasive form. In abrasive wear, small particles of steel are pried loose from wheel and rail, so that these slowly, and usually undesirably, change form. In Archard's theory of friction exactly the same happens, which is the reason that abrasive wear and friction are brought into connection with each other. Indeed, if W is the frictional work and M is the mass of the particles pried loose by wear, then W and M are proportional with a constant C:

$$M = C * W \qquad \text{(linear wear law)} \qquad (1)$$

We will confine ourselves to this linear wear law.

The phenomena due to the linear wear law can be divided into two, *viz.* irregular wear and regular wear.

10 Irregular wear: corrugation [7][8]

In irregular wear the tread of wheel and rail wear irregularly in the rolling direction (corrugation). Ideally, a sinusoidal wave comes into being whose amplitude grows with the number of wheel passages over the spot that is considered. In a more sophisticated model the corrugation is non-harmonic. Corrugation can take place on rail and wheel: in the latter case it is called polygonalisation. Corrugation causes noise, and premature wear of the vehicles. It is combated in practice by grinding the wheels and the rails, a process which has to be repeated at a certain frequency. It is the purpose of the investigation to increase the permitted grinding frequency.

Corrugation has been studied principally by Knothe of the TU Berlin. He started his investigations by considering the birth of the corrugation, that is, he started from a nominally flat rail tread, let (theoretically) run a wheel over it, and investigated whether corrugation would form or not. He used his own Knothe-Kalker-Gross-Thebing non-steady state wheel-rail theory which uses complex sensitivities. This investigation is completed.

As a second entry into the theory of rail corrugation, Kalker used a harmonic corrugation field, let (theoretically) run a wheel over it, and investigated whether the original corrugation field grew or decreased. He used the Vermeulen-Johnson-Kalker wheel-rail theory of steady state rolling without spin creep, and also, more realistically but more slowly, the programme Con93 for non-steady state rolling. This investigation is only partially complete, as a fast computer was not available at the time.

11 Regular wear: profile formation [9]

In regular wear corrugation phenomena are assumed to be entirely absent. The tread of wheel and rail is smooth.

Due to the passage of wheels over a certain spot of the rail (rail profile) or of a fixed wheel over an entire field of rails (wheel profile) the form of rail and wheel change slowly, until after about 300,000 km distance traversed the wheel and the rail are so far deformed that they must be ground.

It is the object of the investigation to start with such a profile that this distance of 300,000 km is increased. To that end we use simulation, as experimentation takes much too long.

A first attempt was made by Kalker in 1984. He used CONTACT, but in a version which was slow, while also the computer used was not very fast. He traversed about 100 km.

A second attempt was made by Kalker and Chudzikiewicz. They abandoned CONTACT for Fastsim, as they found on the basis of the above 100 km that Fastsim could very well be employed, the error being of the order of 10%. Kalker and Chudzikiewicz also did not come very far. A third attempt was made by Kalker and Wiersma. The programme of Kalker and Chuzikiewicz was thoroughly rewritten, and now the 300,000 km boundary was easily attained. This programme and the preceding ones were programmes for the tread; the flange behaviour was still beyond our potentialities. Also, the influence of the motion of the wheel set is still imperfectly understood. At present (1996) the investigation has been taken up by Li and Kalker.

12 Roller and ball bearings: the normal 3D contact problem [2]

It will be recalled that the Hertz solution is used for the normal contact problem in wheel-rail theory was taken. The algorithms which were treated in Part 2 are all such wheel-rail algorithms with the exception of Contact which is more general.

The first normal contact theory dates from 1972 [10]. Shortly thereafter, it was followed by Line contact theory, by which the normal contact for slender and pointed contact areas could be treated (1972, [11]). Then, around 1978, I was approached by the SKF company with the request for a 3D normal contact code. I jumped at the chance, and between 1978 and 1982 I made a sequence of 4 normal contact codes, of which the last smoothly went over into my program Contact.

One unpleasantness marred an otherwise ideal situation: In a roller bearing, the contact area is very long, actually too long to be modelled by the half-space approximation, and this showed in the results. Neither I myself, nor the people of SKF were ever able to put this right. But I still have hopes.

13 Copiers: 2D Layered media [12][13]

In 1984 I was approached by Océ-Nederland with the request to solve the 2D normal and tangential contact problem of a two-layer cylinder pair. Again I was happy to comply, and within a few months I wrote the required code, which I called "Laagrol". A long silence ensued, but in 1988 Dr. Saes of Océ came to me and told me that he had made Laagrol more user friendly, that he had made propaganda for it, and that it was now highly popular at Océ. Did I feel like to extend Laagrol to multilayered, viscoelastic cylinders? It seemed a pleasant challenge to me, and I wrote the blueprint for what was later to become the Multilayer programme [12]. Multilayer was subsequently encoded by my Ph.D. student Gerard Braat[13]. Owing to its superb user-friendliness, Multilayer became a great success at Océ, even when it was no longer supported by Gerard Braat.

Typical of Laagrol and Multilayer was, that the tangential traction influences the normal traction, something that hardly occurs in wheel-rail contact theory. Under those conditions, one starts by setting $\mathbf{X}_T = 0$; then we determine X_N; then we determine $\mathbf{X}_T$; then we determine X_N, as the last X_N, was perturbed; then $\mathbf{X}_T$ again, and so on, till convergence hopefully occurs [14]. This fortunately happens quite often. When it does not, it suffices for convergence to perturb the discretization, and to start again.

14 Conclusion

This paper consists of three parts.

- In the first part, we analyze the contact problem on the basis of three Cartesian coordinate systems that describe the motion of two deformable bodies.

- In the second part, we enumerate a number of algorithms.

- In the third part, we mention three wheel-rail applications which are contact evolutions, and which require very fast contact algorithms. The last two relate the construction of two algorithms which have features that are not usual in wheel-rail theory.

References

[1] Shabane, A.A., Dynamics of Multibody Systems, Wiley, 1989

[2] Kalker, J.J.,Three-Dimensional Elastic Bodies in Rolling Contact, Kluwer Academic Publishers, p. 314, 1990.

[3] Hertz, H., Über die Berührung fester elastischer Körper und über die Härte. In: H. Hertz, Gesammelte Werke, Band 1, Leipzig: J.A. Barth (1895) p. 174-196.

[4] Kalker, J.J., Simplified theory of rolling contact, Delft Progress Report **1**, pp. 1-10, 1973.

[5] Kalker, J.J., A fast algorithm for the simplified theory of rolling contact, Vehicle System Dynamics **11**, pp. 1-13, 1982.

[6] Vermeulen, P.J. and K.L. Johnson, Contact of nonsperical bodies transmitting tangential forces, Journal of Applied Mechanics, **31**, pp. 338-340, 1964.

[7] Knothe, K. and B. Ripke, The effect of the parameters of wheelset, track and running conditions on the growth rate of rail corrugations, pp. 345-356, Institüt für Luft- und Raumfahrt.

[8] Kalker, J.J., Considerations on Rail Corrugation, Vehicle System Dynamics, **23**, pp. 3-28, 1994.

[9] Kalker, J.J., Simulation of the development of a railway wheel profile through wear, Wear, **150**, pp. 355-365, 1991.

[10] Kalker, J.J. and Y. van Randen, A minimum principle for frictionless elastic contact with application to non-Hertzian half-space contact problems, Journal of Engineering Mathematics, **88**, pp. 193-206, 1972.

[11] Kalker, J.J., On elastic line contact, Journal of Applied Mechanics, **39**, pp. 1125-1132, 1972.

[12] Kalker, J.J., Viscoelastic multilayered cylinders rolling with dry friction, Journal of Applied Mechanics, **58**, pp. 666-679, 1991.

[13] Braat, G.F.M., Theory and Experiments on Layered, Viscoelastic Cylinders in Rolling Contact, Thesis Delft University of Technology, pp. 179, 1993.

[14] Panagiotopoulos, P.D., A nonlinear programming approach to the unilateral contact and friction-boundary value problem in the theory of elasticity. Ingenieur-Archiv, **44**, pp. 421-432, 1975.

[15] Kalker, J.J., On the rolling contact of two elastic bodies in the presence of dry friction, Thesis 1967.

[16] Carter F.W., On the action of a locomotive driving wheel, Proceedings Royal Society of London, **A**112, pp. 151,157, 1926.

[17] Kalker, J.J., Book of Tables for the Hertzian Creep-Force Law, Delft University of Technology, Report: 96-61, 1996.

Wear profiles and the dynamical simulation of wheel-rail systems

Martin Arnold*   Helmuth Netter[†,‡]

Abstract

Recently, a quasi-elastic model for the contact between two rigid bodies was introduced, that is tailored to the contact between a wheel with wear profile and a rail ([1]). This new approach to the geometrical description of wheel-rail contact is used successfully in the dynamical simulation of wheel-rail systems that have wheels with wear profiles. In the present paper we discuss the application of the quasi-elastic contact model in the dynamical simulation and the efficient numerical solution of the model equations.

1 Introduction

Simulation tools get more and more important in the design and analysis process of modern advanced railway vehicle developments. Typical applications are parameter studies of the system behaviour in time or frequency domain to get an idea of the sensitivity of the design w. r. t. system parameters.

In this field the multibody system (MBS) approach is well established, i. e. the wheel-rail system is modelled as system of rigid or elastic bodies which are connected by massless joints, springs, dampers or active control elements. MBS models are composed of model components that are more or less self-contained. Among the model components that are specific for wheel-rail systems the most important one consists of conditions for the geometrical contact of wheel and rail and of a model for the friction forces between wheel and rail. Both problems are closely connected.

The traditional description of the wheel-rail contact in a MBS model is based on the geometry of two rigid bodies (wheel and rail) that touch in one single point,

---

*University of Rostock, Department of Mathematics, Postfach, D – 18051 Rostock, Germany, e-mail: arnold@mathematik.uni-rostock.d400.de

[†]Institute for Robotics and System Dynamics, DLR Oberpfaffenhofen, Postfach 1116, D – 82230 Wessling, Germany, e-mail: Helmuth.Netter@dlr.de

[‡]The work of both authors has been supported by the German Federal Minister for Education, Science, Research, and Technology (grant FR7ROK).

in the *contact point*. In real life, however, wheel and rail are elastically deformed in a neighbourhood of this contact point and there is a *contact patch* between wheel and rail. There are various models to compute the friction forces that act in this contact patch (see [3] for a comprehensive study of this subject).

From the numerical point of view a model for the contact of wheel and rail is attractive if

(M1) the condition for geometrical contact between wheel and rail is formulated as algebraic constraint (this avoids high-frequency oscillations in the solution that are typical of pure elastic models) and

(M2) the undeformed surfaces of wheel and rail are locally approximated by paraboloids so that a generalized Hertzian contact model can be applied (in this model the normal stress distribution in the contact patch is given in closed form using elliptic integrals).

If the assumptions (M1) and (M2) are satisfied then the motion of the wheel-rail system is described by the differential-algebraic system

$$\begin{aligned} M(q)\ddot{q} &= f(q,\dot{q},\lambda,t) - G^T(q)\lambda \\ 0 &= g(q) \end{aligned} \quad (1)$$

in the position coordinates $q(t)$ (see e. g. [7]). The Lagrangian multipliers $\lambda(t)$ couple the contact conditions $g(q) = 0$ with the dynamical equations. The constraint forces $-G^T(q)\lambda$ with $G(q) := \frac{\partial}{\partial q} g(q)$ guarantee that the constraints $g = 0$ are always satisfied. The vector $f(q,\dot{q},\lambda,t)$ consists of applied forces including the friction forces that depend on $-G^T(q)\lambda$. $M(q)$ denotes the (symmetric and positive definite) mass matrix.

The classical Hertzian contact model that has been used in various simulation packages results in efficient simulation tools for wheel-rail systems with conic wheels (that are frequently considered in theoretical investigations). However, it has been known since the early eighties (e. g. [4]) that this classical Hertzian contact model may result in large model errors if the wheels have a *wear profile*. These wheels are commonly used by European railway companies.

Recently, an alternative model for wheel-rail contact was proposed that considers the geometry of the (undeformed) surfaces of wheel and rail not only in the contact point but in the whole contact patch ([1]). This *quasi-elastic contact model* satisfies assumptions (M1) and (M2). It can be used both for systems with conic wheels and for systems having wheels with wear profiles. In the present paper we sketch the application of the quasi-elastic contact model in the industrial simulation package SIMPACK. In Section 2 we introduce basic notations and illustrate, why the classical approach fails for wear profiles. The quasi-elastic contact model is defined in Section 3. The numerical integration of the model equations (1) and simulation results for a benchmark of Pascal ([5]) are discussed in Section 4.

2   The geometry of wheel-rail contact

The basic unit of wheel-rail contact in the MBS model is the contact between one single wheel and a rail (Fig. 1). This allows the simulation both for wheel-rail systems with classical rigid wheelsets and for modern advanced railway vehicles without rigid coupling between the wheels of a wheelset. In local systems of coordinates W and R the undeformed surfaces of wheel and rail are given by industrial standards. The wheel is a body of revolution, its surface

$$\{\,(\,F(s)\sin\tau\,,\,s\,,\,F(s)\cos\tau\,)^T\,:\,\tau\in[0,2\pi),\,\underline{s}\leq s\leq\overline{s}\,\}$$

can be defined by a scalar *profile function* F that depends on the wheel coordinate s (in W the s-axis is parallel to the wheel axle). At the rail a system R of cartesian coordinates (u,v,w) is defined such that the u-axis is parallel to the track, the w-axis is oriented downwards and the origin of W lies in the (v,w)-plane (see Fig. 1). In general the rail profile may vary along the track (e. g. in switches), therefore the local geometry of the rail surface depends on the position x that the wheel has reached in its motion along the track. Locally, however, the changes of the profile in u-direction may be neglected and for a given position x the surface of the rail can be approximated by

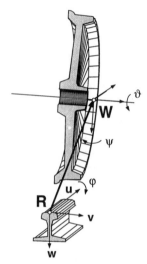

$$\{\,(\,u\,,\,v\,,\,G(v;x)\,)^T\,:\,u\in\mathbb{R},\,\underline{v}\leq v\leq\overline{v}\,\}$$

with a profile function G that is independent of u. In special cases, e. g. in motions along a straight track with non-varying profile, the rail profile does not vary during the dynamical simulation, i. e. $G=G(v)$.

Figure 1: Geometry of wheel and rail.

As rigid body the wheel has 6 degrees of freedom and its relative position to the rail is determined by the coordinates (ξ_u,ξ_v,ξ_w) of the origin of W in R and by 3 angles of rotation (φ,ϑ,ψ) that are defined such that wheel axle, s-axis, and v-axis are parallel if $\varphi=\psi=0$ (R was defined such that $\xi_u=0$, see also Fig. 1). In R the wheel surface is given by

$$\{\,(u,v,w)^T:=(0,\xi_v,\xi_w)^T+A(\varphi,\psi)\cdot(F(s)\sin\tau,s,F(s)\cos\tau)^T\,\} \qquad (2)$$

with a matrix $A(\varphi,\psi)$ that represents the rotations around the u- and the w-axis.

The contact condition determines the vertical displacement $\xi_w=\xi_w(\xi_v,\varphi,\psi;x)$ of the wheel as function of ξ_v, φ, and ψ. To be more specific we assign to each point P_W of the wheel surface a point $P_r(P_W)$ of the rail surface by projection along the w-axis. For a classical contact point P_W^* the points P_W^* and $P_r(P_W^*)$ coincide and the normal vector $n_W(P_W^*)$ to the wheel surface at P_W^* is parallel to the normal

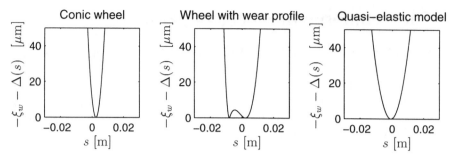

Figure 2: Distance between wheel and rail for points $P_\mathrm{w}(s;\psi) \in \mathfrak{C}$: Rail profile UIC60–ORE, $\xi_v = 9.67\,\mathrm{mm}$, $\varphi = 1.43°$, $\psi = 0°$. left plot: conic wheel, central plot: wheel profile S1002, right plot: distance between the paraboloids that approximate wheel (S1002) and rail in the generalized Hertzian contact model ($\nu = 2 \cdot 10^{-5}$).

vector $n_\mathrm{r}(P_\mathrm{r}(P_\mathrm{w}^*))$ to the rail surface at $P_\mathrm{r}(P_\mathrm{w}^*)$ (otherwise wheel and rail would penetrate each other).

With our assumptions all normal vectors n_r to the *rail* surface are parallel to the (v,w)–plane since G is independent of u. We define the curve $\mathfrak{C}$ as the curve of all points P_W on the *wheel* surface with normal vectors $n_\mathrm{W}(P_\mathrm{W})$ that are also parallel to the (v,w)–plane (i. e. $P_\mathrm{W}^* \in \mathfrak{C}$). In the special case $\psi = 0$ this curve $\mathfrak{C}$ consists of all points P_W that are directly below the centre of the wheel axle. In general $\mathfrak{C}$ depends on ψ and can be parametrized by the wheel coordinate s: $\mathfrak{C} = \mathfrak{C}(\psi) = \{\, P_\mathrm{W}(s;\psi) : \underline{s} \leq s \leq \overline{s}\,\}$. On $\mathfrak{C}$ we define the *height function*

$$\Delta(s;\xi_v,\varphi,\psi,x) := e_3^T \cdot (\, P_\mathrm{W}(s;\psi) - P_\mathrm{r}(P_\mathrm{W}(s;\psi))\,) - \xi_w \tag{3}$$

that depends on ξ_v, φ, ψ, and x but is independent of ξ_w (see (2)).

The distance between $P_\mathrm{W}(s;\psi) \in \mathfrak{C}$ and $P_\mathrm{r}(P_\mathrm{W})$ is given by $\xi_w + \Delta(s;\xi_v,\varphi,\psi,x)$ (the distance is measured along the w-axis that is oriented downwards). Wheel and rail do not penetrate iff $\xi_w + \Delta(s;\xi_v,\varphi,\psi,x) \leq 0$ is satisfied for all $P_\mathrm{W}(s;\psi) \in \mathfrak{C}$. Furthermore the undeformed bodies of wheel and rail are in contact iff

$$\xi_w + \max_s \Delta(s;\xi_v,\varphi,\psi,x) = 0\,. \tag{4}$$

In this classical contact model the contact point P_W^* is a point $P_\mathrm{W}^* = P_\mathrm{W}(s^*;\psi) \in \mathfrak{C}$ with $\Delta(s^*;\xi_v,\varphi,\psi,x) = \max_s \Delta(s;\xi_v,\varphi,\psi,x)$. Since the relative position of wheel and rail is determined by the MBS coordinates q this rigid body contact condition defines an algebraic constraint $g_i(q) = 0$. For a wheel-rail system with l wheels we get in (1) the constraint $0 = g(q) = (g_1(q), \ldots, g_l(q))^T$.

The critical point is the differentiability of g (g has to be at least two times continuously differentiable to get a continuous solution $q(t)$, $\lambda(t)$ of (1)). For conic wheels the height function has only one local maximum (Fig. 2 left) and the position s^* of the contact point varies smoothly with ξ_v, φ, and ψ. Therefore (4) defines in general a sufficiently differentiable constraint.

The central diagram of Fig. 2 shows the corresponding plot for a wheel with wear profile. These profiles are defined on the basis of measured data from wheels that are worn after frequent use. Therefore the surfaces of wheel and rail are approximately parallel over a large area of the tread. For the configuration in Fig. 2 (that is close to the nominal position of a wheel of a rigid wheelset on a straight track with rails inclined 1/40) there are two contact points. During the dynamical simulation jumps of the contact point occur and the constraint of the rigid body contact model is continuous, but only piecewise differentiable.

3 The quasi-elastic contact model

In [1] regularizations of the rigid body contact model (4) were discussed that use the information about the geometry of wheel and rail surface not only in one or more isolated point(s) but over the whole contact patch. Here we consider the *quasi-elastic contact model*

$$\xi_w + \mathrm{smax}_s^{(\nu)} \Delta(s; \xi_v, \varphi, \psi, x) = 0 \tag{5}$$

with

$$\mathrm{smax}_s^{(\nu)} \Delta(s; \xi_v, \varphi, \psi, x) := \nu \ln\Big(\int_{\mathfrak{C}} \exp(\frac{1}{\nu}\Delta(s; \xi_v, \varphi, \psi, x))\, ds \,/ \int_{\mathfrak{C}} ds \,\Big) \tag{6}$$

and a small positive parameter ν. In the limit $\nu \to 0$ function $\mathrm{smax}_s^{(\nu)} \Delta$ converges (pointwise) to the maximum $\max_s \Delta$ but in contrast to (4) condition (5) defines for any $\nu > 0$ a sufficiently differentiable constraint (if Δ is sufficiently differentiable). In MBS coordinates the quasi-elastic contact condition (5) results in an algebraic constraint $g_i^{(\nu)}(q) = 0$, i. e. assumption (M1) of the Introduction is satisfied.

If $\nu > 0$ is small then the parts of $\mathfrak{C}$ that do not belong to the contact patch are negligible in $\int_{\mathfrak{C}} \exp(\frac{1}{\nu}\Delta)\, ds$ since $\exp(\frac{1}{\nu}\Delta)$ grows extremely fast if Δ is increased. We always have $\mathrm{smax}_s^{(\nu)} \Delta(s; \xi_v, \varphi, \psi, x) \leq \max_s \Delta(s; \xi_v, \varphi, \psi, x)$, i. e. (5) defines the vertical displacement ξ_w such that wheel and rail as rigid bodies would penetrate each other. The parameter ν in (6) is chosen so that $\max_s \Delta - \mathrm{smax}_s^{(\nu)} \Delta$ is of the same size as the elastic approach δ in a pure elastic model. (δ depends on the axle load, typical values are in the range of $10 \ldots 100\,\mu$m.) For profiles S1002 (wheel) and UIC60–ORE (rail) appropriate values of ν are in the range $10^{-5} \ldots 5 \cdot 10^{-5}$, in the actual computations we used $\nu = 2 \cdot 10^{-5}$. Then ξ_w in (5) is close to the vertical displacement of the wheel in a pure elastic contact model ("quasi-elastic" contact model).

In a rigid body contact model it is natural to suppose that constraint forces and friction forces act in the contact point P_{W}^*. The straightforward generalization to the quasi-elastic contact model considers constraint forces and friction forces that act in a point $P_{\mathrm{W}}(\tilde{s}; \psi) \in \mathfrak{C}$ where $\tilde{s}$ is defined as weighted mean value of s along $\mathfrak{C}$ with weights $\exp(\frac{1}{\nu}\Delta(s; \xi_v, \varphi, \psi, x))$:

$$\tilde{s} := \int_{\mathfrak{C}} s \cdot \exp(\frac{1}{\nu}\Delta(s;\xi_v,\varphi,\psi,x))\,ds \Big/ \int_{\mathfrak{C}} \exp(\frac{1}{\nu}\Delta(s;\xi_v,\varphi,\psi,x))\,ds\,. \qquad (7)$$

To get the friction forces we need an efficient method to compute the contact patch and the normal stress distribution in this contact patch (see (M2)). For conic wheels the classical Hertzian contact model can be used with paraboloids that approximate the surfaces of wheel and rail in a neighbourhood of the contact point. This is roughly illustrated by the left plot of Fig. 2 since the height function Δ coincides approximately with a parabola that has the contact point as apex. The central plot of Fig. 2 proves that the classical Hertzian contact model may result in large errors if the wheel has a wear profile. Here paraboloids with a contact point P_W^* as apex give only a very poor approximation of the surfaces of wheel and rail (see e. g. [4] for a more detailed discussion of non-Hertzian contact between a wheel with wear profile and a rail).

Eq. (7) motivates a generalization of the classical Hertzian contact model: the surfaces are approximated by paraboloids, but now with $P_W(\tilde{s};\psi)$ as apex. The curvatures of the paraboloids are defined as weighted mean values of the curvatures of wheel surface and rail surface, respectively. Similar to (7) these mean values are computed along $\mathfrak{C}$ with weights $\exp(\frac{1}{\nu}\Delta)$. With this approach a reasonable approximation of the surfaces of wheel and rail in the contact patch is obtained. This is illustrated by the right plot of Fig. 2 that shows for the same configuration as in the central plot the height function if the surfaces of wheel and rail are substituted by these paraboloids ($\nu = 2 \cdot 10^{-5}$).

This generalized Hertzian contact model defines an *equivalent Hertzian contact patch* that can be obtained in the same way as in the classical Hertzian theory since assumption (M2) of the Introduction is satisfied. Software that has been developed originally for classical Hertzian contact patches (e. g. the code FASTSIM of Kalker [3]) can be carried over straightforwardly to compute the friction forces with an accuracy that is acceptable for most applications. Extensions to more complicated contact patches are currently under development.

4 The numerical solution of the model equations

To solve the model equations (1) efficiently the specific structure of the wheel-rail contact model in typical applications has to be exploited. Formulating the quasi-elastic contact model three essential points have been considered a priori: the assumptions (M1) and (M2) from the Introduction and the transition from the two dimensional surfaces of wheel and rail to the one dimensional curve $\mathfrak{C}$.

Eqs. (1) form a differential-algebraic system of index 3. Theory, numerical analysis and software for (1) are well developed (see [7] and the references therein). In the dynamical simulation the integral in (6) is discretized by the trapezoidal rule on a grid that is kept fixed during integration. With $\nu = 2 \cdot 10^{-5}$ the stepsize h_i

of the trapezoidal rule can be set to $h_i \approx 1\,\mathrm{mm}$ at the tread and $h_i \approx 0.2\,\mathrm{mm}$ at the flange of a wheel with wear profile S1002. Approximately 200 evaluations of Δ are necessary to compute one function value of $\mathrm{smax}_s^{(\nu)}$. The same discretization scheme is used to get the weighted mean values $\tilde{s}$, ...

This basic implementation of the quasi-elastic model can be used in a wide range of applications including rail profiles that vary along the track (e. g. in switches) and wheelsets without rigid coupling of the wheels. The numerical effort to evaluate (5) is, however, much larger than in the classical rigid contact model (4).

In the simulation package SIMPACK this basic implementation is used only, if the rail profile really varies along the track, i. e. if G depends on x. However, in most of the industrial applications simulations have to be performed for non-varying rail profiles (straight tracks, curved tracks with fixed curvature). In this important special case the functions in (5) and (7) depend on ξ_v, φ, and ψ, only. They can be approximated by polynomial 3D–tensorproduct splines such that during the dynamical simulation the cpu-time for the evaluation of the contact condition is reduced drastically. Spline grid and spline order are adapted to the geometrical data of wheel-rail contact ([2]).

For given profile functions F and G the spline coefficients have to be computed only once, the splines can then be used in the simulation again and again. In SIMPACK the spline coefficients are computed during the model setup in a pre-processing step, i. e. before starting the simulation (for details see [2]). The spline approximation of the contact conditions reduces the overall simulation time by the factor 5...12, it is used as default whenever the rail profile does not vary along the track. We refer to [6] for the discussion of typical industrial applications of SIMPACK.

To illustrate the benefits of the quasi-elastic contact model and the importance of efficient numerical methods we end this section with simulation results for a model problem:

Example 1 Pascal ([5]) proposed as benchmark problem the motion of a rigid wheelset along a straight track with a force $F_y = 20\,\mathrm{kN}$ that acts in lateral direction and shifts the wheelset to the right. In (1) the wheelset is described by 6 position coordinates ($q(t) \in \mathbb{R}^6$) and 2 constraints (right/left wheel), the profile functions are S1002 (wheel) and UIC60 (rail). The motion tends to a quasi-stationary state ($y \approx 5.0\,\mathrm{mm}$) that is at the right wheel close to a configuration where the contact point P_W^* jumps. Fig. 3 shows that the rigid body contact model fails in this benchmark. The non-differentiability of (4) introduces discontinuities in $\lambda(t)$ and $\dot{q}(t)$ and artificial oscillations in $q(t)$. For the quasi-elastic contact model the quasi-stationary state is reached. The trajectory that is obtained using the basic implementation coincides with the one that is based on the (3D–)approximation of the geometrical data. The spline approximation reduces, however, the cpu-time for the dynamical simulation from 497.1 s to 42.4 s on a SUN Sparc5 workstation (i. e. by the factor 12).

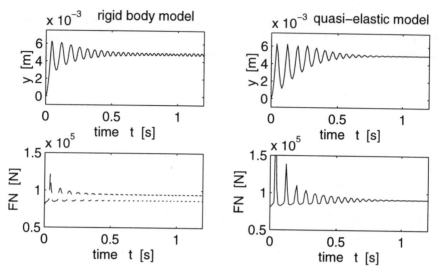

Figure 3: Rigid body contact model and quasi-elastic contact model applied to a benchmark of Pascal ([5]): lateral displacement y of the wheelset, constraint force FN at the right wheel.

References

[1] M. Arnold, K. Frischmuth, M. Hänler, and H. Netter. Differentialgleichungen und singuläre Mannigfaltigkeiten in der dynamischen Simulation von Rad–Schiene– Systemen. In K.-H. Hoffmann et al., editors, *Mathematik –Schlüsseltechnologie für die Zukunft*, Springer–Verlag, Berlin Heidelberg New York, (in press).

[2] M. Arnold and H. Netter. The approximation of contact conditions in the dynamical simulation of wheel-rail systems. Technical Report IB 515–96–08, German Aerospace Research Establishment, 1996.

[3] J.J. Kalker. *Three-Dimensional Elastic Bodies in Rolling Contact*. Kluwer Academic Publishers, Dordrecht Boston London, 1990.

[4] K. Knothe and H. Le The. A contribution to the calculation of the contact stress distribution between two elastic bodies of revolution with non-elliptic contact area. *Computers & Structures*, 18:1025–1033, 1984.

[5] J.P. Pascal. Benchmark to test wheel/rail contact forces. Technical report, INRETS Paris, 1990.

[6] W. Rulka, A. Haigermoser, L. Mauer, and H. Netter. Anwendung moderner Auslegungsstrategien für Schienenfahrzeuge durch Einsatz des Simulationsprogramms SIMPACK. VDI Berichte Nr. 1219, Düsseldorf, 1995.

[7] B. Simeon, C. Führer, and P. Rentrop. Differential-algebraic equations in vehicle system dynamics. *Surveys on Mathematics for Industry*, 1:1–37, 1991.

Dynamic response of a periodically supported railway track in case of a moving complex phasor excitation

István Zobory and Vilmos Zoller
Department of Railway Vehicles, Technical University of Budapest
H-1521 Budapest, Hungary, E-mail: `vilmos@kmf.hu`

Abstract

Most of the known solution methods of periodically supported beam problems involve application of infinite sums (see eg. [2] and references there). This fact makes it hard to investigate qualitative properties. In 1995 Droździel, Sowiński and Żochowski [1] proposed a new method in case of fixed loads, that involves only finite sums for obtaining the solution of the problem. Independently, the present authors and Zoltán Zábori [3] gave a closed–form solution in case of finitely many, altough inhomogeneous in position, stiffness and damping, supports. Combining the last two methods we give here a finite, closed–form solution for the periodically supported infinite Bernoulli–Euler beam problem, that works even in the case of moving time–dependent loads represented by moving complex phasors.

1 Introduction

The partial differential equation of the periodically supported infinite Bernoulli–Euler beam under moving load has the form

$$EI\frac{\partial^4 z}{\partial x^4} + \rho A\frac{\partial^2 z}{\partial t^2} + \sum_{j=-\infty}^{\infty}(m_s\ddot{Z}_j + k_b\dot{Z}_j + s_bZ_j)\delta(x-jL) = F_0 e^{wt}\delta(x-vt). \quad (1)$$

Equation (1) is coupled with the system of ordinary differential equations

$$m_s\ddot{Z}_j + k_b\dot{Z}_j + s_bZ_j = k_p\left(\frac{\partial}{\partial t}z(jL,t) - \dot{Z}_j\right) + s_p(z(jL,t) - Z_j), \quad j \in \mathbf{Z} \quad (2)$$

representing the pad/sleeper/ballast elements, where $z(x,t)$ is the vertical displacement of the beam, $Z_j(t)$ is the vertical position of the jth sleeper located at

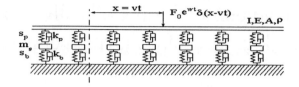

Figure 1: The periodically supported track model

$x = jL$, and w is a complex number generating the complex phasor excitation e^{wt}. Distributional system (1-2) must satisfy boundary condition

$$\lim_{|x|\to\infty} z(x,t) = 0. \tag{3}$$

The exterior excitation of the track is coming from the load, i.e. from solutions of the unsupported beam equation

$$EI\frac{\partial^4 z}{\partial x^4} + \rho A \frac{\partial^2 z}{\partial t^2} = F_0 e^{wt}\delta(x - vt).$$

If we have a fixed load $F_0 e^{wt}$ at $x = l_0$, then solutions of the unsupported beam equation take the form $e^{\epsilon_j \lambda x + wt}$, where $\epsilon_j := i^{j-1}$ is a fourth complex unit root for $j = 1,...,4$, while $\epsilon_j \lambda$ is any root of equation $EI\lambda^4 + \rho A w^2 = 0$, see e.g.[3]. The time response of the sleepers to it, and the response of the sleepers to each other are both of form e^{wt}, and the full response of the system over any of the free parts of the beam (between two consecutive sleepers or between the load and a neighbouring sleeper) is a linear combination of the four functions $e^{\epsilon_j \lambda x + wt}$, $j = 1,...,4$.

If the load is moving, then solutions of the unsupported beam equation have the form $e^{\lambda_i x + (w - \lambda_i v)t}$, time excitation of the track is of form $e^{(w - \lambda_i v)t}$, where the λ_i's are roots of polynomial

$$P(\lambda_i) := EI\lambda_i^4 + \rho A(w - \lambda_i v)^2 \tag{4}$$

for $i = 1,...,4$, see e.g.[4]. It means, that the excitation has the same form as if we had a superposition of four fixed loads $F_0 e^{(w - \lambda_i v t)}$. Solutions of equation $EI\lambda^4 + \rho A(w - \lambda_i v)^2 = 0$ are complex numbers $\epsilon_j \lambda_i$, $j = 1,...,4$. The dynamic response of the full system over any free regions of the beam is then a linear combination of the 16 functions $e^{\epsilon_j \lambda_i x + (w - \lambda_i v)t}$, $i,j = 1,...,4$.

In the following sections we fit together in a sufficiently smooth way these free part solutions into finite, closed-form solutions of the periodically supported beam problem (1-3) on the whole plane in case of static, fixed and moving loads, respectively.

2 Static load case

In this section we give another formulation and proof for a result from [1], and build up our method in the case when conditions $v = 0$ and $w = 0$ are satisfied, i.e. we have a static load F_0 positioned at $x = l_0 \in [0, L)$ on the periodically supported beam. Then the right-hand side of equation (1) has the form $F_0\delta(x - l_0)$. In this case the solution of problem (1-3) does not depend on time t, i.e. $z(x,t) = z(x)$ and $Z_j \in \mathbb{R}$, $j \in \mathbb{Z}$. After substitution we obtain formula

$$Z_j = \frac{s_p}{s_b + s_p} z(jL), \quad j \in \mathbb{Z} \tag{5}$$

and we get ordinary differential equation

$$z^{IV} + 6c \sum_{j=-\infty}^{\infty} \delta(x - jL)z = \frac{F_0}{EI}\delta(x - l_0) \tag{6}$$

with $c = s_b s_p / (6EI(s_b + s_p))$.

The general solution of the unsupported equation $EIz^{IV} = F_0\delta(x - l_0)$ determines the form of the free parts of the unloaded beam as $z(nL + x) = A_n x^3 + B_n x^2 + C_n x + D_n$, where $x \in [0, L]$ and $A_n, B_n, C_n, D_n \in \mathbb{R}$. If we would like to be transferred in a sufficiently smooth way from the nth part $[nL, (n+1)L]$ of the unloaded, periodically supported beam to the $(n + 1)$st part, we obtain transition matrix $\mathbf{A}$ with entries

$$\begin{bmatrix} A_{n+1} \\ B_{n+1} \\ C_{n+1} \\ D_{n+1} \end{bmatrix} = \begin{bmatrix} 1 - cL^3 & -cL^2 & -cL & -c \\ 3L & 1 & 0 & 0 \\ 3L^2 & 2L & 1 & 0 \\ L^3 & L^2 & L & 1 \end{bmatrix} \begin{bmatrix} A_n \\ B_n \\ C_n \\ D_n \end{bmatrix}.$$

Characteristic polynomial of matrix $\mathbf{A}$ has the form $z^4 - 2az^3 + 2bz^2 - 2az + 1$, which implies that if ξ is an eigenvalue of $\mathbf{A}$, then $1/\xi$ is also an eigenvalue. In our concrete case $a = 2 - cL^3/2$ and $b = 3 + 2cL^3$ hold. Let ξ_k, $k = 1, ..., 4$ denote the eigenvalues of $\mathbf{A}$, while $\mathbf{U}$ stands for the matrix composed of the eigenvectors, that exist in the consequence of the forthcoming propositions, in the way $\sum_{i=1}^{4} a_{ji}u_{ik} = \xi_k u_{jk}$.

Lemma . Complex polynomial $p(z) = z^4 - 2az^3 + 2bz^2 - 2az + 1$ has a multiple root if and only if $b = -1 \pm 2a$ or $b = \frac{1}{2}a^2 + 1$. If $a, b \in \mathbb{R}$, then p has a root z with $|z| = 1$ if and only if $2|a| - 1 \leq b \leq \frac{1}{2}a^2 + 1 \leq 3$. If a and b are complex numbers, then p has a root of module 1 if and only if $\text{Im}(b)^2 - 2a\text{Im}(a)\text{Im}(b) + 2(b - 1)\text{Im}(a)^2 = 0$ and $|\text{Im}(b)| \leq 2|\text{Im}(a)|$ are satisfied.

Corollary . In the static case the characteristic polynomial of matrix $\mathbf{A}$ can have neither multiple roots nor roots laying on the unit circle, because in this case we have $b = 11 - 4a$ with $a < 2$, and this halfline does not intersects the point set determined by the conditions of Lemma.

Solutions of the homogeneous equation corresponding to (6) can be written into the form $z_{\text{hom}}(nL + x) = \sum_{j,k=1}^{4} x^{4-j}(\mathbf{A}^n)_{jk}c_k$, where $n \in \mathbf{Z}$, $x \in [0, L)$ and c_k is arbitrary constant for $k = 1, ..., 4$. A particular solution of equation (6) for $x \in (0, L)$ can be given by methods used in [4] as $z_{\text{part}}(x) = \frac{F_0}{6EI}(x - l_0)^3 \text{H}(x - l_0)$, where H stands for Heaviside's unit jump function. This solution extends in a sufficiently smooth way to the whole real line as

$$z_{\text{part}}(nL + x) = \frac{F_0}{6EI} \sum_{j,k=1}^{4} x^{4-j}(\mathbf{A}^n)_{jk} \binom{3}{k-1}(-l_0)^{k-1} \text{H}(nL + x - l_0).$$

If, in order to produce the general solution of inhomogeneous equation (6), we sum up homogeneous and inhomogeneous solutions obtained above, then we are able to choose constants c_k, $k = 1, ..., 4$ in such a way, that boundary condition (3) is satisfied. Actually, the correct choice can be determined as follows.

Let $\mathbf{c}$ denote the vector composed of constants c_k, and let vector $\mathbf{f}$ be given by its coordinates $f_k := \frac{F_0}{6EI}\binom{3}{k-1}(-l_0)^{k-1}$, $k = 1, ...4$. Boundary condition (3) means $\mathbf{A}^n(\mathbf{c} + \text{H}(nL + x - l_0)\mathbf{f}) \to 0$ if $n \to \pm\infty$. Let $\mathbf{u}_1$ and $\mathbf{u}_2$ denote the eigenvectors of $\mathbf{A}$ corresponding to the two eigenvalues of module less than 1. Such eigenvalues always exist in consequence of Corollary. Let $\mathbf{u}_3$ and $\mathbf{u}_4$ stand for the other two eigenvectors. Now let us decompose vector $\mathbf{f}$ by the four eigenvectors as $\mathbf{f} = \sum_{l=1}^{4} v_l \mathbf{u}_l$, $l = 1, ..., 4$, where vector $\mathbf{v} := [v_1, ..., v_4]^T$ can be determined by $\mathbf{v} = \mathbf{U}^{-1}\mathbf{f}$. In case $n \to \infty$ we have $\text{H}(nL + x - l_0) \to 1$, and, in order to satisfy boundary condition (3), $\mathbf{c} + \mathbf{f}$ must lay in the linear subspace generated by eigenvectors $\mathbf{u}_1$ and $\mathbf{u}_2$. For $n \to -\infty$ vector $\mathbf{c}$ must lay in the subspace spanned by $\mathbf{u}_3$ and $\mathbf{u}_4$. Because of linear independence one can conclude

$$\mathbf{c} = -v_3\mathbf{u}_3 - v_4\mathbf{u}_4.$$

Hence we obtain $\mathbf{c} + \text{H}(nL + x - l_0)\mathbf{f} = \sum_{i,l=1}^{4} \sigma_l(\mathbf{U}^{-1})_{li}f_i\text{H}(\sigma_l(nL + x - l_0))\mathbf{u}_l$, if $nL + x - l_0 \neq 0$ is satisfied, since $\text{H}(s) - 1 = -\text{H}(-s)$ for $s \neq 0$. Here σ_l denotes the sign of $1 - |\xi_l|$. On the other hand we have $(\mathbf{A}^n\mathbf{u}_l)_j = \xi_l^n(\mathbf{U})_{jl}$. This way we are able to formulate the solution of problem (1-3) in the static case.

Theorem 1. The solution of the static beam problem has the form

$$z(nL + x) = \frac{F_0}{6EI} \sum_{i,j,k=1}^{4} x^{4-j}\binom{3}{i-1}(-l_0)^{i-1}\sigma_k\xi_k^n(\mathbf{U})_{jk}(\mathbf{U}^{-1})_{ki}\text{H}(\sigma_k(nL + x - l_0))$$

in case when $nL + x - l_0 \neq 0$. If $nL + x - l_0 = 0$, then we can take the limit from one side of the previous formula. Constants Z_j can be calculated by formula (5). Here $n \in \mathbf{Z}$ and $x \in [0, L)$. The ξ_k's are the eigenvalues and $\mathbf{U}$ is the matrix composed of the eigenvectors of matrix $\mathbf{A}$, while $\sigma_k := \text{sgn}(1 - |\xi_k|)$.

3 Fixed load case

In this section we generalize results of [1] for the case $v = 0$ and $w \neq 0 \in \mathbb{C}$, i.e. when complex phasor $F_0 e^{wt}$ is fixed at longitudinal position $x = l_0 \in [0, L)$. In this case the right-hand side of partial differential equation (1) transforms into $F_0 e^{wt}\delta(x - l_0)$. The form of the solution is $z(x,t) = u(x)e^{wt}$, $Z_j(t) = \beta_j e^{wt}$, $\beta_j \in \mathbb{C}$, $j \in \mathbb{Z}$. After substitution we obtain formula

$$Z_j(t) = \frac{(k_p w + s_p)z(jL,t)}{(m_s w^2 + k_b w + s_b) + (k_p w + s_p)}, \quad j \in \mathbb{Z}, \tag{7}$$

and ordinary differential equation

$$EI u^{IV} + (\rho A w^2 + \gamma \sum_{j=-\infty}^{\infty} \delta(x - jL))u = F_0 \delta(x - l_0) \tag{8}$$

with $\frac{1}{\gamma} = \frac{1}{m_s w^2 + k_b w + s_b} + \frac{1}{k_p w + s_p}$.

We can look for solutions of the homogeneous equation corresponding to (8) in the form $u_{\text{hom}}(nL+x) = \sum_{j=1}^{4} u_{nj}e^{\lambda \epsilon_j x}$, where $EI\lambda^4 + \rho A w^2 = 0$. Utilizing methods from [4] we obtain correspondence $u_{nj} = -\frac{\gamma \epsilon_j}{4EI\lambda^3} \sum_{k=-\infty}^{n} u_{\text{hom}}(kL)e^{-\lambda \epsilon_j kL} + a_j$, where a_j is arbitrary constant. From this we can deduce recurrence formula $u_{n+1,j} = \sum_{k=1}^{4} e^{\lambda \epsilon_k L}(\delta_{jk} - \frac{\gamma \epsilon_j}{4EI\lambda^3})u_{nk}$. The entries of the transition matrix have form

$$(\mathbf{A})_{jk} := e^{\lambda \epsilon_k L}\left(\delta_{jk} - \frac{\gamma \epsilon_j}{4EI\lambda^3}\right). \tag{9}$$

Characteristic polynomial of matrix (9) is again of form $z^4 - 2az^3 + 2bz^2 - 2az + 1$. Its coefficients have been determined in [1] as

$$a = \cos(\lambda L) + \cosh(\lambda L) + \frac{\gamma}{4EI\lambda^3}(\sin(\lambda L) - \sinh(\lambda L)),$$

$$b = 1 + 2\cos(\lambda L)\cosh(\lambda L) + \frac{\gamma}{2EI\lambda^3}(\sin(\lambda L)\cosh(\lambda L) - \cos(\lambda L)\sinh(\lambda L)).$$

Solutions of the homogeneous equation can be written into the form $u_{\text{hom}}(nL+x) = \sum_{j,k=1}^{4} e^{\lambda \epsilon_j x}(\mathbf{A}^n)_{jk}c_k$ with arbitrary constants c_k, $k = 1,...,4$. A particular solution of inhomogeneous equation (8) on the interval $(0, L)$ can be given by $u_{\text{part}}(x) = \frac{F_0}{4EI\lambda^3}\sum_{j=1}^{4}\epsilon_j e^{\lambda \epsilon_j(x-l_0)} H(x-l_0)$. This solution extends to $\mathbb{R}$ as

$$u_{\text{part}}(nL+x) = \frac{F_0}{4EI\lambda^3} \sum_{j,k=1}^{4} \epsilon_k e^{\lambda(\epsilon_j x - \epsilon_k l_0)}(\mathbf{A}^n)_{jk}H(nl+x-l_0).$$

Now we apply the method of Section 2 to determine constants c_k, $k = 1,...,4$. After defining $f_k := \frac{F_0 \epsilon_k}{4EI\lambda^3}e^{-\lambda \epsilon_k l_0}$ we can make the same steps as in the static case, altough the coefficients in the characteristic polynomial of $\mathbf{A}$ look more complicated. The only known case, when they satisfy conditions of Lemma is that of *pinned-pinned resonance* in [1] with $w = \pm i\sqrt{\frac{EI}{\rho A}}(\frac{N\pi}{L})^2, N \in \mathbb{Z}$.

Theorem 2. *The solution of the beam problem under fixed load, provided $w \neq 0, mL + x - l_0 \neq 0$ and that the characteristic coefficients of matrix (9) do not satisfy conditions of Lemma, has the form*

$$z(nL+x,t) = \frac{F_0 e^{wt}}{4EI\lambda^3} \sum_{i,j,k=1}^{4} e^{\lambda(\epsilon_j x - \epsilon_i l_0)} \epsilon_i \sigma_k \xi_k^n (\mathbf{U})_{jk} (\mathbf{U}^{-1})_{ki} H(\sigma_k(nL+x-l_0)),$$

with $n \in \mathbb{Z}$, $0 \leq x < L$, $EI\lambda^4 + \rho A w^2 = 0$, and $\epsilon_j = i^{j-1}$, where ξ_k and $(\mathbf{U})_{jk} = \epsilon_j / (e^{\lambda \epsilon_j L} - \xi_k)$ *are the eigenvalues and eigenvectors of matrix* (9), *respectively, while* $\sigma_k := \mathrm{sgn}(1 - |\xi_k|)$. *Functions Z_j can be reckoned by formula* (7). *If $nL + x - l_0 = 0$, then limits from one side provide the solution.*

4 Moving load case

The solution can be written into the form $z(x,t) = u(x,t)e^{wt}$, $Z_j(t) = U_j(t)e^{wt}$, $j \in \mathbb{Z}$, where functions u and U_j possess Floquet symmetry property

$$u(x+L, t+L/v) = u(x,t), \quad U_{j+1}(t+L/v) = U_j(t), \quad \text{for any } (x,t) \in \mathbb{R}^2. \quad (10)$$

The time excitation coming from the load in case $w \neq 0$ is $e^{(w-\lambda_i v)t}$, $i = 1, ..., 4$, where the λ_i's are the roots of polynomial (4). In region $\mathbb{R} \times (0, L/v)$ solutions of the homogeneous equation corresponding to (1) can be constructed from

$$u_{\mathrm{hom}}(nL+x,t) = \sum_{i=1}^{4} e^{-\lambda_i v t} \sum_{j,k=1}^{4} e^{\lambda_i \epsilon_j x} (\mathbf{A}_i^n)_{jk} c_{ki} = \sum_{i,j,k=1}^{4} e^{\lambda_i(\epsilon_j x - vt)} (\mathbf{A}_i^n)_{jk} c_{ki},$$

with arbitrary constants $c_{ki} \in \mathbb{C}$, $i, k = 1, ..., 4$. Transition matrices are given by

$$(\mathbf{A}_i)_{jk} := e^{\lambda_i \epsilon_k L} \left(\delta_{jk} - \frac{\gamma_i \epsilon_j}{4EI\lambda_i^3} \right), \quad i,j,k = 1, ..., 4, \quad (11)$$

with

$$\frac{1}{\gamma_i} = \frac{1}{m_s(w-\lambda_i v)^2 + k_b(w-\lambda_i v) + s_b} + \frac{1}{k_p(w-\lambda_i v) + s_p}, \quad i = 1, ..., 4.$$

Utilizing the fact, that solution of the unsupported problem has the same form as solution of the homogeneous, periodically supported equation with $\epsilon_j = 1$, i.e. $j = 1$, a particular solution of inhomogeneous equation (1) can be built up by

$$u_{\mathrm{part}}(nL+x,t) = F_0 \sum_{i,j=1}^{4} \frac{1}{P'(\lambda_i)} e^{\lambda_i(\epsilon_j x - vt)} (\mathbf{A}_i^n)_{j1} H(nL+x-vt),$$

where $0 \leq t < L/v$ and P' is the derivative of polynomial (4), see e.g.[4]. Applying the arguments of Section 2 for vector $\mathbf{f} := [F_0/P'(\lambda_i), 0, 0, 0]^T$ and matrix $\mathbf{A} := \mathbf{A}_i$

for any particular $i = 1, ..., 4$ separately, now we are able to choose coefficients c_{ki}, $i, k = 1, ..., 4$ to satisfy boundary condition (3) on $\mathbb{R} \times (0, L/v)$. With the help of symmetry property (10) we can extend our solution to the whole plane, and this way we obtain the main result of the paper.

Theorem 3. If conditions of the Lemma do not hold for the characteristic coefficients of matrices (11), $(n-m)L + x - vt \neq 0$ is satisfied, and the roots $\lambda_i, i = 1, ..., 4$ of polynomial (4) satisfy $\lambda_i^4 \neq \lambda j^4$ for $i \neq j$, then the solution of beam problem (1-3) under moving load can be given by

$$z(nL + x, mL/v + t) =$$

$$F_0 e^{w(mL/v+t)} \sum_{i,j,k=1}^{4} e^{\lambda_i(\epsilon_j x - vt)} \frac{1}{P'(\lambda_i)} \sigma_{ik} \xi_{ik}^{n-m} (\mathbf{U}_i)_{jk} (\mathbf{U}_i^{-1})_{k1} H(\sigma_{ik}((n-m)L + x - vt))$$

with $m, n \in \mathbb{Z}$, $0 \leq x < L$, $0 \leq t < L/v$ and $\epsilon_j := i^{j-1}$, $j = 1, ..., 4$. Here P' is the derivative of polynomial (4), while ξ_{ik} and $\mathbf{U}_i$ stand for the eigenvalues and the eigenvector matrices of matrices (11), respectively, and $\sigma_{ik} := \mathrm{sgn}(1 - |\xi_{ik}|)$ holds. For any $n \in \mathbb{Z}$ functions Z_n have the form

$$Z_n(mL/v+t) = F_0 e^{w(mL/v+t)} \sum_{i,j,k=1}^{4} \frac{(k_p(w-\lambda_i v)+s_p)e^{-\lambda_i vt}}{m_s(w-\lambda_i v)^2 + (k_b+k_p)(w-\lambda_i v) + (s_b+s_p)} \times$$

$$\frac{1}{P'(\lambda_i)} \sigma_{ik} \xi_{ik}^{n-m} (\mathbf{U}_i)_{jk} (\mathbf{U}_i^{-1})_{k1} H(\sigma_{ik}((n-m)L - vt))).$$

If $(n-m)L + x - vt = 0$, then limits from one side provide the solutions.

Remark. $\lambda_1 = i\lambda_2$ appears if $w = \pm i\sqrt{\frac{\rho A}{EI} \frac{v^2}{2}}$. $\lambda_1 = \lambda_2$ takes place if $w = \pm i\sqrt{\frac{\rho A}{EI} \frac{v^2}{4}}$ or $w = 0$. In these resonance cases suitable modifications of the previous method result in the corresponding solutions of problem (1-3).

Fig.2 shows the motion form of a periodically supported system in the case when we take the dynamic part of the moving load into consideration with data $F_0 = 2 \times 10^4$ N, $w = -1 + 10i \frac{1}{s}$, $v = 10$ km/h $k_p = 6.3 \times 10^4$ Ns/m $s_p = 2.6 \times 10^8$ N/m, $m_s = 145$ kg, $k_b = 8.2 \times 10^4$ Ns/m, $s_b = 1.8 \times 10^8$ N/m, $I = 3.052 \times 10^{-5}$ m^4, $E = 2.1 \times 10^{11}$ N/m^2, $\rho A = 60.31$ kg/m, $L = 0.6$ m.

5 Conclusions

The developed method of analyzing the dynamic behaviour of railway tracks makes it possible to obtain in finitely many steps the actual motion forms and loading conditions of the beam and the supporting pad/sleeper/ballast elements in any periodically supported cases. Further developments can be: generalizations of the results to inhomogeneous in sleeper position, pad and ballast stiffness and damping, cases and characterization of the combined motion/loading conditions of the track/wheel system interconnected by the usual Hertzian springs and dampers, tractated in [4].

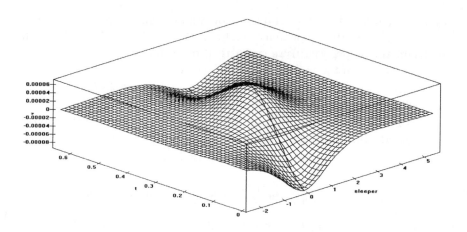

Figure 2: Responses for dynamic load $F_0 e^{-t}\cos(10t)\delta(x-vt)$

References

[1] Droździel, J./Sowiński, B./Żochowski, A.: Methods of solving steady–state and transient vibrations of discretely supported track, Machine Dynamics Problems 11 (1995), pp. 19-38.

[2] Krzyżyński, T.: On continuous subsystem modelling in the dynamic interaction problem of a train–track–system, Vehicle System Dynamics 24, S (1995), pp. 311-324.

[3] Zobory, I./Zoller, V./Zábori, Z.: Time domain analysis of a railway vehicle running on a discretely supported continuous rail model at a constant velocity, Z. angew. Math. Mech. 76, S4 (1996), pp. 169-172.

[4] Zobory, I./Zoller, V./Zibolen, E.: Theoretical investigations into the dynamical properties of railway tracks using a continuous beam model on elastic foundation, Periodica Polytechnica, Ser. Transp. Eng. 22, 1 (1994), pp. 35-54.

A mathematical treatise on periodic structures under travelling loads with an application to railway tracks

Tomasz Krzyżyński

Inst. Fund. Technological Res. of the Polish Academy of Sciences
Świętokrzyska 21, PL–00-049 Warsaw, Poland,
E–mail: tkrzyz@ippt.gov.pl

and

Karl Popp

Institute of Mechanics, University of Hannover
Appelstr. 11, D–30167 Hannover, Germany,
E–mail: popp@ifm.uni-hannover.de

Abstract

The paper deals with the vertical dynamics of a railway track. In the system under consideration a single rail is modelled as a Timoshenko beam. The rails are coupled by means of periodically spaced sleepers which are modelled as rigid bodies with two degrees–of–freedom. The railway track forms a typical periodic system consisting of a number of identical flexible elements (cells) which are coupled in an identical way (by means of the sleepers). The solution method applied in the paper consists in the direct application of Floquet's theorem to the equations of motion of the beam structure. Arranging the periodic boundary conditions for the whole infinite system makes it possible to reduce the analysis to one cell of the periodic structure. There are two modes of travelling waves propagating in the two–dimensional periodic structure. The first mode corresponds to the in–phase propagation of waves in the two rails. The second mode represents the case of a half–wave–length phase difference between the waves. The solution for the system under moving harmonic forces consists of the sum of these two modes.

1 Introduction

A conventional railway track is composed of rails mounted to sleepers which rest on the ballast, with pads between the rail and the sleepers. Usually, the track is a system of longitudinal symmetry, which generally is not the case for loads acting

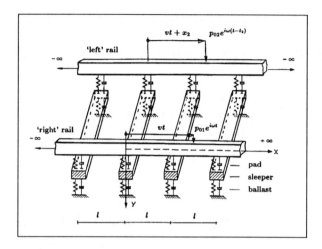

Figure 1: System model.

on it. In this paper the track response to travelling loads in form of harmonic forces is investigated. In the system under consideration a single rail is modelled as a Timoshenko beam. The rails are coupled by means of periodically spaced sleepers which are modelled as rigid bodies with two degrees–of–freedom, i.e. vertical displacement and rotation. Since the railway track forms a typical periodic system which consists of a number of identical flexible elements (cells) coupled in an identical way (by means of the sleepers), its dynamical analysis can be based on Floquet's theorem. The method which we use consists in a direct application of this theorem to the equations of motion of a beam structure.

2 Formulation of the problem

The system under consideration, Figure 1, is composed of two parallel infinite subsystems (rails) which are coupled together by means of equally spaced elements (sleepers). Assuming longitudinal symmetry of the track we model the rails as Timoshenko beams with equal parameters. In this paper we refer to the 'right' and 'left' rail by subscripts 'j=1' and 'j=2', respectively. The equations of motion of a Timoshenko beam resting on a viscoelastic foundation, which is subjected to the load $p_j(x,t)$, are taken in the following form:

$$
\begin{aligned}
K\frac{\partial}{\partial x}\left(\frac{\partial w_j}{\partial x}-\psi_j\right)-\rho A\frac{\partial^2 w_j}{\partial t^2}-\eta\frac{\partial w_j}{\partial t}-qw_j+p_j(x,t)=0,\\
EI\frac{\partial^2 \psi_j}{\partial x^2}+K\left(\frac{\partial w_j}{\partial x}-\psi_j\right)-\rho I\frac{\partial^2 \psi_j}{\partial t^2}=0,
\end{aligned}
\quad (1)
$$

where $w_j = w_j(x,t)$ is the displacement, $\psi_j = \psi_j(x,t)$ is the rotation of the beam cross–section due to bending moments, both of them are functions of the spatial variable x and the time t. In Eqs. (1) $K = \kappa GA$ is the shear stiffness, EI is the flexural stiffness, ρA is the beam mass per unit length, ρI is the rotatory inertia of the beam cross–section, q and η denote the coefficients of elasticity and damping of the foundation, respectively.

The periodicity of the railway track, Figure 1, results in the following boundary conditions for the functions w_j and ψ_j:

$$w_j(nl_-,t) = w_j(nl_+,t), \qquad \psi_j(nl_-,t) = \psi_j(nl_+,t),$$
$$-EI\frac{\partial \psi_j}{\partial x}(nl_-,t) + EI\frac{\partial \psi_j}{\partial x}(nl_+,t) = 0, \qquad (2)$$
$$-K\left[\frac{\partial w_j}{\partial x}(nl_-,t) - \psi_j(nl_-,t)\right] + K\left[\frac{\partial w_j}{\partial x}(nl_+,t) - \psi_j(nl_-,t)\right] + R_j(nl,t) = 0.$$

In Eqs. (2) l is the spacing of the sleepers, n is the sleeper number ($n \in \{-\infty,..,-1,0,1,...,+\infty\}$). Eqs. (2) represent the continuity of displacement, rotation, bending moment and the equilibrium of shearing forces, respectively, for the n-th periodic support. The sleeper–rail interaction force $R_j = R_j(nl,t)$ can be determined by means of the dynamics of the sleepers.

The sleeper is modeled as a rigid body with two degrees of freedom, resting on a viscoelastic foundation, which is mounted to the rails by means of the pads modeled as viscoelastic elements. The equations of motion of the n–th sleeper read

$$(\tilde{\Delta}_I + \tilde{\Delta}_{II})u_1 + (\tilde{\Delta}_I - \tilde{\Delta}_{II})u_2 = R_1,$$
$$(\tilde{\Delta}_I - \tilde{\Delta}_{II})u_1 + (\tilde{\Delta}_I + \tilde{\Delta}_{II})u_2 = R_2, \qquad (3)$$

where $u_j = u_j(nl,t)$ denotes the displacement of the n–th sleeper at the mounting point $j = 1,2$, and

$$R_j = R_j(nl,t) = \tilde{\Delta}_P \cdot (w_j - u_j), \qquad \tilde{\Delta}_P(t) = (q_P + \eta_P \frac{d}{dt}) \qquad (4)$$

is the force in the pad. In Eqs. (3) we have

$$\tilde{\Delta}_I(t) = \frac{1}{4}\left(m_S\frac{d^2}{dt^2} + \eta_I\frac{d}{dt} + q_I\right), \qquad \tilde{\Delta}_{II}(t) = \frac{1}{l_p^2}\left(M_S\frac{d^2}{dt^2} + \eta_{II}\frac{d}{dt} + q_{II}\right), \qquad (5)$$

with the nomenclature given in Table 1.

For the steady–state motion with the frequency ω, the interaction force $R = R(nL,\tau)$ determined for the n-th sleeper reads

$$R_1(nl,t) = \delta_1(\omega) \cdot w_1(nl,t) + \delta_2(\omega) \cdot w_2(nl,t),$$
$$R_2(nl,t) = \delta_1(\omega) \cdot w_2(nl,t) + \delta_2(\omega) \cdot w_1(nl,t), \qquad (6)$$

$m_S = \mu_S l_S$	l_S – sleeper length
$q_I = q_B l_S$	μ_S – sleeper mass per unit length
$\eta_I = \eta_B l_S$	q_B – ballast elasticity coefficient
$M_S = \frac{1}{12}\mu_S l_S^3$	η_B – ballast damping coefficient
$q_{II} = \frac{1}{12}l_S^3 q_B$	l_P – spacing of pads on the sleeper
$\eta_{II} = \frac{1}{12}l_S^3 \eta_B$	q_P – pad elasticity coefficient
	η_P – pad damping coefficient

Table 1: Nomenclature for the sleeper, the ballast, and the pad.

where $\delta_1(\omega)$ and $\delta_2(\omega)$ are certain coefficients depending on the parameters of sleepers and pads. For the following analysis we use the non–dimensional quantities given in Table 2. The non–dimensional form of the equations of motion, Eqs. (1), reads

$$\frac{1}{\alpha}\frac{\partial}{\partial X}\left(\frac{\partial W_j}{\partial X} - \Psi_j\right) - \frac{\partial^2 W_j}{\partial \tau^2} - N\frac{\partial W_j}{\partial \tau} - QW_j + P_j(X,\tau) = 0,$$
$$\frac{\partial^2 \Psi_j}{\partial X^2} + \frac{1}{\alpha}\left(\frac{\partial W_j}{\partial X} - \Psi_j\right) - \beta\frac{\partial^2 \Psi_j}{\partial \tau^2} = 0. \qquad (7)$$

The boundary conditions given by Eqs. (2) take now the following form:

$$W_j(nL_-,\tau) = W_j(nL_+,\tau), \qquad \Psi_j(nL_-,\tau) = \Psi_j(nL_+,\tau),$$
$$-\frac{\partial \Psi_j}{\partial X}(nL_-,\tau) + \frac{\partial \Psi_j}{\partial X}(nL_+,\tau) = 0, \qquad (8)$$
$$-\frac{1}{\alpha}\left(\frac{\partial W_j}{\partial X} - \Psi_j\right)_{nL_-} + \frac{1}{\alpha}\left(\frac{\partial W_j}{\partial X} - \Psi_j\right)_{nL_+} - R_j(nL,\tau) = 0.$$

In the next section of the paper we determine the solution of Eqs. (7) with the conditions of Eqs. (8) for the system under travelling disturbance sources.

$X = x a_0$	$\tau = t\omega_0$	$W(X,\tau) = w/w_0$
$\Psi = \psi/(a_0 w_0)$	$\Omega = \omega/\omega_0$	$V = v/v_0$
$N = \eta/\eta_0$	$Q = q/E$	$\alpha = \mu/k\, v_0^2$
$\beta = \rho/E\, v_0^2$	$K = \kappa G A$	$L = l a_0$
$\eta_0 = \sqrt{E\mu}$	$a_0 = \sqrt[4]{1/I}$	$\omega_0 = \sqrt{E/\mu}$
$w_0 = p_{01} a_0/E$	$p^* = p_{02}/p_{01}$	$v_0 = \omega_0/a_0$

Table 2: Nomenclature for the rail.

3 Dynamic system response to moving forces

We consider the case of a load in the form of travelling harmonic forces

$$p_1(x,t) = p_{01}\delta(x - vt)e^{(i\omega t)}, \qquad p_2(x,t) = p_{02}\delta(x - x_2 - vt)e^{[i\omega(t-t_2)]}, \qquad (9)$$

where v is the load velocity, ω is the load frequency, p_{01} and p_{02} are magnitudes of the load travelling on the 'right' and 'left' rail, respectively. In Eqs. (9) x_2 and t_2 represent the space and time shifts of the load p_2 in reference to the load p_1, respectively. The non-dimensional form of Eqs. (9), which has to be introduced into Eq. (7), read

$$P_1(X,t) = \delta(X - V\tau)e^{(i\Omega\tau)}, \qquad P_2(X,t) = p_0^*\delta(X - X_2 - V\tau)e^{[i\Omega(\tau-\tau_2)]}. \qquad (10)$$

The solutions of Eqs. (7) can be written in the following form:

$$W_j(X,\tau) = \frac{1}{2\pi}\int_{-\infty}^{\infty} A_{w_j}(X,\lambda)\exp[i\lambda(X - V\tau) + i\Omega\tau]d\lambda,$$

$$\Psi_j(X,\tau) = \frac{1}{2\pi}\int_{-\infty}^{\infty} A_{\psi_j}(X,\lambda)\exp[i\lambda(X - V\tau) + i\Omega\tau]d\lambda. \qquad (11)$$

We assume that the functions $A_{w_j} = A_{w_j}(X,\lambda)$, $A_{\psi_j} = A_{\psi_j}(X,\lambda)$ describe dynamically admissible displacement fields. According to Floquet's theorem we suppose they are periodic functions, i.e. functions which are independent from the choice of a cell in the periodic structure, $A_{w_j}(X+L,\lambda) = A_{w_j}(X,\lambda)$, $A_{\psi_j}(X+L,\lambda) = A_{\psi_j}(X,\lambda)$. Introducing Eqs. (10) and performing the Fourier transform yields

$$\frac{1}{\alpha}D_A\left(D_A A_{w_j} - A_{\psi_j}\right) + \overline{\Omega}^2 A_{w_j} - iN A_{w_j} - Q A_{w_j} = P_j(\lambda),$$

$$D_A^2 A_{\psi_j} + \frac{1}{\alpha}\left(D_A A_{w_j} - A_{\psi_j}\right) + \beta\overline{\Omega}^2 A_{\psi_j} = 0, \qquad D_A = i\lambda + \frac{d}{dX}, \qquad (12)$$

where

$$P_1(\lambda) = -1, \quad P_2(\lambda) = -p^*e^\phi, \quad \phi = -i(\lambda X_2 + \Omega\tau_2). \qquad (13)$$

The frequency $\overline{\Omega} = -\lambda V + \Omega$ can be considered as the 'forced' frequency of the waves. The conditions for the functions $A_{w_j} = A_{w_j}(X,\lambda)$ and $A_{\psi_j} = A_{\psi_j}(X,\lambda)$, which follow from the boundary conditions given by Eqs. (8), read

$$A_{w_1}[(n+1)L] = A_{w_1}(nL), \quad A_{\psi_1}[(n+1)L] = A_{\psi_1}(nL),$$
$$-D_A A_{\psi_1}[(n+1)L] + D_A A_{\psi_1}(nL) = 0, \qquad (14)$$
$$(D_A^2 + \beta\overline{\Omega}^2)A_{\psi_1}[(n+1)L] - (D_A^2 + \beta\overline{\Omega}^2)A_{\psi_1}(nL) + \delta_1 A_{w_1}(nL) + \delta_2 A_{w_2}(nL) = 0,$$

$$A_{w_2}[(n+1)L] = A_{w_2}(nL), \quad A_{\psi_2}[(n+1)L] = A_{\psi_2}(nL),$$
$$-D_A A_{\psi_2}[(n+1)L] + D_A A_{\psi_2}(nL) = 0, \qquad (15)$$
$$(D_A^2 + \beta\overline{\Omega}^2)A_{\psi_2}[(n+1)L] - (D_A^2 + \beta\overline{\Omega}^2)A_{\psi_2}(nL) + \delta_1 A_{w_2}(nL) + \delta_2 A_{w_1}(nL) = 0.$$

It should be noted that the conditions for the 'right' rail (Eqs. (14)) and for the 'left' rail (Eqs. (15)) are coupled together by means of the term $\delta_2 A_{w_2}(nL)$ and $\delta_2 A_{w_1}(nL)$.

The solutions of Eq. (13), which satisfy the boundary conditions given by Eqs. (14) and (15), for

$$\xi = X - nL, \quad X \in <nL, (n+1)L>, \quad \xi \in <0, L>, \tag{16}$$

can be written in the following form:

$$A_{\psi_1}(\xi) = (\delta_1 w_{p1} + \delta_2 A^0_{w_2})\frac{f_\psi(\xi)}{|\mathcal{M}|} + \psi_{p1}, \quad A_{w_1}(\xi) = (\delta_1 w_{p1} + \delta_2 A^0_{w_2})\frac{f_w(\xi)}{|\mathcal{M}|} + w_{p1},$$

$$A_{\psi_2}(\xi) = (\delta_1 w_{p2} p^* e^\phi + \delta_2 A^0_{w_1})\frac{f_\psi(\xi)}{|\mathcal{M}|} + \psi_{p2} p^* e^\phi, \tag{17}$$

$$A_{w_2}(\xi) = (\delta_1 w_{p2} p^* e^\phi + \delta_2 A^0_{w_1})\frac{f_w(\xi)}{|\mathcal{M}|} + w_{p2} p^* e^\phi,$$

where $\psi_{pj}, w_{pj}, j = 1, 2$, are certain coefficients depending on the system parameters. In Eq. (17) we have also $A^0_{w_j} = A_{w_j}(\xi = 0)$, $A^0_{\psi_j} = A_{\psi_j}(\xi = 0)$. The term $|\mathcal{M}|$ in Eq. (17) has the same form as the dispersion relation for waves propagating in a Timoshenko beam resting on periodic flexible supports, cf [1]. Similary, the function $f_w(\xi)$ in Eq. (17) can be interpreted as a displacement field due to the array of harmonic interaction forces $R_j(nL, \tau)$, whereas the function $f_\psi(\xi)$ can be interpreted as a rotation field due to the array of these forces which excite the one–dimensional periodic system, cf [1].

Writing Eqs. (17) for $\xi = 0$ we obtain the following relation:

$$\begin{bmatrix} |\mathcal{M}| & 0 & 0 & -\delta_2 f^0_\psi \\ 0 & |\mathcal{M}| & -\delta_2 f^0_\psi & 0 \\ 0 & 0 & |\mathcal{M}| & -\delta_2 f^0_w \\ 0 & 0 & -\delta_2 f^0_w & |\mathcal{M}| \end{bmatrix} \begin{bmatrix} A^0_{\psi_1} \\ A^0_{\psi_2} \\ A^0_{w_1} \\ A^0_{w_2} \end{bmatrix} = \begin{bmatrix} \tilde{\psi}_1 \\ \tilde{\psi}_2 \\ \tilde{w}_1 \\ \tilde{w}_2 \end{bmatrix}, \tag{18}$$

where $\tilde{\psi}_j, \tilde{w}_j, j = 1, 2$, are certain coefficients depending on the system parameters. We can determine the unknown functions $A^0_{\psi_1}, A^0_{\psi_2}, A^0_{w_1}, A^0_{w_2}$ from Eqs. (18) and introduce them into Eqs. (17), and then into Eqs. (11). Solving Eq. (11) by means of Cauchy's theorem and taking into account a certain finite number of poles of the integrands yields the solution for cells ahead of or behind the load, which can be written in the following form:

$$W_j(X, \tau) = \sum_{N=1}^{2} \sum_{M_N=1}^{2} \sum_{k=1}^{K_{M_N}} Q_j^{(NM_Nk)} \cdot f_w(\xi, \lambda_{NM_Nk}) \cdot e^{i(\lambda_{NM_Nk} \cdot nL + \overline{\Omega}_{NM_Nk} \cdot \tau)}. \tag{19}$$

The solution obtained is a superposition of waves travelling in the periodic system. There are two forms of the wave motion: for $N = 1$ we have 'in–phase motion', i.e. in-phase propagation of waves in the two rails; for $N = 2$ we have 'out-of-phase motion', i.e. the half-wave-length phase difference between the propagating waves. For each of the two forms there are two modes of the wave motion, i.e. one wave frequency corresponds to two wavenumbers, cf [2]. We have the first mode for $M_N = 1$, and the second mode for $M_N = 2$. The wavenumber λ_{NM_Nk} is determined by means of the system dispersion relations depending on the 'forced' frequency $\overline{\Omega}_{NM_Nk} = -\lambda_{NM_Nk}V + \Omega$:

$$|\mathcal{M}(\lambda_{NM_Nk}, \overline{\Omega}_{NM_Nk})| \mp \delta_2(\overline{\Omega}_{NM_Nk})f_w^0(\lambda_{NM_Nk}, \overline{\Omega}_{NM_Nk}) = 0, \qquad (20)$$

where we have the minus sign for $N = 1$ and the plus sign for $N = 2$.
The wave 'shape' is determined by the term $f_\psi(\xi, \lambda_{NM_Nk})$ or $f_w(\xi, \lambda_{NM_Nk})$. Additionally, in Eq. (19) we have $Q_j^{(N\,M_N\,k)}(V, \Omega)$ – the wave amplitudes , $K_{M_N} = K_{M_N(A)}$ – number of waves ahead of the load $[nL \geq V\tau]$, $K_{M_N} = K_{M_N(B)}$ – number of waves behind the load $[(n+1)L \leq V\tau]$, for a given mode M_N. For amplitudes of waves corresponding to the in-phase form we have $Q_1^{(1\,M_1\,k)} = Q_2^{(1\,M_1\,k)}$, whereas for out-of-phase case we have $Q_1^{(2\,M_2\,k)} = -Q_2^{(2\,M_2\,k)}$. The solution for the actually loaded cell $[nL \leq V\tau \leq (n+1)L]$ is composed of terms where λ satisfies Eq. (20) and terms where λ satisfies the characteristic equation of a continuous and non-periodic beam under a moving harmonic load, cf [3].

Numerical calculations have been carried out for the system parameters presented in Table 3. In Figure 2 the amplitudes of the rail displacements and rotations at $x = 0.4\,m$ due to two harmonic forces ($v = 0$) are shown. One force acts at the first rail at $x = 0.4\,m$, and the other force at the second rail at $x = -0.2\,m$.

In the pure elastic system ($\eta_P = 0, \eta_B = 0, \eta = 0$) for certain load velocities and frequencies the wave amplitudes increase infinitely in time. This corresponds to the case where the load travels with a group velocity of the wave generated, cf [3].

4 Conclusions

The method applied in the paper consists in the direct application of Floquet's theorem to equations of motion of a beam structure. Arranging the periodic boundary conditions for the whole infinite system makes it possible to reduce the analysis to one cell of the periodic structure. There are two modes of travelling waves propagation in the two-dimensional periodic structure. The first mode corresponds to the in-phase propagation of waves in the two rails. The second mode represents the case of the half-wave-length phase difference between the waves. The solution for the system under moving harmonic forces consists of a sum of these two modes.

5 Acknowledgements

The paper contains results of investigations supported by the Polish State Committee for Scientific Research (KBN grant T07B01410) and the cooperative research

$E = 2.1 \cdot 10^{11} \frac{N}{m^2}$	$I = 3.052 \cdot 10^{-5} m^4$	$\mu = 60.31 \frac{kg}{m}$
$K = 2.62 \cdot 10^8 N$	$l_S = 2.36 m$	$l_P = 1.435 m$
$q_P = 2.6 \cdot 10^8 \frac{N}{m}$	$\eta_P = 6.3 \cdot 10^4 \frac{Ns}{m}$	$\mu_S = 122.88 \frac{kg}{m}$
$q_B = 1.525 \cdot 10^8 \frac{N}{m^2}$	$\eta_B = 6.95 \cdot 10^4 \frac{Ns}{m^2}$	$l = 0.6 m$
$\eta = 4 \cdot 10^4 \frac{Ns}{m^2}$	$q \to 0$	

Table 3: System parameters.

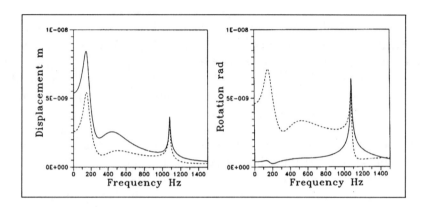

Figure 2: Amplitude–frequency relations for $v = 0$, $p_{01} = 1 N$, $p_{02} = 1 N$, $x_2 = -0.6 m$, $\omega t_2 = \pi$; continuous line for the 'right' rail, dashed line for the 'left' rail.

project "Dynamics of Periodic Structures with Mistuning" granted by the Volkswagen Foundation, Hannover, Germany. The authors express their thanks to these institutions.

References

[1] D. J. Mead: A new method of analyzing wave propagation in periodic structures; Applications to periodic Timoshenko beams and stiffened plates, J. Sound Vib., 104, 1 (1986), pp. 9–27.

[2] T. Krzyżyński: On continuous subsystem modelling in the dynamic interaction problem of a train–track–system, Supplement to Vehicle System Dynamics, 24 (1995), pp. 311–324.

[3] R. Bogacz / T. Krzyżyński / K. Popp: On the generalization of Mathews problem of the vibrations of a beam on elastic foundation, Z. angew. Math. Mech., 69, 8 (1989), pp. 243–252.

The Bifurcation Behaviour of Periodic Solutions of Impact Oscillators

J.P. Meijaard
Laboratory for Engineering Mechanics, Delft University of Technology,
Mekelweg 2, NL-2628 CD Delft, The Netherlands

Abstract

The dynamics in a neighbourhood of a periodic solution of an autonomous or periodically forced impact oscillator is investigated for the case that a zero-velocity impact occurs. Bifurcations of the periodic solution are determined and estimates of the maximal Liapunov exponent are made by means of an approximate one-dimensional map.

1 Problem description

The dynamics are considered of mechanical systems in which impacts may occur. Especially bifurcation phenomena when parts start to impact at low velocities are investigated. These systems may have a periodic solution, either if the system is autonomous and a supercritical Hopf bifurcation has occurred, or if the system is periodically forced. As some parameter of the system is changed continuously, the amplitude of the periodic solution may grow until some part starts to hit a stop. Examples can be found in the lateral dynamics of railway vehicles. The lateral motion is limited by the wheel flanges, which can impact against the rails [10, 4, 8]. Other examples comprise the flow-induced vibrations in heat exchangers and rotors rubbing their casing.

It has been observed that at or very close to the parameter value at which some part of the structure starts to hit some other part, chaotic motion sets in, apparently without passing through any of the known transition routes to chaos. In [9], the mechanism by which this transition occurs in autonomous systems near a Hopf bifurcation is explained. In this paper, additional results for the bifurcations of general periodic solutions are discussed. In the next section, the dynamics near the point at which a zero-velocity impact occurs is investigated. For the case that the motion is trapped in a small region near the unstable periodic motion, the dynamics are approximated by a one-dimensional map, whose behaviour is analysed.

2 Dynamics at low-velocity impacts

2.1 Differential equations and associated Poincaré maps

Systems whose behaviour is described by a set of ordinary differential equations are considered. These systems may be autonomous,

$$\dot{\mathbf{x}} = \mathbf{f}(\mathbf{x}, \lambda), \tag{1}$$

or periodically forced,

$$\dot{\mathbf{x}} = \mathbf{f}(\mathbf{x}, t, \lambda), \qquad \mathbf{f}(\mathbf{x}, t + T, \lambda) = \mathbf{f}(\mathbf{x}, t, \lambda). \tag{2}$$

Here, $\mathbf{x}$ is the n-dimensional vector of state variables, which are partitioned in mechanical configuration and velocity variables, and possibly some other state variables which arise from internal variables in constitutive equations or control systems. Furthermore, λ is the parameter that can be varied, t is the time and T is the minimal period of the periodic forcing.

If the motion is restrained by stops, there will be one or more boundaries in the phase space which cannot be crossed. These boundaries can be defined by surfaces S_i in the phase space,

$$S_i : g_i(\mathbf{x}) = 0, \tag{3}$$

where the admissible region is given by $g_i \leq 0$. In mechanical systems, the functions g_i only depend on the configuration variables. If a trajectory in the phase space reaches a boundary, an impact occurs, where the state after the impact is related to the state before the impact by a transition condition, $\mathbf{x}^+ = \mathbf{T}_{i0}(\mathbf{x}^-)$. (Here and in the sequel, a superscript minus sign denotes a quantity just before an impact under consideration and a superscript plus denotes the same quantity just after this impact.) From now on, it is assumed that only one specific boundary can become active, and the index i is dropped.

In a neighbourhood of a periodic solution, the dynamics of the system as described by the system of ordinary differential equations (1) or (2) can be reduced to the dynamics of a time-independent Poincaré map. For the autonomous case, an $(n-1)$-dimensional local surface of section that is transversal to the phase flow is selected. The Poincaré map is defined as the mapping from this surface of section into itself that sends a point to the first point of return. For the periodically forced case, the Poincaré map is simply defined as the discrete time-T phase flow for a fixed phase of the forcing. In the surface of section, local coordinates $\mathbf{y}$ are introduced, which may be identical to $\mathbf{x}$ in the case of a periodically forced system, and the Poincaré map can be written as

$$P : \mathbf{y} \mapsto \mathbf{P}(\mathbf{y}). \tag{4}$$

In the surface of section, one can distinguish a region with points that will not reach the impact boundary S before a return to the section, and a complementary region of points that will reach this boundary. These two regions are separated by a boundary, which is obtained by mapping back the part of S with $\dot{g} = \partial g/\partial \mathbf{x} \cdot \mathbf{f} = 0$ (Note that this is the boundary between the directly reachable and not directly reachable parts of S) to the surface of section by the inverse phase flow. This boundary in the surface of section, which will be denoted by Σ_0, can be described by an equation of the form $g_p(\mathbf{y}, \lambda) = 0$, where the part $g_p < 0$ is the region without impact. The image of Σ_0 will be denoted by Σ_1. Note that Σ_0 and Σ_1 depend on the parameter λ.

In following a periodic solution if the parameter is changed, it can happen that the corresponding fixed point of the Poincaré map reaches the boundary Σ_0. At this instant, the orbit touches the boundary S with zero velocity and a bifurcation takes place, which belongs to the class of the so-called border-collision bifurcations [13]. By a shift of values of the parameter, one can assume that this bifurcation occurs at $\lambda = 0$, while $\lambda < 0$ corresponds to the situation in which the fixed point is in the region without impact, $g_p < 0$.

2.2 Expansion of the dynamics near a zero-velocity impact bifurcation

In the Poincaré section, a local curvilinear coordinate system can be chosen with coordinate lines of the first coordinate perpendicular to the level surfaces of g_p and the other coordinate lines in the level surfaces. Note that these coordinates depend on λ. The origin is chosen at the border $g_p = 0$, on the coordinate line that corresponds to the fixed point if it is in the region without impact or otherwise a virtual fixed point. One can define a virtual fixed point as the fixed point for a map that is equal to the Poincaré map in the non-impacting region, while it is extended in some determined smooth way on the impacting region, for instance by analytic continuation.

The impact at S gives rise to a jump in the velocities, while there is no jump in the configuration variables or other variables. This yields a square-root singularity in the Poincaré map. An initial condition at $y_1 > 0$ would lead to an indentation of the surface of impact proportional in first order to this value y_1. As this indentation is generically parabolic in first approximation, the velocity of approach at the impact is proportional in first order to $\sqrt{y_1}$. This gives rise to a perturbation in the state that is in first approximation proportional to $\sqrt{y_1}$, which is smoothly mapped back to the surface of section by the phase flow. This gives rise to the square-root singularity. The direction of maximal stretching must be along the image of Σ_0, Σ_1, because the Jacobian determinant of the Poincaré map remains finite and the mapping from Σ_0 to Σ_1 is smooth. Now we can make an expansion of the dynamics in the neighbourhood of the fixed point near the bifurcation,

$$\mathbf{y} \mapsto \begin{cases} \mathbf{A}\mathbf{y} + \mathbf{b}\lambda + o(|\mathbf{y}|, |\lambda|) & (y_1 \leq 0), \\ \mathbf{A}\mathbf{y} + \mathbf{b}\lambda + \mathbf{A}\mathbf{e}_2\sqrt{y_1} + \mathbf{c}y_1 + o(|\mathbf{y}|, |\lambda|) & (y_1 > 0). \end{cases} \quad (5)$$

Here, $\mathbf{A}$ is the Jacobian matrix of the Poincaré map at the fixed point at $\lambda = 0$ and $\mathbf{e}_2$ is a vector tangent to Σ_0, not necessarily of unit length, along which the maximal stretching occurs. Though the mapping is not differentiable at Σ_0, it is continuous and invertible.

2.3 Trapping conditions and continuation of periodic solutions

First we study the dynamics at the bifurcation, $\lambda = 0$, for the case that the fixed point is stable for $\lambda < 0$. As has been pointed out by Koiter [5] for static stability, the behaviour at the bifurcation determines the behaviour in a neighbourhood of the bifurcation. The fixed point will be conditionally stable if an orbit can approach the fixed point without the occurrence of impacts. This gives the conditions for the domain of attraction,

$$\mathbf{e}_1^T \mathbf{A}^k \mathbf{y} < 0 \quad (k = 0, 1, 2, \ldots), \tag{6}$$

where $\mathbf{e}_1$ is a vector with its first component equal to one and the other equal to zero. A necessary condition that the domain of attraction has positive measure is that the eigenvalue with maximal absolute value is real and positive. All initial points close to the fixed point will be attracted by it if the direction of maximal stretching $\mathbf{A}\mathbf{e}_2$, tangential to Σ_1, is in the domain defined by Eq. (6); this is the condition for the existence of a trapping region near the fixed point for λ close to 0. Indeed, if the fixed point is attracting at the bifurcation, then it will have a small neighbourhood that is mapped into itself by some iterate of the Poincaré map. As the dependence on parameters is continuous (though not differentiably continuous), this domain will remain attracting for $\lambda \neq 0$. The condition on the largest eigenvalue will generally be met in autonomous systems if the periodic solution originates from a supercritical Hopf bifurcation, but in periodically forced mechanical systems with low damping, this eigenvalue will be complex in most cases.

Now the evolution of the fixed point near the bifurcation is investigated for the general case and an algorithm for finding fixed points and periodic orbits is developed. First, a periodic orbit which is a fixed point for the Poincaré map with one impact is determined. The residual vector $\mathbf{r}$ is defined as

$$\mathbf{r}(\mathbf{y}) = \mathbf{P}(\mathbf{y}) - \mathbf{y} = -(\mathbf{I} - \mathbf{A})\mathbf{y} + \mathbf{b}\lambda + \mathbf{A}\mathbf{e}_2\sqrt{y_1} + \mathbf{c}y_1 + o(|\mathbf{y}|, |\lambda|), \tag{7}$$

which has to be the zero vector for a periodic orbit. The correction for y_1 is determined by premultiplying the residual by $\mathbf{e}_1^T(\mathbf{I} - \mathbf{A})^{-1}$,

$$\mathbf{e}_1^T(\mathbf{I} - \mathbf{A})^{-1}\mathbf{r} + \mathbf{e}_1^T(\mathbf{I} - \mathbf{A})^{-1}\mathbf{A}\mathbf{e}_2\Delta(\sqrt{y_1}) = 0. \tag{8}$$

From this equation, $\Delta(\sqrt{y_1})$, and hence Δy_1, can be calculated. The corrections for the other components of $\mathbf{y}$ are calculated as

$$\Delta \mathbf{y} = (\mathbf{I} - \mathbf{A})^{-1}[\mathbf{r} + \mathbf{A}\mathbf{e}_2\Delta(\sqrt{y_1})]. \tag{9}$$

If we start the iteration procedure with the zero vector, a first approximation for the fixed point is obtained as

$$\sqrt{y_1} = [\mathbf{e}_1^T(\mathbf{I} - \mathbf{A})^{-1}\mathbf{b}\lambda]/[-\mathbf{e}_1^T(\mathbf{I} - \mathbf{A})^{-1}\mathbf{A}\mathbf{e}_2]. \tag{10}$$

In order that this solution exists, the numerator and the denominator of the fraction on the right-hand side must have the same sign. If there is a trapping region, the denominator is positive, and an impacting periodic solution exists if $\lambda > 0$, because the numerator is an approximation for the first component of the virtual fixed point of the Poincaré map, which is positive for $\lambda > 0$ by assumption. If there is no trapping region, the fixed point exists for one of the signs of λ. If it exists for $\lambda < 0$, we have a kind of fold. The convergence of the iteration procedure for sufficiently small values of $|\lambda|$ can now be proved by Banach's contraction mapping theorem.

Next, the existence of periodic orbits with a higher period, say period-k, is investigated. If it is assumed that there is an impact in the first iteration and no impacts at the next $k-1$ iterations of the Poincaré map, the residual $\mathbf{r}^k$ is given by

$$\mathbf{r}^k(\mathbf{y}) = \mathbf{P}^k(\mathbf{y}) - \mathbf{y} = -(\mathbf{I} - \mathbf{A}^k)\mathbf{y} + (\sum_{i=0}^{k-1} \mathbf{A}^i)\mathbf{b}\lambda + \mathbf{A}^k \mathbf{e}_2 \sqrt{y_1} + \mathbf{A}^{k-1} \mathbf{c} y_1 + o(|\mathbf{y}|, |\lambda|). \tag{11}$$

An iteration procedure similar to the one given above can be applied. The first approximation now yields

$$\sqrt{y_1} = [\mathbf{e}_1^T (\mathbf{I} - \mathbf{A})^{-1} \mathbf{b}\lambda] / [-\mathbf{e}_1^T (\mathbf{I} - \mathbf{A}^k)^{-1} \mathbf{A}^k \mathbf{e}_2]. \tag{12}$$

Again, if there is a trapping region, the period-k solution exists for sufficiently small positive values of λ.

2.4 Dynamics of the corresponding one-dimensional map

For a better understanding of the behaviour of the impacting system, its dynamics is approximated by a one-dimensional map. This reduction is simply done by projecting the state on the eigenvector $\mathbf{u}_1$ corresponding to the largest real eigenvalue of the matrix $\mathbf{A}$. This approximation is justified if there are relatively long intervals without impacts between impacts, so the state has time to converge to the direction corresponding to the least stable eigenvalue. This condition is fulfilled if there is a trapping region and λ is small. If we premultiply the map by the adjoint eigenvector, $\mathbf{u}_1^*$, and denote the projection of $\mathbf{y}$ by u, the approximate mapping becomes

$$u \mapsto \begin{cases} \mu u + b\lambda & (u \leq 0), \\ \mu u + b\lambda - a\sqrt{u} + cu & (u > 0). \end{cases} \tag{13}$$

Here, μ is the largest real eigenvalue, and the other parameters are given by

$$a = -\mathbf{u}_1^* \mathbf{A} \mathbf{e}_2 \sqrt{u_{11}}, \quad b = \mathbf{u}_1^* \mathbf{b}, \quad c = \mathbf{u}_1^* \mathbf{c} u_{11}, \tag{14}$$

where u_{11} is the first component of the eigenvector $\mathbf{u}_1$. In practical situations, the value of c is difficult to calculate, and does not affect the dynamics in a sensitive way. Because $\sqrt{u} \gg u$, the one-dimensional map may be further simplified to

$$u \mapsto \begin{cases} \mu u + b\lambda & (u \leq 0), \\ b\lambda - a\sqrt{u} & (u > 0). \end{cases} \tag{15}$$

This map and related one- and two-dimensional maps have already been analysed in several papers [11, 12, 1, 6]. Some of the results are reviewed here, and some additional results are given.

The qualitative behaviour strongly depends on the value of μ; four cases of interest can be distinguished: $\mu = 1$, $2/3 < \mu < 1$, $1/4 < \mu < 2/3$ and $\mu < 1/4$. The simplest case is the first, $\mu = 1$, which has been analysed previously by the author [7, 8]. As soon as the parameter λ becomes positive, all motion becomes unstable, because at the impact for $u > 0$, nearby trajectories move away from each other, which is not compensated by a contraction at $u < 0$. Hence, a chaotic solution is obtained. Period-k orbits with one impact exist as long as

$$\lambda \leq \frac{a^2}{b}\frac{1}{(k-1)^2}, \quad \text{with} \quad u = \frac{1}{4}\left(-a + \sqrt{a^2 + 4kb\lambda}\right)^2. \tag{16}$$

For the case $2/3 < \mu < 1$, the situation is similar to the previous case, although the analysis is a little more involved. Period-k solutions with one impact exist if

$$\lambda \leq \frac{a^2}{b}\left[\frac{\mu^{k-2}(1-\mu)}{1-\mu^{k-1}}\right]^2, \quad \text{with} \quad u = \left(-\frac{1}{2}a\mu^{k-1} + \sqrt{\frac{1}{4}a^2\mu^{2k-2} + \frac{1-\mu^k}{1-\mu}b\lambda}\right)^2. \tag{17}$$

Also in this case, all solutions are unstable, which can be seen as follows. If k is the highest period for which solutions given by Eq. (17) exist, there can be at most $(k - 1)$ iterations with $u \leq 0$ between iterations with $u > 0$, so small deviations are multiplied by at least $\mu^{k-1}a/(2\sqrt{b\lambda})$, which is larger than one for large k. Note that the case for $\mu = 1$ can be obtained by taking the limit.

For the cases with $\mu < 2/3$, the periodic solution given by Eq. (17) can be stable for large values of k as long as

$$\lambda > [3a^2(1-\mu)\mu^{k-2}]/[4b(1-\mu^k)]. \tag{18}$$

All period-doubling bifurcations at the stability boundary given by Eq. (18) are subcritical because of the positive Schwarzian derivative of the map. (The Schwarzian derivative of a one-dimensional map f is defined as $f'''/f' - (3/2)(f''/f')^2$ [2].) For increasing positive λ, if $1/4 < \mu < 2/3$, one has alternating parameter regions with chaotic motion and stable periodic motion, which follows a decreasing sequence for the period k. For $\mu < 1/4$, the intervals in which stable periodic solutions exists overlap, and the intervals with chaotic motion disappear.

2.5 Estimation of the maximal Liapunov exponent

For the sake of simplifying the notation, the following auxiliary quantities are introduced for the map given by Eq. (15). The maximal value is denoted by u_{max}, $u_{max} = b\lambda$, the minimal reachable value, denoted by u_{min}, is the image of u_{max}, $u_{min} = b\lambda - a\sqrt{b\lambda}$. In order to quantify the sensitivity for the initial conditions in the chaotic regions, the Liapunov exponent will be approximated. This exponent is most easily obtained from an invariant measure, denoted by $p(u)$, which describes the probability density of finding a point of an orbit in some place between u_{min} and u_{max}. This measure can be obtained approximately by making some assumptions about its nature. First, it is assumed that this measure is a smooth function of u; second, its value is considered constant ($p = p_0$) for $0 \leq u \leq u_{max}$; finally, the exact boundary conditions near $u = u_{min}$ and $u = 0$ are ignored. Under these assumptions, the functional equation for the invariant measure can be formulated as (Perron-Frobenius)

$$p(u) = \frac{1}{\mu}p[u - \frac{1-\mu}{\mu}(u_0 - u)] - \frac{2u}{a^2}p_0 + \frac{2b\lambda}{a^2}p_0, \quad u_{min} \leq u \leq u_{max}, \qquad (19)$$

where $u_0 = u_{max}/(1-\mu)$ is the virtual fixed point of the one-dimensional map. The meaning of this equation is that the measure at a point is equal to the sum of the measures of all of its direct pre-images, each divided by the absolute value of the derivative of the mapping in the pre-images. The homogeneous part of this functional equation is of the form of an infinite-order Euler differential equation, which admits a solution of the form

$$p(u) = p_0(C_1 u + C_2 + C_3/(u_0 - u)), \quad (u \leq 0) \qquad (20)$$

$$C_1 = \frac{2\mu^2}{a^2(1-\mu^2)}, \quad C_2 = \frac{2b\lambda\mu^3}{a^2(1-\mu)^2(1+\mu)}, \quad C_3 = (1-C_2)\frac{b\lambda}{1-\mu}.$$

The value of p_0 is determined by the condition that the total measure is equal to one, which yields

$$p_0 = 1/[-\frac{1}{2}C_1 u_{min}^2 - C_2 u_{min} - C_3 \ln(u_0/(u_0 - u_{min})) + u_{max}]. \qquad (21)$$

For the case that $\mu = 1$, these formulae are not valid, and the invariant measure can be approximated by the alternative formula

$$p(u) = p_0(-\frac{u^2}{a^2 b\lambda} + \frac{u}{a^2} + 1), \quad (u \leq 0) \qquad (22)$$

where

$$p_0 = 1/[\frac{u_{min}^3}{3a^2 b\lambda} - \frac{u_{min}^2}{2a^2} - u_{min} + u_{max}]. \qquad (23)$$

From the invariant measure one can calculate the Liapunov exponent Λ according to the ergodic theorem [3] as

$$\Lambda = \lim_{n \to \infty} \frac{1}{n} \sum_{k=0}^{n-1} \ln|f'(u_k)| = \int p(u) \ln|f'(u)| du, \qquad (24)$$

$$\Lambda = \int_{u_{min}}^{0} p(u) \ln \mu \, du + \int_{0}^{u_{max}} p_0 \ln(\frac{a}{2\sqrt{u}}) \, du = \qquad (25)$$

$$\ln \mu + p_0 u_{max}(\frac{1}{2} - \ln 2 - \ln \mu - \frac{1}{2}\ln(u_{max}/a^2)).$$

The Liapunov exponent is an increasing function of the variable parameter λ, starting from a value of zero at $\lambda = 0$. From Eq. (25), it appears that the Liapunov exponent does not start from zero, which is due to the approximations. The approximations become more accurate as μ is closer to one and λ closer to zero.

3 Conclusions

With the analysis we have gained insight in a quite general mechanism by which chaotic vibrations may set in when a periodic solution of a mechanical systems starts to hit a rigid stop. The conditions for the existence of a trapping region near the periodic solution have been derived and the existence and stability conditions for several periodic solutions were investigated. Some quantitative predictions on periods and Liapunov exponents can be made. It appears that the behaviour of systems with many degrees of freedom is comparable to systems with one or two degrees of freedom.

Several open questions remain to be investigated. More reliable estimates of the Liapunov exponents are still needed; the behaviour of systems with an infinite number of degrees of freedom may differ considerably from corresponding finite-dimensional models; special cases in a multiparameter context are to be considered.

A comparison between analytical predictions and numerical results will be published elsewhere [9].

References

[1] W. Chin; E. Ott; H.E. Nusse; C. Grebogi: "Grazing bifurcations in impact oscillators," *Physical Review* **E 50** (1994), pp. 4427–4444.

[2] P. Collet; J.-P. Eckmann: *Iterated Maps on the Interval as Dynamical Systems*, Birkhäuser, Boston, 1980.

[3] J.-P. Eckmann; D. Ruelle: "Ergodic theory of chaos and strange attractors," *Reviews of Modern Physics* **57** (1985), pp. 617–656.

[4] C. Knudsen; R. Feldberg; H. True: "Bifurcations and chaos in a model of a rolling railway wheelset," *Philosophical Transactions of the Royal Society of London* **A 338** (1992), pp. 455-469.

[5] W.T. Koiter: *Over de Stabiliteit van het Elastisch Evenwicht* (Dissertation, Delft University of Technology), H.J. Paris, Amsterdam, 1945.

[6] H. Lamba; C.J. Budd: "Scaling of Lyapunov exponents at nonsmooth bifurcations," *Physical Review* **E 50** (1994), pp. 84–90.

[7] J.P. Meijaard: *Dynamics of Mechanical Systems, Algorithms for a Numerical Investigation of the Behaviour of Non-Linear Discrete Models* (Dissertation, Delft University of Technology), Private publication, Delft, 1991.

[8] J.P. Meijaard: "Continuous and discontinuous modelling of the contact between wheel flange and rail," in J.F. Dijksman; F.T.M. Nieuwstadt (eds): *Topics in Applied Mechanics, Integration of Theory and Applications in Applied Mechanics*, Kluwer Academic Publishers, Dordrecht, 1993, pp. 119–126.

[9] J.P. Meijaard: "A mechanism for the onset of chaos in mechanical systems with motion-limiting stops," *Chaos, Solitons & Fractals* **7** (1996), pp. 1649–1658.

[10] J.P. Meijaard; A.D. de Pater: "Railway vehicle systems dynamics and chaotic vibrations," *International Journal of Non-Linear Mechanics* **24** (1989), pp. 1–17.

[11] A.B. Nordmark: "Non-periodic motion caused by grazing incidence in an impact oscillator," *Journal of Sound and Vibration* **145** (1991), pp. 279–297.

[12] H.E. Nusse; E. Ott; J.A. Yorke: "Border-collision bifurcations: an explanation for observed bifurcation phenomena," *Physical Review* **E 49** (1994), pp. 1073–1076.

[13] H.E. Nusse; J.A. Yorke: "Border-collision bifurcations including "period two to period three" for piecewise smooth systems," *Physica* **D 57** (1992), pp. 39–57.

Industrial processes

MODELLING INDUSTRIAL PROCESSES INVOLVING INFILTRATION IN DEFORMABLE POROUS MEDIA

D. Ambrosi* and L. Preziosi**

* CRS4: Centro Ricerche e Sviluppo Studi Superiori Sardegna
Via Nazario Sauro 10, Cagliari, 09123, Italy
** Dipartimento di Matematica, Politecnico
Corso Duca degli Abruzzi 24, Torino, 10129, Italy

Abstract

The paper deduces a new model to simulate industrial processes involving infiltration in deformable porous media. If the porous material is not rigid, then it can deform under the action of the forces associated with the flow, both in the infiltrated and in the uninfiltrated region. The coupled flow/deformation problem in the two interfaced regions is then formulated with the proper evolution equations for the boundaries delimiting the two domains and with the relative boundary and interface conditions. The unsteady problem is solved numerically, putting in evidence the influence of the deformations on the bulk flow and on the propagation of the advancing front and allowing as a by-product a prediction of the stress and deformation states.

1 Introduction

Many technological processes involve deformable porous media. In some of them, e.g. fabric dyeing, coffee brewing, composite materials manufacturing, and in general in all processes involving infiltration in sponge-like materials, there is a need of modelling the coupled flow/deformation problem in two interfaced domains, the region which has been already infiltrated and the one which has not.

This paper will consider the just stated problem in one dimension and in the slug-flow approximation, i.e. assuming that a sharp front divides the three domains, which is reasonable when the applied pressure is much larger than the capillary pressure.

Another non-trivial problem which is often overlooked and is discussed here in detail is the formulation of the boundary and interface conditions that need be

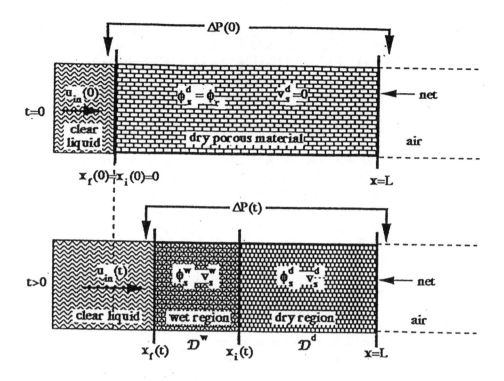

Figure 1: Geometrical schematization of the infiltration problem.

joined to the deformable porous media model and the deduction of the evolution equations for the interfaces. The model then presents features that encourage the introduction of domain decomposition methods. The coupled flow/deformation problem is then solved mainly by finite differences methods using upwind methods based on characteristic decomposition.

The evolution of the system is very different from that obtained when the coupling between fluid flow and deformation of the porous medium is absent. In particular, the stationary configuration is non-uniform. In fact, when a fluid flows through a deformable porous medium the forces associated with the flow deform the porous material. In turn, the deformation of the porous medium influences the flow. In addition, in some industrial processes, e.g. composite materials manufacturing, it is important to monitor deformation and stress states to identify in advance possible inhomogeneities and damages in the reinforcing network, which can be obtained as a by-product of the integration of the present model.

2 Infiltration in Deformable Porous Media

Assume that the deformable porous medium undergoes one-dimensional infiltration and deformation along a principal direction of its permeability tensor.

For $t < 0$ the porous material is dry, occupies the region $x \in [0, L]$ and is compressed at a given volume ratio $\phi_s(x, t = 0) = \phi_0(x)$, or possibly it is in its undeformed configuration $\phi_s(x, t = 0) = \phi_r$. Its right border is constrained by a net, so that it can not be pushed beyond $x = L$. At the same time, however, both air and liquid may freely pass through the net which does not offer any resistance to the flow.

Referring to Fig. 1, the liquid flows in the positive direction either pushed at a given velocity $u_{in}(t)$, or forced by an imposed pressure difference $\Delta P(t)$ between the extrema of the porous medium. At $t = 0$, the liquid touches the free border x_f of the porous material which compresses while the liquid starts infiltrating. Therefore, while the right border of the sponge is fixed at $x = L$, the other one moves to $x = x_f(t)$, and part of the porous material wets, say up to $x = x_i(t)$.

In doing this we are implicitly simplifying capillary phenomena assuming the existence of a sharp flat front, which divides the fully saturated porous medium $\mathcal{D}^w = [x_f(t), x_i(t)]$ from the remaining uninfiltrated portion $\mathcal{D}^d = [x_i(t), L]$. This assumption, often called slug-flow approximation, is valid when the applied pressure is much larger than the capillary pressure.

As infiltration proceeds both the dry and the wet porous material compress or expand back, according to the flow conditions. In [1-3] it is suggested that the wet porous material be modelled as an anelastic solid or as a Voigt-Kelvin solid. Unfortunately, the literature lacks of experimental results in this direction. Hence, as a first step, we here assume that the deformable porous medium respond elastically to deformations and in a way independent on its being dry or wet.

Following [3], we can write the following model

$$\begin{cases} \dfrac{\partial \phi_s^w}{\partial t} + v_s^w \dfrac{\partial \phi_s^w}{\partial x} + \phi_s^w \dfrac{\partial v_s^w}{\partial x} = 0, \\ \\ \rho_s \phi_s^w \left(\dfrac{\partial v_s^w}{\partial t} + v_s^w \dfrac{\partial v_s^w}{\partial x} \right) + \mu \dfrac{v_s^w - u_{in}(t)}{K(\phi_s^w)} + \Sigma'(\phi_s^w) \dfrac{\partial \phi_s^w}{\partial x} = 0, \end{cases} \quad x \in \mathcal{D}^w = [x_f(t), x_i(t)]$$

$$\begin{cases} \dfrac{\partial \phi_s^d}{\partial t} + v_s^d \dfrac{\partial \phi_s^d}{\partial x} + \phi_s^d \dfrac{\partial v_s^d}{\partial x} = 0, \\ \\ \rho_s \phi_s^d \left(\dfrac{\partial v_s^d}{\partial t} + v_s^d \dfrac{\partial v_s^d}{\partial x} \right) + \Sigma'(\phi_s^d) \dfrac{\partial \phi_s^d}{\partial x} = 0, \end{cases} \quad x \in \mathcal{D}^d = [x_i(t), L]$$

(1)

where the superscripts w and d stand, respectively, for *wet* and *dry*. In Eq.(1) $K(\phi_s^w)$ is the permeability and $\Sigma'(\phi_s)$ is the derivative with respect to the volume ratio ϕ_s of the strictly increasing function relating the excess stress (taken positive in compression) with the volume ratio. As already stated, the same relation is used for both the dry and the wet porous medium. Finally, the quantity v_s is the velocity of the solid and μ is the dynamic viscosity of the liquid.

As far as the evolution equations determining the position of the sponge boundary $x_f(t)$ and the infiltration front $x_i(t)$ are concerned, we can observe that $x_f(t)$ is a material interface fixed on the solid phase, while $x_i(t)$ is a material interface fixed on the liquid phase. Therefore, they have to move respectively with the velocity of the solid and of the liquid at the relative interface. This gives

$$\begin{cases} \dfrac{dx_f}{dt} = v_s^w(x_f(t), t) ,\\[6pt] \dfrac{dx_i}{dt} = \dfrac{u_{in}(t) - \phi_s^w(x_i(t), t) v_s^w(x_i(t), t)}{1 - \phi_s^w(x_i(t), t)} . \end{cases} \qquad (2)$$

The boundary conditions are obtained observing that one end of the porous medium is stress-free, while the other one is fixed, which gives

$$\begin{cases} \phi_s^w(x_f(t), t) = \phi_r , \\[4pt] v_s^d(L, t) = 0 , \end{cases} \qquad (3)$$

where ϕ_r is defined by $\Sigma(\phi_r) = 0$.

Furthermore, following [3,4], one has that at the infiltration front $x_i(t)$ both the volume ratio and the velocity of the solid constituent are continuous

$$\begin{cases} v_s^w(x_i(t), t) = v_s^d(x_i(t), t) , \\[4pt] \phi_s^w(x_i(t), t) = \phi_s^d(x_i(t), t) . \end{cases} \qquad (4)$$

Finally, the initial conditions are

$$\begin{cases} \phi_s^d(x, 0) = \phi_r & \text{for } 0 \leq x \leq L, \\[4pt] v_s^d(x, 0) = 0 & \text{for } 0 \leq x \leq L, \\[4pt] x_f(0) = 0 , \\[4pt] x_i(0) = 0 . \end{cases} \qquad (5)$$

When the porous material is fully infiltrated, i.e. $\mathcal{D}^d = \emptyset$, then the integration of the initial boundary value problem continues with Eqs.(1a,b) with the evolution equation (2a), and the boundary conditions (3a) and

$$v_s^w(L, t) = 0 . \qquad (6)$$

The quantity $u_{in}(t)$ which appears in the initial boundary value problem depends on how the liquid is injected in the porous material. If it is pushed by a piston at a given velocity, then $u_{in}(t)$ is actually a given function of time, e.g. $u_{in} = \text{const}$. If, instead, the flow is driven by a prescribed pressure difference

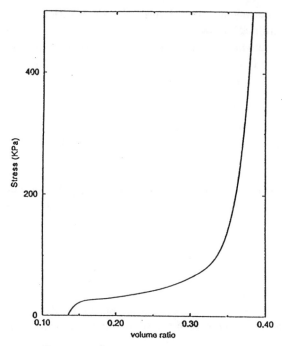

Figure 2: Stress-volume ratio relation.

$\Delta P_\ell(t) = P_\ell(x_f(t), t) - P_\ell(L, t)$ between the extrema of the porous media (say, constant), then the inflow velocity is determined by

$$u_{in}(t) = \frac{\dfrac{\Delta P_\ell(t)}{\mu} + \displaystyle\int_{x_f(t)}^{x_i(t)} \dfrac{v_s^w(x,t)\,dx}{K(\phi_s^w(t,x))} + \dfrac{\mu_{air}}{\mu} \displaystyle\int_{x_i(t)}^{L} \dfrac{v_s^d(x,t)\,dx}{K(\phi_s^d(t,x))}}{\displaystyle\int_{x_f(t)}^{x_i(t)} \dfrac{dx}{K(\phi_s^w(t,x))} + \dfrac{\mu_{air}}{\mu} \displaystyle\int_{x_i(t)}^{L} \dfrac{dx}{K(\phi_s^d(t,x))}}, \qquad (7)$$

(see [3] for more details).

3 Simulation Method and Results

The differential system (1) has hyperbolic nature and can therefore yield discontinous solutions, depending on the initial and boundary conditions. It has been discretized by using an upwind scheme of Godunov type, based on characteristic decomposition of the equations. This approach ensures sharp capturing of the discontinuities and accurate computation of the solution in the regions with smooth behavior. The movement of the boundary $x_f(t)$ has been taken into account by a coordinate-free discretization of the fluxes, so that the nodes of the computational grid can move into the domain remaining equi-spaced as the solid compresses and relaxes. The interface boundary $x_i(t)$ is treated as a marker, indicating where the former and the latter equations of system (1) are to be solved.

In the simulations which follow we use, as a starting point, a stress-volume ratio dependence similar to the one given in Fig. 2 which refers to Sommer and Mortensen's [5] experiments on a polyurethane sponge (TF-5070-10). The permeability-volume ratio relation is $K(\phi_s) = K_r 10^{-6.9654(\phi_s - \phi_r)}$. The values of ϕ_r, K_r and $\Sigma'(\phi_r)$ relative to Sommer and Mortensen's polyurethane sponge are, respectively, $\phi_r = 0.135175$, $\Sigma'(\phi_r) = 2.522532$ MPa and $K_r = 1.6851 \cdot 10^{-11}$ m^2.

Figure 3 presents a simulation obtained for given u_{in}. In particular, Figure 3a gives the evolution of the volume ratio versus x/L at different dimensionless times $\tau_i = u_{in} t_i / L$, while Figure 3b gives the position of the infiltration front $x_i(\tau)/L$ and of the boundary of the porous material $x_f(\tau)/L$. The sponge gradually compresses while the position of its left boundary monotonically increases. The infiltration front, which can be easily identified in Fig. 3a locating the discontinuity in the slope of the graph, travels at a nearly constant speed. It has to be remarked that the volume ratio in the dry region is nearly space independent, but depends on time. After the infiltration front x_i has reached the other end of the sponge, which, in dimensionless terms, occurs for $\tau = 1 - \phi_r$, as analitically predicted, the solution tends to the stationary configuration determined in [3], which is clearly not uniformly compressed. The sponge may completely relax back only if inflow is stopped and further spontaneous imbibition is allowed (see Figs. 3 and 4 of [2]).

In Fig. 4 the flow is driven by a constant pressure difference between the extrema of the porous material. At the very beginning, the applied pressure difference generates a very high inflow velocity (say, up to 10 m/s), due to the smallness of air viscosity and of the infiltrated layer (see Eq.(7)). Therefore, the elastic porous material suddenly compresses. Due to the hyperbolic character of the evolution equation, which is in turn related to the elastic nature of the porous material, a compression front propagates into the sponge at a speed of the order of $\sqrt{\Sigma'(\phi_s)/\rho_s}$. This is evident at very small times (see [3]). Due to its elasticity, after compressing the porous material expands back, generating oscillations in the location of the interfaces having period related to the time needed for the compression-rarefaction wave to travel back and forth between the borders of the porous material (see Fig. 4b). Referring to [1–3] for a more detailed discussion, it is possible to state that these effects will be strongly smoothed out if the porous material were more properly modelled as a Voigt-Kelvin solid or as an anelastic solid. In this last case the system of equations would still be hyperbolic, but a stronger attenuation factor would be present. In the mean the border of the porous material moves forward and then partially backwards, so that the sponge over-compresses and then slowly expands tending to its steady configuration. The time needed to achieve full infiltration can be identified locating the discontinuity in slope in Figs. 3b and 4b, i.e. when the graph reaches $x/L = 1$. Comparing Figs. 3a and 4a, it has to be remarked that in Fig. 4a the volume ratio in the dry porous material becomes nearly space independent only after a transient.

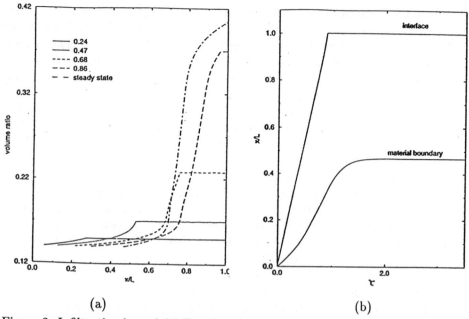

Figure 3: Infiltration in an initially relaxed sponge-like material due to a constant inflow velocity corresponding to $\mathcal{P} = \frac{\rho_s u_{in} K_r}{\mu L} = 4 \cdot 10^{-4}$ and $\mathcal{S} = \frac{\Sigma'(\phi_r) K_r}{\mu u_{in} L} = 25$. (a) Volume ratio versus x/L at different dimensionless times $\tau_i = u_{in} t_i / L$. The discontinuity in slope locates the advancing front separating the wet and the dry region. (b) Temporal evolution of the infiltration front and of the sponge border.

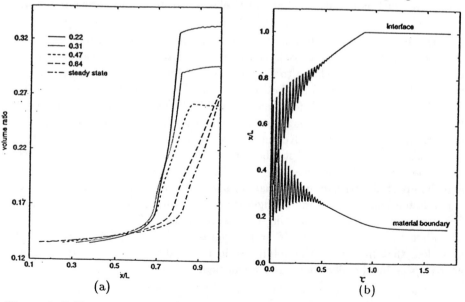

Figure 4: Infiltration in an initially relaxed sponge-like material due to a constant pressure drop corresponding to $\mathcal{P} = \rho_s \Delta P_\ell \left[\frac{K_r}{\mu L}\right]^2 = 4 \cdot 10^{-4}$ and $\mathcal{S}_P = \frac{\Sigma'(\phi_r)}{\Delta P_\ell} = 50$.

zu Figure 4:

(a) Volume ratio versus x/L at different dimensionless times $\tau_i = u_{in}t_i/L$. The initial sudden compression at the preform end generates an initial over-compression of the sponge. The discontinuity in slope locates the advancing front separating the wet and the dry region. (b) Temporal evolution of the infiltration front and of the sponge border. The oscillations of the interfaces are not numerical effects, but are due to the elasticity of the porous material.

4 Acknowledgements

The authors are grateful to the Italian National Research Council (C.N.R.), to the Italian Ministry for the University and Scientific Research (M.U.R.S.T.) and to the Sardinian Regional Authorities for funding the present research.

References

1. Preziosi, L., The theory of deformable porous media and its application to composite material manufacturing, Surveys in Mathematics for Industry, **6**, (1996), pp.167–214.

2. Preziosi, L., Joseph, D.D., and Beavers, G.S., Infiltration in initially dry, deformable porous media, Int. J. Multiphase Flows, **22**, (1996), pp.1205–1222.

3. Ambrosi, D., and Preziosi, L., Modelling matrix injection through elastic porous preform, *Composites A*, To appear.

4. Müller, I., Thermodynamics of mixtures of fluids, J. Mécanique, **14**, (1975), pp.267–303.

5. Sommer, J.L., and Mortensen, A., Forced unidirectional infiltration of deformable porous media, J. Fluid Mech., **311**, (1996), 193–215.

Modeling the Thermal Decomposition of Chlorinated Hydrocarbons in an Ideal Turbulent Incinerator

Markus Kraft, Harald Fey,
Fachbereich Chemie, Universität Kaiserslautern,
Erwin Schrödinger Straße, D 67663 Kaiserslautern, FRG
mkraft@rhrk.uni-kl.de, fey@mathematik.uni-kl.de

Carlo Procaccini, John P. Longwell, Adel F. Sarofim
Department of Chemical Engineering,
Massachusetts Institute of Technology, Cambridge, MA 02139, USA
cpro@mit.edu

Henning Bockhorn
Institut für Chemische Technik, Universität Karlsruhe
Kaiserstraße 12, D 76128 Karlsruhe, FRG
bockhorn@ict.uni-karlsruhe.de

Keywords *reactor model, stochastic differential equations, turbulent mixing, chlorinated hydrocarbons, incineration*

1 Introduction

Combustion of wastes is an effective disposal technology. Although there are many examples of application in industry, severe environmental concerns are raised regarding the emissons from incinerators. Thermal treatment plants are usually comprised of a primary combustion chamber followed by a second combustion chamber, as schematically described in Fig. 1. During the oxidation of solid and liquid wastes in the primary chamber the mixing and the reaction conditions are inhomogeneous. Thus, the wastes undergo incomplete combustion and the gases traveling to the secondary stage of the reactor are rich in potentially hazardous products of incomplete combustion (PICs) together with unreacted wastes. For instance, incineration of chlorinated hydrocarbons can lead to highly toxic chemical species like polychlorinated dibenzo-p-dioxins (PCDDs) and polychlorinated dibenzofurans (PCDFs). The secondary chamber is the component of the plant which most strongly influences the air quality of the exhaust. Additionally, fuel (usually a waste) is injected to raise the temperature, promote better mixing of the reactants, and thus complete the oxidation of wastes. Formation of large droplets, quenching due to the existence of cold zones, and poor turbulent mixing of the gaseous reactants are examples of faults which may lead to unwanted emissions. In Germany, for instance, the BImSchG law (Bundes-Immissionsschutzgesetz) requires a residence time of $2s$ at a temperature of 1200°C for the combustion of chlorinated hydrocarbons to achieve their complete conversion. Nonetheless, the current regulation is not based on an adequate knowledge of the oxidation of halogenated species in the incineration systems [3].

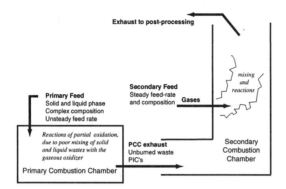

Figure 1: Conversion processes in thermal treatment plants

This paper focuses on the effects of turbulent mixing on the formation of trace, potentially toxic, byproducts in the secondary combustion chambers of incinerators.

2 The Partially Stirred Plug Flow Reactor

The physical and chemical processes in the secondary combustion chamber of an incinerator can be described in terms of the Navier-Stokes equations and the transport equations for all involved chemical species. Since we are interested in trace byproducts there is a need to employ a detailed chemical model which leads to a high number of chemical species which have to be considered. In this work we use the detailed chemical model given in [1] by Barat et al. This model contains 50 chemical species and 224 reactions. Thus, the resulting system of transport equation is very large. Direct numerical simulations of a typical turbulent high Reynolds number flows exceed the present computational capacities. Therefore, there is a need to use a simpler model. In the present work, a stochastic chemical reactor model, the Partially Stirred Plug Flow Reactor (PaSPFR) [5], is used. This is a modification of the classical Plug Flow Reactor (PFR) scheme, similar to the Partially Stirred Reactor (PaSR) in [2].

2.1 Governing Equations

We consider the composition vector in the PaSPFR $\underline{\phi}_t = (\phi_t^1, \ldots, \phi_t^L)^T = (Y_t^1, \ldots, Y_t^{L-1}, h^*)^T$ to be a stochastic process. The pressure is assumed to be constant at 1 atm and the reactor is assumed to be adiabatic. $Y^l \in [0,1]$ $l =$

$1,\ldots, L-1$ are mass fractions of $L-1$ chemical species and $h^* = \int_{T_0}^{T} c_p\, dT$, $h^* \in \mathbb{R}$ is the enthalpy contribution due to the heat of the gas mixture in the volume element of the reactor. The time evolution of the stochastic process $\{\phi_t, t \leq T\}$ is given by the vector stochastic differential equation (SDE) with initial condition $\underline{\psi}_{t_0} = \underline{\psi}_0$. The l-th component the SDE is given by (see Ref. [7])

$$d\psi_t^l = -\left(A^l(\underline{\psi}_t) + S^l(\underline{\psi}_t)\right) dt + \sum_{m=1}^{L} \sqrt{B^{lm}(\underline{\psi}_t)} dWbin_t^l, \qquad (1)$$

where $Wbin = \{Wbin_t, t \geq 0\}$ is L- dimensional normalized binomial process with components $Wbin_t^1, \ldots, Wbin_t^L$ which are independent scalar binomial processes with respect to a common family of $\sigma-$ algebras. Then $A : \mathbb{R}^L \mapsto \mathbb{R}^L$, $S : \mathbb{R}^L \mapsto \mathbb{R}^L$, and $B : \mathbb{R}^L \mapsto \mathbb{R}^{L \times L}$ are given by:

$$A^l(\underline{\psi}) = -\frac{1}{2}\left[1 + k\left(1 - \frac{D(\phi_l)}{\psi_{l*}^2}\right)\right]\frac{C_\phi}{\tau}(\psi_l - E(\phi_l)) \qquad (2)$$

$$B^{lm}(\underline{\psi}) = \begin{cases} k\left[1 - \frac{(\psi_l - E(\phi_l))^2}{\psi_{l*}^2}\right]\frac{C_\phi}{\tau}D(\phi_l) & : l = m \\ 0 & : l \neq m \end{cases} \qquad (3)$$

$$\psi_{l*} = \begin{cases} (\psi_l^{\max} - E(\phi_l)) & : (\psi_l - E(\phi_l)) \geq 0 \\ (\psi_l^{\min} - E(\phi_l)) & : (\psi_l - E(\phi_l)) < 0 \end{cases} \qquad (4)$$

The empirical constant k is set to be 0.1 and $C_\phi = 2.0$ as suggested in, e.g., Ref. [6]. The chemical source term is $S^l = \rho^{-1}\dot{\omega}_l(\underline{\psi})W_l$ ($l = 1, \ldots, L-1$) where $\dot{\omega}_l(\underline{\psi})$ is the molar production rate and W_l is the molar weight of the l-th species. Source term for the enthalpy is $S^L = \rho^{-1}\sum_{l=1}^{L-1} h_{0l}W_l\dot{\omega}_l(\underline{\psi})$ where h_{0l} is the formation enthalpy of the l -th species. To obtain statistical moments of the stochastic process information on its probability density, $f(\underline{\psi}, t)$, is required. Here, the mean of the random variable ϕ_l is denoted as $E(\phi_l)$ and its variance as $D(\phi_l)$. In a variable density flow the quantity which is usually studied is the mass probability density function (MDF) $\mathcal{F}(\underline{\psi}; t) = \rho(\underline{\psi})f(\underline{\psi}, t)$. To calculate $\mathcal{F}(\underline{\psi}; t)$ at time t we approximate the initial condition MDF $\mathcal{F}(\underline{\psi}; 0) \approx \sum \Delta m\, \delta(\underline{\psi} - \underline{\psi}^n)$ by N particles with weight Δm. Each particle serves as initial condition with $\underline{\psi}_{t_0}^n = \underline{\psi}_0^n$ of equation (1). The ensemble of particles at time t, $\underline{\psi}_t^n$, approximate the MDF $\mathcal{F}(\underline{\psi}, t)$. Details of the numerical procedure are given in Ref. [5].

2.2 Initial Conditions

The PaSPFR model accounts for two of the processes which take place in a combustion chamber. The first one is a set of chemical reactions, the second one is the mixing of chemical species due to turbulence. The PaSPFR is a tubular reactor in which a statistically stationary, isotropic, homogeneous turbulent flow exists. The evolution of a small volume of gas, moving downstream at a speed equal to the mean velocity, can be described through the spatial changes of the relevant physical quantities along the reactor axis. The relaxation constant τ represents different degrees of turbulent mixing intensity. For $\tau = 0$, ideal mixing takes place inside the reactor, and the PaSPFR model is equivalent to a homogeneous PFR model. For $\tau = \infty$, no mixing occurs. In the present study, we assume that two initially separated streams, V_1 and V_2, mix and react inside the PaSPFR. V_2 represents the hot products of fuel-lean combustion, coming from the primary chamber. V_1 is a stream of CH_3Cl, chosen as a typical waste surrogate. The ratio between the volume flow rates, $\dot{V}_1/\dot{V}_2$, is fixed at 0.05. The only parameter varied in this study is the characteristic turbulent mixing time τ. Numerical simulations are performed for $\tau = 0s, 0.025s, 0.05s$, and $0.1s$. For $\tau = 0.1s$, the characteristic mixing and chemical times are of the same order of magnitude. The perfectly premixed inlet composition and temperature, when $\tau = 0$, are shown in table 1 as "PFR" inlet conditions. As initial condition for the MDF transport equation, we have chosen the joint composition mass density function at time $t = t_0$.

$$\mathcal{F}(\underline{\psi}; t_0) = \mathcal{F}_0(\underline{\psi}) = \frac{1}{2}\rho(\underline{\psi})\left(\delta(\underline{\psi} - \underline{\psi}_0^{(1)}) + \delta(\underline{\psi} - \underline{\psi}_0^{(2)})\right) \quad (5)$$

In equation (5) the vectors $\underline{\psi}_0^{(1)}$ and $\underline{\psi}_0^{(2)}$ are the mass fraction and enthalpy composition vectors of stream V_1 and stream V_2 respectively.

	Waste $\dot{V}_1$	Exhaust $\dot{V}_2$	PFR
$X(H_2O)$	0.000	0.064	0.062
$X(O_2)$	0.000	0.062	0.060
$X(CO_2)$	0.000	0.064	0.062
$X(CH_3Cl)$	1.000	0.000	0.023
$X(N_2)$	0.000	0.810	0.793
Temp.	1100 K	1350 K	1334 K

Table 1: Inlet conditions to the PaSPFR ($\dot{V}_1,\dot{V}_2$) and the PFR as used in this study. $X(k)$ are mole fractions of the chemical species and the ratio of the volume streams between V_1 and V_2 is set to be $\dot{V}_1/\dot{V}_2 = 0.05$.

3 Results and Discussion

In order to investigate the influence of different degrees of mixing (i.e.different values of τ) on the chemical reaction pathways during the ignition process, we have first studied the perfectly premixed case ($\tau = 0$). Here, the PFR model can be used because both streams V_1 and V_2 are homogeneously mixed at the time $t = 0s$. Later, different degrees of mixing intensity, represented by different values of the parameter τ, have been studied in the model. A special focus of the study is on the effect of unmixedness on the mechanism of chlorine inhibition. Chlorine is know to inhibit the hydrocarbon oxidation by depleting the H radical, used to produce HCl, and consequently causing a decrease in OH available for the burnout of CO [1].

3.1 Chemical System

Figure 2 schematically represents, the final stage of the hydrocarbon combustion, i.e. the CO oxidation. The reaction numbers refer to the reactions in Ref. [1]. The oxidation of CO takes place mainly through reaction No. 62, which is driven by the chainbranching reaction No. 133. For this reaction as well as others (Nos. 135,147,137) the H radical is an important starting species.

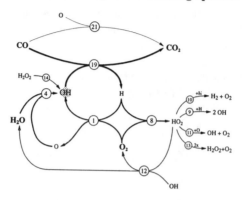

Figure 2: Reaction paths for the oxidation of CO.

In Figure 3 the formation paths of HCl are displayed. The numbers on the arrows represent the net specific mole fluxes of the corresponding species which are converted to HCl. Each path is comprised of a certain number of elementary reactions. These reactions are listed in the same Figure on the right hand side. The percentage of how much each reaction contributes to the formation of HCl are also given. The reactions which are responsible for the H abstraction are mainly Nos. 152 and 178.

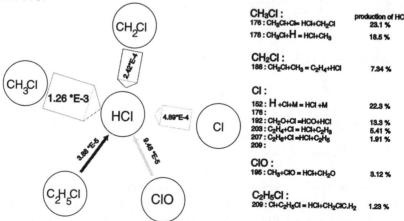

Figure 3: Reaction paths for the HCL production

3.2 Influence of Mixing

In Figures 4 and 5 the time evolution of mean and standard deviation for the OH radical and HCl is shown. The distribution of the mean mass fraction of OH radical is strongly influenced by the mixing intensity. A rapid increase in the OH concentration signals that ignition takes place. Due to the reduction of the mixing intensity (higher τ), the presence of OH radical is shifted to an earlier point in time. Moreover, the sharp peak present in the diagram of the mean OH concentration for $\tau = 0$ is broadened, meaning that the ignition process is distributed over a longer interval.

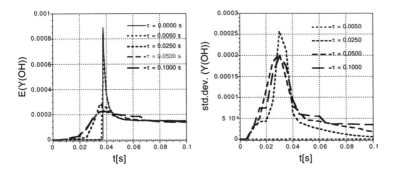

Figure 4: Time evolution of estimated mean and standard dev. of OH mass fraction in the PFR and PaSPFR for different turbulent times.

At $t = 0.07s$ the ignition phase is finished an the system can be considered to be in equilibrium. The turbulence time $\tau = 0.005s$ in the PaSPFR corresponds to the PFR because mixing is completed before the chemical reactions can take place. The time evolution of the standard deviation for $\tau = 0.005s$ illustrates this point. The different ignition process leads to a slight acceleration of CH_3Cl degradation and HCl formation. As indicated by the mean mass fraction of HCl in Figure 5.

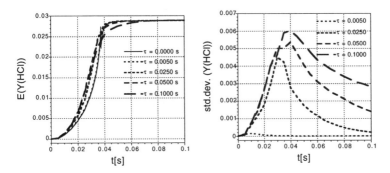

Figure 5: Time evolution of estimated mean and standard dev. of HCl mass fraction in the PFR and PaSPFR for different turbulent times.

The time evolution of the standard deviation indicates that two main processes superimpose. Chemical reactions lead to a strong increase of the standard deviation of the mass fraction of OH and HCl and molecular mixing reduces this variance because of the drift term in equation 1. The less the mixing the slower the decrease in the variance. Since the decay of the variance is slowed down, radicals like OH and H are present for a longer time interval, but at lower concentrations. This change of the radical mass fraction influences the reaction paths.

Figure 6 shows the estimated mean specific conversion of reactions that consume H radicals. Reactions 135 and 133, the chain branching reactions, are important for the oxidation of CO and reactions 152 and 178 contribute to the formation of HCl. These two reactions are responsible for approximately 45 % of the HCl production and because of the consumption of the H radical, they slow down the overall CO oxidation. Figure 6 shows that the reactions 152 and 178 decrease their mean specific molar production as τ increases. On the other hand, reaction 135 and 133 increase the production of H radical.

The H radical contributes to the oxidation of CO rather than to the formation of HCl. As a consequence, the chlorine inhibition is less strong and the degradation of methyl chloride is accelerated as exhibited in the mean mass fraction profiles. The faster formation of hydrogen chloride indicates that the formation paths, as displayed in Figure 3, are shifted. Figure 7 illustrates how the mean of the specific production of HCl by the main contributing reactions varies according to different turbulence mixing times τ.

The overall production of HCl remains constant over the time interval considered because after $0.1s$ the oxidation is fully completed.

As indicated in Figure 6, the contribution of reactions 152 and 178 is not as strong as in the case of perfect mixing ($\tau = 0s$). Instead, reactions 176 and 192 increase their production. These reactions do not compete for the H radical but capture additional chlorine radicals and accelerate the formation of HCl.

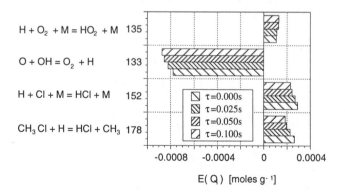

Figure 6: Estimated mean specific conversion of reactions 135, 133, 152, and 178 for varying turbulence times τ. The integration time is $0.1s$.

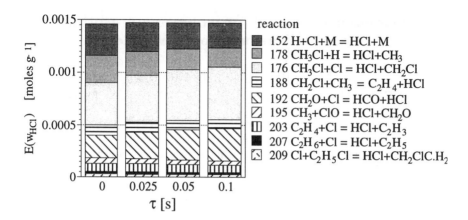

Figure 7: Estimated mean of the specific overall production of HCl for different turbulence mixing times τ. The integrated time interval is set to be [0,0.1]s.

References

[1] Barat R.B., Sarofim A.F., Longwell J.P., Bozelli J.W. : Inhibition of a Fuel Lean Ethylene/Air Flame in a Jet Stirred Combustor of Methyl Chloride: Experimental and Mechanistic Analysis *Combust. Sci. and Tech.*, Vol.74, pp. 361-378, 1990

[2] Chen J.-Y. : Stochastic Modeling of Partially Stirred Reactors *submitted to Combust. Sci. and Tech.* 3/1995

[3] Christill M., Kolb Th., Seifert H., Leukel W., Kufferath A. : Untersuchungen zum termischen Abbauverhalten chlorierter Kohlenwasserstoffe *VDI Berichte* Nr. 1193, pp. 381-388, 1995

[4] Kee R.J., Rupley F.M., Miller J.A.: Chemkin-II: A Fortran Chemical Kinetics Package for the Analysis of Gas Phase Chemical Kinetics *Sandia Report SAND89-8009B UC-706*, 1993

[5] Kraft M., Fey H., Schlegel A., Chen J.-Y., Bockhorn H.: A Numerical Study on the Influence of Mixing Intensity on NOx Formation. *3rd Workshop on Modelling of Chemical Reaction Systems* Proceedings, to appear 1997

[6] Pope S. B.: PDF Methods for Turbulent Reactive Flows *Progress in Energy and Combustion Science*, Vol.11, pp. 119-192, 1985

[7] Valiño L., Dopazo C.: A binomial Langevin model for turbulent mixing *Phys. Fluids*, A3(12), pp.3034-3037, 1991

Mathematical model for isobaric non-isothermal crystallization of polypropylene

S. Mazzullo, R. Corrieri, C. De Luigi
MONTELL Italia S.p.A. - Centro Ricerche "G. Natta" - Ferrara - Italy

Abstract

The paper describes a mathematical model for an experiment in which a cylindrical sample of molten polypropylene is cooled under constant pressure. During cooling, crystallization takes place and density increases.

The model describes the transition of the polymer from the liquid to the solid state, including the process of crystallization.

The governing equations are: the energy balance, the Fourier constitutive equation, the crystallization kinetics and the state equation for density.

Pressure affects the crystallization process mainly through the rise of the glass transition temperature Tg and the equilibrium melting temperature $T°_m$.

Input data to the model are kinetic constants obtained from isothermal crystallization at atmospheric pressure.

Output is the prediction of the pressure-volume-temperature (PVT) diagram of polypropylene.

1 Introduction

One of the most interesting features of polymeric semicrystalline materials is that they melt at relatively low temperature. The polymer in the molten state is a viscous liquid that can be injected into a cold mould to obtain objects of different shapes, after cooling. The technological process of injection moulding requires special machines that operate at high pressure and high speed. During the moulding process, the molten polymer undergoes an intense stress field which interferes with the crystallization process due to cooling in the mould.

In order to facilitate the analysis and the understanding of the complex phenomena that take place during injection moulding, it is necessary to design simplified experiments capable of simulating the individual stages of the whole process.

This paper concerns the final stage of injection moulding: the so-called "packing stage", in which the polymer cools and crystallize under constant pressure. We shall give evidence, both experimentally and theoretically, of the thermodynamic and kinetic effects caused by an external pressure over a polymer, initially in the molten state, during its non-isothermal crystallization.

2 Apparatus and experiments

The apparatus, which is based on a self modified Instrom Capillary Rheometer, consists of a metal cylinder with a cylindrical hole along its axis (Fig. 1). At the bottom, the inner cylinder is sealed off by a copper tool plug sorrounded by a teflon ring, while at the top, a steel piston is inserted, so that a chamber is formed. In the experiment the polymer is loaded into this sample chamber, which has a diameter $\varnothing = 9$ mm. and an initial length $L = 7$ cm.

The thick aluminum wall contains an electrical heating device as well as a network of holes through which air-cooling is possible. The tool plug, at the bottom of the sample chamber, contains a thermocouple that measures the temperature of the sample at the centre. In addition a constant pressure can be applied to the sample by means of the steel piston. Attached to the piston there is an indicator giving the actual position of the piston during the experiment. After stabilization of the system at the operating temperature and pressure, the wall temperature is decreased to initiate the cooling of the system. During the experiment the following variables are measured as a function of time:
- the displacement of the piston
- the temperature at the centre of the sample.

Furthermore the pressure is held at a fixed value throughout the experiment.

The molten polymer crystallizes during cooling, giving rise to spherulites which are birifringent structures, if observed under polarized light. Fig. 2 shows a cross section of the polymeric sample after crystallization, close to the tip of the thermocouple (shown as a semicircle at the bottom). The number and size of spherulites are strongly influenced by the thermomechanical history of the polymer. The spherulites, in turn, influence the ultimate mechanical properties of the manufactured articles.

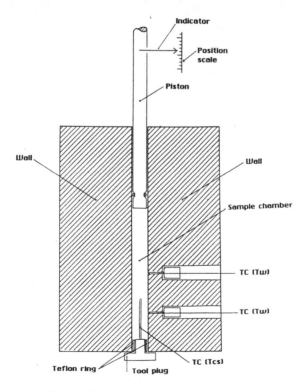

Fig. 1. Schematic of the apparatus

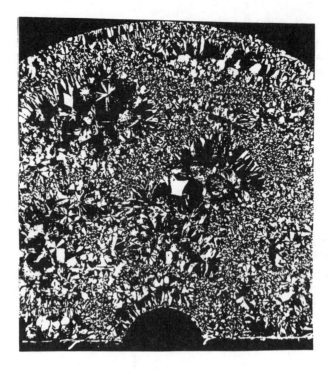

Fig. 2. Cross section of the sample at the tip of the thermocouple, observed under polarized light.

From the dimensions of the chamber and the time dependency of the displacement of the piston, the sample volume as a function of time can easily be calculated. Since no polymer leaks out of the sample chamber, the density as a function of time is also known. Density increase during the cooling experiment is almost 25%.

Eliminating the time variable from the plots of sample temperature and density the PVT diagram of Fig. 4 is obtained where density varies as a function of temperature, while the pressure is a parametric variable. From the PVT diagram, the effect of pressure is clearly detectable. The sharp increase in the derivative of each curve is due to the beginning of crystallization. At higher pressures, the temperatures of incipient crystallization increases as does the final density. A mathematical model was developed to describe and simulate such kinds of experimental evidence.

3 Model equations

The polymeric material in the sample chamber is slowly displaced due to the combined action of cooling and applied pressure. To give some figures, the total displacement of the piston is almost 10 mm. and the cooling experiment takes almost 1^h consequently the average velocity of the piston is $V = 10$ mm/h. For simplycity, we have not taken into account convective and viscosity contributions in the energy balance equation.

The main variables of the model are therefore temperature T, the heat flux q, the degree of crystallinity w and the polymer density ϱ.
The 4 governing equations of such variables are:
the energy balance equation

$$\varrho c_p \frac{\partial T}{\partial t} + div\, q = \varrho \Delta H \frac{\partial w}{\partial t} \qquad (1)$$

the Fourier conduction law

$$q = - K\, grad\, T \qquad (2)$$

the Tobin crystallization kinetic equation

$$\frac{dw}{dt} = k\,(T)\, w^\alpha\, (w_{eq} - w)^\beta\, w^{1-\alpha-\beta}; \quad w_{eq} = 0.6 \qquad (3)$$

the state equation for the density

$$\varrho(P,T) = w\, \varrho_c(T) + (1-w) \frac{B}{B-P} \varrho_a(T) \qquad (4)$$

The observed density ϱ is a weighed average between the density of the crystalline ϱ_c and the amorphous part ϱ_a of the polymer. The weighing function is the degree of crystallinity w. It is assumed that ϱ_c and ϱ_a are functions of temperature [2]:

$$1/\varrho_a = v_a = 1.134 + 9.28E - 4 \times T\,(°C) \qquad (5)$$
$$1/\varrho_c = v_c = 1.059 + 4.25E - 4 \times T\,(°C) \qquad (6)$$

For the amorphous part of the polymer, it is introduced a last correction for the pressure P through the elastic modulus of compressibility B [3]. Consequently, the density of the amorphous phase reversibly increases according to the applied pressure.

However, the largest effect of pressure is on the glass transition temperature (T_g) and the equilibrium melting temperature ($T°_m$). The change of the T_g can be estimate from the firs law of Ehrenfest [4]:

$$\frac{dT_g}{dP} = V_g \frac{\Delta \alpha}{\Delta c_p} \cong 0.26\, \frac{K}{Mpa}\, ; \quad T_g(0) = T^*_g \qquad (7)$$

The variation of the $T°_m$ can be estimate from the Clausius-Clapeyron equation:

$$\frac{d T°_m}{dP} = \frac{\Delta V}{\Delta H}\, T°_m \cong 0.33\, \frac{K}{MPa}\, ; T°_m(0) = T^*_m \qquad (8)$$

The two temperatures influence the kinetics of crystallization (in the sense to anticipate the crystallization when the pressure is increased) through the Hoffman's kinetic constant of overall crystallization $k\,(T)$ [5]:

$$k\,(T) = k_o \exp\left(\frac{u/R}{T-T_g}\right) \exp\left(\frac{K_g(T+T°_m)}{2T^2(T°_m-T)}\right) \quad (9)$$

From a phenomenological point of view, the temperature range in which crystallization takes place, lies between Tg and $T°_m$. The effect of pressure is that it induces a shift toward higher values of both T_g and T_m. Such a shifting effect can be better appreciated by plotting the behaviour of the overall rate constant $k\,(T)$ as a function of temperature, the pressure being a parameter, Fig.3. At higher pressures, the curves shift toward higer temperature values. This implies that, for a given crystallization temperature, the rate of phase change increases as pressure increases.

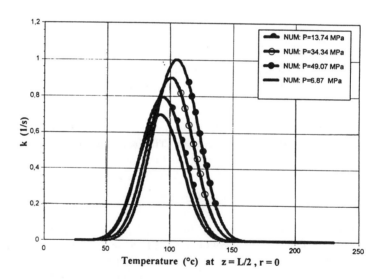

Fig. 3. Overall rate constant as function of temperature and pressure

4 Numerical simulation and results

The natural coordinates to work out the model are cylindrical coordinates with axial symmetry. This imphies that the model is naturally formulated in radial and axial space coordinates and in time coordinate.

Initial and boundary conditions can be easily formulated from the experimental conditions. They are prescribed as follows:

Initial conditions:
$$T(r,z,0) = T_o \quad (10)$$
$$w(r,z,0) = 0 \quad (11)$$

Boundary conditions:
$$\frac{\partial T}{\partial r} = 0 \qquad r = 0 \quad (12)$$

$$\frac{\partial T}{\partial z} = \frac{H}{K}(T - T_e) \qquad z = 0 \quad (13)$$

$$\frac{\partial T}{\partial r} = \frac{-H}{K}(T - T_e) \qquad r = b \quad (14)$$

$$\frac{\partial T}{\partial z} = \frac{-H}{K}(T - T_e) \qquad z = L \quad (15)$$

An extra equation was added to the model to take into account the conservation of the polymer mass, m. This equation is consistent with the choise of disregarding the velocity of polymer contraction. <u>The macroscopic mass conservation</u> reads:

$$m = \frac{\pi}{4} \varrho^2 L \varrho_0 = \frac{\pi}{4} \varrho^2 z \varrho \quad (16)$$

The meaning of this equation may be better seen from the Navier-Stokes equation in the limit case of motion occurring only in the axial z-direction:

$$\varrho \left(\frac{\partial v_z}{\partial t} + v_z \frac{\partial v_z}{\partial z} \right) = -\frac{\partial P}{\partial z} + \frac{1}{Re} \frac{\partial^2 v_z}{\partial z^2} \quad (17)$$

The Reynolds number is very low due to the low velocity of the piston and the very high viscosity of the polymer:

$$Re = \frac{L V \varrho_0}{\mu} \le 3 \times 10^{-8} \quad (18)$$

This implies that, as a good approximation, the motion equation reduces to the solution of Laplace equation, which is immediately integrable in explicit form giving :

$$v_z = -\frac{V}{L} z \quad (19)$$

Consistent with this approximation, the continuity equation should be studied in the steady state approximation:

$$\frac{\partial}{\partial z} \rho v_z = 0 \qquad (20)$$

This equation too is integrable in closed form giving rise to <u>macroscopic mass conservation</u> which link the actual height z of the polymer sample to its actual density ρ:

$$z = L \frac{\rho_o}{\rho} \qquad (21)$$

At each time step, the conservation of the mass in the chamber allows the volume accupied by the contracting polymer to be calculated and, consequently, the displacement of the piston.

$$z = \frac{m}{\sum_{ij} S_{ij} \rho_{ij}} \qquad (22)$$

Where S_{ij} ρ_{ij} are the surface area and the density in the discretized domain of integration.

The model was numerically solved by finite differences using an implicit alternate direction A.D.I. method which is inconditionally stable. For the non linear part of the model it was used the Newton Raphson method.

The model uses as primary information the kinetics of crystallization at atmospheric pressure, obtained by DSC calorimetry operating in isothermal conditions [1].

The input parameters are summarized in table 1.

k_o (s^{-1})	K_g (K^2)	u/R (K)	$T°_g$(K)	$T°_m$ (K)	ΔH (J Kg^{-1})
1.6 E 9	4.87 E 5	6280/8.314	261.2	471.2	1.0 E 5

B (MPa)	K (W m^{-1}K^{-1})	c_p (J Kg^{-1} K^{-1})	α	β
860.0	0.16	2093.0	2/3	0.7650

Tab. 1.Thermodynamic parameters of the model.

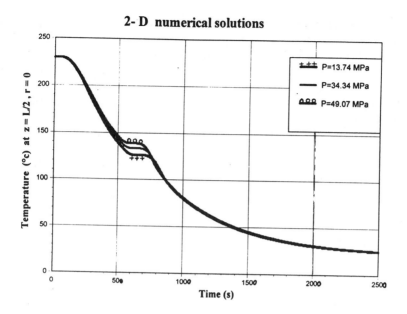

Fig. 4. Calculated temperature at the sample centre.

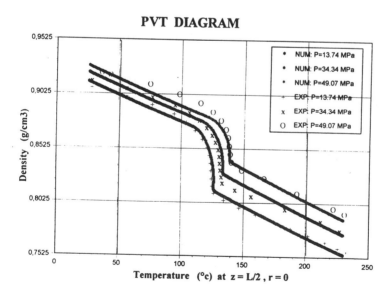

Fig. 5. Experimental and calculated PVT diagram for polypropylene.

The computed temperature profile (Fig. 4) in the middle of the axis of the polymeric sample agrees with the measured profile:
- In particular the inflection point, which is a signal of the beginning of phase transition, is located at the right temperature. Crystallization temperature increases with increasing pressure.

Similarly, the calculated piston displacement is consistent with the observed experimental ones.

The calculated PVT diagram (Fig. 5) is obtained by eliminating the time variable between the two above calculated functions.

Symbols are experimental points and continuous lines are calculated values.

The agreement looks satisfactory and confirms that the simplified assumption of the model are not as rough as they may appear to be at a first sight.

5 *Acknowledgments*

We wish to thank P. Sgarzi for optical microscopy observations.

6 References

[1] R. Caselli, S. Mazzullo, M. Paolini, C. Verdi: "Models, experiments and numerical simulation of isothermal crystallization of polymers"" ECMI VII (A. Fasano, M. Primicerio eds)   Teubner, Stuttgart, pp 167-174 (1993)

[2] W.R. Krigbaum, I. Uematsu : "Heat and entropy of fusion of isotactic polypropylene" J. Polym.Sci.A, 3, pp 767-776 (1965)

[3] S. Timonshenko, J.N. Goodier: "Theory of elasticity", M.C. Graw Hill, New-York (1951)

[4] J.P. Trotignon, J. Verdu: "Skin core structure fatigue behaviour relationships for injection molded parts of polypropylene. I: Influence of molecular weight and injection conditions on the morphology" - J.Appl.Polym.Sci. 34, pp 1-18 (1987)

[5] E.J. Clark, J.D. Hoffman: "Regime III crystallization in polypropylene" Macromolecules, 17, pp 878-885 (1984)

Stochastic modelling and morphological features of polymer crystallization processes

V. Capasso   I. Gialdini   A. Micheletti

Abstract

Crystallization of polymers is modelled as a spatially structured stochastic process consisting of a nucleation phase and a growth phase. A counting process approach together with methods of stochastic geometry lead us to the evolution equations of the relevant quantities of the process.

Estimation of the relevant parameters of the process are obtained via a nonparametric method based on the estimation of the hitting function of the associated Boolean Model, and also by a semi-parametric method based on the estimation of the intensity of the spatial point process known as "Lower tangent points" process.

The theory of random Johnson-Mehl tessellation is then applied to describe the main morphological properties of the final crystalline structure.

1 Introduction

There is a great variety of polymers and their applications are incredibly diversified, including the production of components destined to sophisticated technologies; a growing interest is in their biomedical applications, e.g. for fixing metal prostheses to bones. It is therefore clear how crucial it is to obtain products with well defined chemical and mechanical properties. Schematically, the crystals are assumed to be

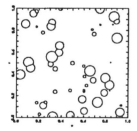 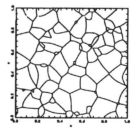

Figure 1: An intermediate instant of a simulation of a 2-D crystallization process and the final generated Johnson-Mehl Tessellation.

balls in the appropriate n-dimensional space, $n = 1, 2, 3$. The crystals (spherulites) appear (or "nucleate") in the amorphous phase randomly either in time or in space and grow with a radial speed $\dot{R}(t) > 0$. The functional form of $\dot{R}(t)$ depends mainly on the temperature, so that in an isothermal crystallization process it may be assumed time invariant. We define V as the available volume and $\tilde{\omega}(t)$ as the crystalline stochastic fraction of the volume at time $t \geq 0$; hence $0 \leq \tilde{\omega}(t) \leq 1$, and it is assumed $\tilde{\omega}(0) = 0$, i.e. that at the initial time the polymer is entirely amorphous. When two growing spherulites hit each other, they stop growing at the contact surface; this phenomenon is called impingement. Its slowing effect on the overall crystallization is a relevant feature of the process, causing the graph of $\omega(t) := E(\tilde{\omega}(t))$ to exhibit the typical sigmoid shape. The geometrical pattern of the crystals at the end of the crystallization process is known as Johnson–Mehl (random) tessellation [8] (see Fig. 1).

If we are not interested in the shape properties of the final tessellation, but only in the time evolution of the stochastic occupied volume fraction, we can neglect the effects of impingement and use the theory of Boolean models to describe the crystallization process.

In Section 2 we will give an account of this second type of modelling; in Section 3 we will discuss two methods, still based on stochastic geometry, to estimate the typical growth parameters in the constant case; in Section 4 we will use the theory of random Johnson-Mehl tessellations to describe the morphological properties of the final crystalline structure.

2 The dynamic Boolean model

In order to model the crystallization process we introduce a basic concept in stochastic geometry: the Boolean model [9]. Let $\phi = \{x_1, x_2, \ldots\}$ be a stationary spatial Poisson process with intensity Λ, where Λ is a σ-finite measure on $\mathbb{R}^d$. Let Θ_i, $i \in \mathbb{N}$ be a sequence of independent identically distributed Random Closed Sets (RACS's), all equivalent to a RACS Θ_0 called the primary grain of the Boolean model. The quantity $\Theta = \bigcup_{n=1}^{\infty} (\Theta_n + x_n)$. is known as a Boolean model. Experimental results and numerical simulations (see Fig. 1) show that, if we neglect the effects of impingement, at a fixed instant t of observation, the crystallization process can be represented by a Boolean model, where the grains are d-dimensional spheres. The only difficulty that arises is that the Boolean model is static, while the crystallization process is dynamic; so we have introduced a "dynamic Boolean model", that is a time-dependent reformulation of the classical Boolean model. In our modelling we will assume that the crystallization is isothermal and that the nucleation rate $\dot{N}_0$ and the growth rate $\dot{R}_0$ are constant. It is also assumed that at the instant $t = 0$ there is no crystal in the amorphous mass. We shall make use of the two following models [2, 5].

Model 1 ("Dynamic" Boolean model) A spatially marked Poisson process $\Phi = \{x_{s_1}, x_{s_2}, ...\}$ is given, with intensity-kernel $\Lambda_s(ds) = \mu_s(.)ds$ where $\mu_s(.)$ is a σ-finite measure on $\mathbb{R}^d$, absolutely continuous with respect to the Lebesgue measure $\nu_d(.)$. At any fixed observation time $t > 0$, a family of independent RACS is associated with Φ, for all $s_i \leq t$, $\{\Theta^t_{s_1}, ..., \Theta^t_{s_n}\}$. The crystallization process is modelled at time $t > 0$ by the stochastic set $\Theta^t = \bigcup_{i=1}^{\infty}(\Theta^t_{s_i} + x_{s_i})$. In the constant case ($\dot{N}_0 = const.$) we assume $\mu_s(.) = \dot{N}_0 \nu_d(.), \forall s \in \mathbb{R}_+$; the mark x_{s_i} represents the location of the germ born at time s_i. Conditional to the birth times $s_1, ..., s_n$, the grains $\Theta^t_{s_i}$ are d-dimensional spheres of deterministic radius given by $R^t_i = \dot{R}_0(t - s_i)$.

Equivalently, in the case $\dot{N}_0, \dot{R}_0$ constant, the following model applies.

Model 2 ("Static" Boolean Model)
The crystallization process at time t is represented as a standard Boolean model with a random spherical primary grain $\Theta^t = \bigcup_{i=1}^{\infty}(\Theta^t_{s_i} + x^t_i)$, where the germs are points of a simple spatial Poisson process $\Phi^t = \{x^t_1, x^t_2, ...\}$, with intensity $\lambda(t)$, that, in the constant case, is $\lambda(t) = \dot{N}_0 t$. The grains $\Theta^t_{s_i}$ are d-dimensional spheres with random radius $R^t_i = \dot{R}_0(t - s_i)$, where s_i is a random variable uniformly distributed in $[0, t]$.

It can be shown [2] that under our assumptions, the two models lead to the same expression for the occupied volume fraction at time t. Due to the Choquet Capacity Theorem [9], the probability law of the process is completely characterized by the "hitting function" which is defined as

$$T_{\Theta^t}(K) = P(\Theta^t \cap K \neq \emptyset) \qquad (1)$$

for every compact K in $\mathbb{R}^d$.

3 Parameter estimation

The aim of this section is to provide "good" estimators for the parameters of the crystallization process in isothermal conditions, that is for the nucleation and the growth rates, which are assumed to be constant ($\dot{N}(t) = \dot{N}_0, \dot{R}(t) = \dot{R}_0$). First we give an account of a non-parametric estimator, based on the estimation of the hitting function of a Boolean model. This allows the estimation of both the nucleation rate $\dot{N}_0$ and of the growth rate $\dot{R}_0$. In this case, while the consistency is obtained as a consequence of the consistency of the estimator of the hitting function, we have no rigorous result about their asymptotic normality. A second semi-parametric estimator is also provided for the sole nucleation rate (assuming the growth rate as known), based on the estimate of the intensity of the Lower

Tangent Point Process, whose consistency and asymptotic normality is provided by a theorem due to I.Molchanov and D.Stoyan [6].

3.1 Nonparametric estimators of the growth parameters

We may estimate our growth parameters $\dot{N}_0$ and $\dot{R}_0$ via a nonparametric estimate of the hitting function, coupled with the concept of *causal cone* [2, 4] (for details about the procedure see [2, 5]). The estimators $\widehat{\dot{N}_0}$ and $\widehat{\dot{R}_0}$ are obtained by substituting a nonparametric estimator of the hitting function, $\widehat{T}_s$ (see [2]), in a non-linear system of two equations in two unknowns. The estimate procedure is based on the appropriate choice of two test sets, which must be neither "too big", nor "too small". If the choice is well done, consistency of the two estimators is obtained, when the available volume V tends to infinity.

As a numerical simulator of the crystallization process [5], up to 3-D, was available, we have used the data provided by this simulator to perform the estimation procedure in order to test the convergence rate of $\widehat{\dot{R}_0}$ and $\widehat{\dot{N}_0}$. A Kolmogorov-Smirnov test shows that $\widehat{\dot{R}_0}$ and $\widehat{\dot{N}_0}$ have an approximate normal distribution [4, 5]; so it is possible to build 95% confidence intervals for the estimators. Experimental results (see [2, 4]) show that in the 1-D case the estimators are evidently biased for small values of the available volume ($V = 10$), but the bias is rapidly reduced by increasing the value of the available volume V, thanks to consistency. Instead, in the 2-D case, the confidence bands show a strong bias, which is not sensibly reduced even increasing the value of V. This fact shows that, in the 2-D case, the rate of convergence of the estimators to their true values, when $V \to \infty$, is slower than in the 1-D case. To improve the estimation procedure, it is necessary to use some *edge correction* tecniques. The results are reported in Fig. 2. With the correction, the true value of the parameter is almost always included in the confidence band, showing so a better agreement between the estimated values and the true ones.

3.2 A semiparametric estimator of the nucleation rate

In many real experiments, the growth rate is easily measurable, while the nucleation rate is more difficult to be computed, because it is not easy to observe the actual number of crystals at each time. A feature that is more easily observable, from photographs or films of real transformations, is the intensity of the Lower Positive Tangent Points Process (LPP), a particular spatial point process (for the definition see [6, 7]). which we will denote by N_{lp}. The intensity kernel of N_{lp} is given by $\lambda_{lp}(t) = \dot{N}_0(1-\widetilde{\omega}(t))\nu_d(.)$. A consistent and unbiased estimator of $\lambda_{lp}(t)$ is $\widehat{\lambda}_r(t) = \frac{card(N_{lp}(t) \cap W_r)}{\nu_2(W_r)}$, where $W_r = rW$ is a window of observation, scaled by the factor r, and $\nu_2(W_r)$ is its area. A consistent and unbiased estimator of $\widetilde{\omega}(t)$, $\widehat{\omega}(t)$, can be obtained, for example, via Montecarlo methods; then a consistent

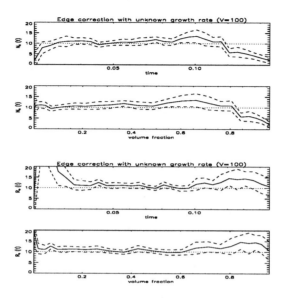

Figure 2: Edge corrected confidence bands for the nucleation and growth rates, with respect to time and volume fraction. In every figure the continuous line is the mean estimated value of the parameter, the dashed lines are the confidence bounds, and the dotted straight line is the real value of the parameter (= 10 for both $\dot{R}_0$ and $\dot{N}_0$).

and unbiased estimator of the nucleation rate is $\widehat{N_{0,r}} = \frac{\widehat{\lambda_r(t)}}{1-\widehat{\omega(t)}}$. The asymptotic normality of this estimator, when $r \to \infty$, is provided by Theorem 5.5 in [6]. This fact allows the construction of confidence bands for the estimated values. Also in this case we have tested the estimation procedure by using data provided by numerical simulations of the crystallization process. The results are reported in Fig. 3. Note that either in presence or in absence of edge correction the confidence bands always include the true value of the parameter.

4 The morphology of the final tessellation

As many physical and chemical properties of the polymeric matherial depend mainly on the crystalline microstructure (that is the number, the dimensions and the shape of the final crystals), it is very relevant the study of the mean geometrical features of the final Johnson-Mehl tessellation [8]. To this aim we have to use a counting process approach to model the crystallization process.

A tessellation of the d-dimensional Euclidean space R^d is a subdivision of R^d into d-dimensional subsets C_i (called cells or crystals) with disjoint topological interiors and so that $C_i = cl(intC_i)$, where int and cl denote topological interior

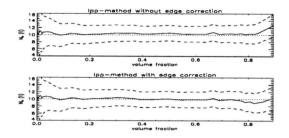

Figure 3: Confidence bands for the nucleation rate (LPP method), with respect to volume fraction. First figure: without edge correction; second figure: with edge correction. In every figure the continuous line is the mean estimated value of the parameter, the dashed lines are the confidence bounds, and the dotted straight line is the real value of the parameter (= 10).

and closure respectively. A Johnson-Mehl Tessellation (see Fig. 1) can be regarded as the result of a birth-growth process of nuclei which form a space-time Poisson point process

$$\Phi = \{a_i = (t_i, \xi_i) \in [0, +\infty) \times R^d\}. \tag{2}$$

The intensity kernel of this process is $\Lambda \times \nu_d$, where ν_d again denotes the Lebesgue measure on R^d and Λ is an arbitrary Radon measure on R^d. Note that the number of nuclei born up to time t in a borel set B is $\Lambda([0,t])\nu_d(B)$. Furthermore the process Φ is spatially stationary.

Suppose now to observe the process when the Johnson-Mehl Tessellation is already completed. Let $C((\xi_i, t_i)|\Phi) = C(a_i|\Phi) := \{y \in R^d : \forall j \neq i \ T_i(y) \leq T_j(y)\}$ be the cell generated by the nucleus ξ_i born at time t_i. Here $T_k(y)$ is the time when the growing crystal of nucleus ξ_k reaches the point y.

Definition. We call *n-facet* a non empty intersection $F_n := F(a_0, \ldots, a_m|\Phi) = \bigcap_{i=0}^{m} C(a_i|\Phi)$, with $m = d - n \geq 0$. The connected components of F_n are called *n-interfaces*.

Let $L_k \subseteq R^d$ be a k-dimensional affine subspace, $0 < k \leq d$. Consider the k-dimensional tessellation generated by the intersection of the Johnson-Mehl Tessellation with L_k. The mean (k-m)-content of n-interfaces in a Borel set B$\subseteq L_k$, $\mu_{k,k-m}(B)$ is invariant under translations for the spatial isotropy of the process, so it holds $\mu_{k,k-m}(B) = \mu_{k,k-m}\nu_k(B)$. A theorem, due to Møller [8], allows the computation of $\mu_{k,k-m}$, called *density* of $(k-m)$-interfaces, for $1 \leq k \leq 3$ and $0 \leq m \leq k$ when the growth speed of crystals is constant. The results are reported in Tables 1, 2, 3.

By coupling the theory of Boolean models with the theory of random tessellations, it is possible to reinterpret the formula which has been found by Møller in the constant case, in such a way that it doesn't depend on the particular functional

	General case
mean number of crystals for unit volume	$\lambda = \int_0^\infty p(t)\Lambda(dt)$
mean number of vertexes for unit volume	$\mu_{30} = (\dot{R}_0)^9 \frac{256}{105}\pi^5 \int_0^\infty \left[\int_0^t (t-s)^2 \Lambda(ds)\right]^4 p(t)dt$
mean length of edges for unit volume	$\mu_{31} = (\dot{R}_0)^7 \frac{64}{15}\pi^4 \int_0^\infty \left[\int_0^t (t-s)^2 \Lambda(ds)\right]^3 p(t)dt$
mean area of facets per unit volume	$\mu_{32} = (\dot{R}_0)^5 \frac{32}{3}\pi^2 \int_0^\infty \left[\int_0^t (t-s)^2 \Lambda(ds)\right]^2 p(t)dt$

Table 1: Mean contents of interfaces in $\mathbb{R}^3$. Here and in the following tables, the quantity $p(t)$ is given by $p(t) = \exp[-(\dot{R}_0)^2 \pi \int_0^t (t-s)^2 \Lambda(ds)]$ and represents the mean free volume fraction at time t.

	General case
mean number of vertexes for unit surface	$\mu_{20} = (\dot{R}_0)^7 \frac{32}{15}\pi^4 \int_0^\infty \left[\int_0^t (t-s)^2 \Lambda(ds)\right]^3 p(t)dt$
mean length of edges for unit surface	$\mu_{21} = (\dot{R}_0)^5 \frac{16}{3}\pi^3 \int_0^\infty \left[\int_0^t (t-s)^2 \Lambda(ds)\right]^2 p(t)dt$

Table 2: Mean contents of interfaces in a 2-dimensional section of $\mathbb{R}^3$

	General case
mean number of crystals intersected for unit length	$\mu_{10} = (\dot{R}_0)^5 \frac{16}{3}\pi^2 \int_0^\infty \left[\int_0^t (t-s)^2 \Lambda(ds)\right]^2 p(t)dt$

Table 3: Mean contents of interfaces in a 1-dimensional section of $\mathbb{R}^3$

form of the growth rate. The new form is given by [3]

$$\mu_{k,k-m}(B) = \bar{c}_{kn}\mathbf{E}_T\left([S_{ex}(T)]^m\right)\nu_k(B), \tag{3}$$

where T is the random stopping time at which the origin is covered for the first time by the crystalline phase, $\mathbf{E}_T$ denotes the expectation with respect to T, $\bar{c}_{kn}$ is a geometrical constant and $S_{ex}(T)$ is the *extended surface* of the crystals at time T, that is the surface of the crystals computed in the absence of impingement. We conjecture that formula (3) holds also in the non-constant case, so that it is more general of the one found by Møller.

5 Conclusions

In this paper we have provided some tecniques to solve either direct or inverse problems related to the morphology of polymer crystallization, mainly in the isothermal and constant case. The conjecture stated at the end of the previous paragraph (for which a proof is being studying) could be an attempt to extend some results to more general transformations, with a temperature dependence.

6 Acknowledgements.

Useful discussions and a continuous exchange of information are acknowledged with G.C. Alfonso (Genoa), A. Fasano (Florence), S. Mazzullo (Montell, Ferrara), C. Verdi (Milan) and G. Eder (Linz). This work has been performed with partial support of the CNR (National Research Council) Special Programme "Mathematics for Technology and Society" (1995), an ASI (Italian Space Agency) contract (1995) and the MURST 40% programme on Nonlinear Systems and Applications.

References

[1] V.Capasso, M.De Giosa, R. Mininni, Asymptotyc properties of the maximum likelihood estimators of parameters of a spatial counting process modelling crystallization of polymers, *Stoch. Anal. and Appl.*,**13**, 279-294, (1995).

[2] V.Capasso, A.Micheletti, M.De Giosa, R.Mininni, Stochastic modelling and statistics of polymer crystallization processes, *Surv. Math. Ind.*, **6**, (1996), pp.109-132.

[3] I. Gialdini, A. Micheletti, Geometrical interpretation of the mean content of n-interfaces in a Johnson-Mehl tessellation of $\mathbb{R}^d$, Proceedings of the Seconda Conferenza Internazionale di Geometria Stocastica, Corpi Convessi e Misure Empiriche, *Rendiconti del Circolo Matematico di Palermo*. Submitted.

[4] A.Micheletti, V.Capasso, Stochastic models for polymer crystallization. Parameter estimation, *Quaderno N.38/1996, Dip. di Matematica, Universitá di Milano*. In press on the Proceedings of HERMIS '96, 1996.

[5] A. Micheletti, V.Capasso, The stochastic geometry of polymer crystallization process, *Stoch. Anal. and Appl*,**15**, (1997). To appear.

[6] I.S.Molchanov, D.Stoyan, Asymptotic properties of estimators for parameters of the Boolean model. *Adv. in Appl. Probab.* **26**, 301–323 (1994).

[7] I.S.Molchanov, Statistics of the Boolean model: from the estimation of means to the estimation of distributions. *Adv. in Appl. Probab.* **27**: 63–86 (1995).

[8] J.Møller, Random Johnson-Mehl tessellations, *Adv. in Appl. Probab.*, **24**, 814–844 (1992).

[9] D.Stoyan, W.S.Kendall, J.Mecke, *Stochastic Geometry and its Applications*, Akademie-Verlag, Berlin, Second edition, 1995.

Mathematical Model of Deep–Bed Grain Layer Drying by Ventilation

Aivars Aboltins
Department of Mathematics, Latvia University of Agriculture,
Liela Str.2, Jelgava LV-3001, Latvia
E-mail *aivars@inka.cs.llu.lv*

Abstract

The research work is devoted to matematical modeling the deep-bed grain drying process by ventilation. The purpose of research is to determine the thicknessof the first and the next stew upon grain layers dependent on different grain moistures . The mathematical model encloses a set of four partial differential equations which contain grain and air temperature, grain moisture and air humidity. Initial and boundary conditions for the system are constant values. The system with initial and boundary conditions is solved numerically by weighted time and space finite differences. The permissible thickness of the first and the next strew upon grain layers for various grain moistures that doesn't provoke grain deterioration is fixed. It is 1.5–2 times greater as that is used in production.

Key words: pde numerics, grain drying, dps model.

1 Introduction

Grain has a special place in food production. Therefore production of grain in sufficient amount and good quality is one of the main problem of every state. An important stage in grain production is its primary processing. This process includes a complex of technological operations. The most important of them is drying.

In the Republic of Latvia approximalety 25% of yield is dried by ventilation. This type of drying becomes particularly important when prices of energy are high. In order to investigate conditions for a rational use of energy, it is necessary to simulate the drying process.

Latvia University of Agriculture worked out new technology of grain drying strew upon grain layers [1]. A point of the technology is the new grain layer strewed when drying front reaches the top of the previous layer. Our aim is to determine the thickness of the first and the next grain layers dependent on different grain moistures for more effective use of the ventilation equipment.

2 Mathematical model

Drying of any substance is based on heat-mass transfer processes. Grain drying is based on heat-mass transfer between grain and intergrain space. There are many heat–mass transfer process mathematical models of thin layer grain drying [2]. A solution of these models provides a temperature history of the kernel. We propose mathematical model which contains temperature and moisture functions of the grain and intergrai space (air) [3].

To describe the kinetics of the deep-bed grain drying process we assume the following:

- water is liquid in grain,
- water vaporization in grain proceeds according to the Dalton law,
- heat transfer between grain and drying agent (air) goes on by convection,
- drying agent velocity is constant in the grain layer,
- inner temperature gradient for a single grain is very small and has not been considered.

Using basic laws of physics, i.e. the mass transfer law between grain and drying agent, the law of substance concervation, the law of heat transfer between grain and drying agent and law of energy conservation we obtain the following system of four partial differential equations which contain grain and air temperature ($\Theta(x,t)$ and $T(x,t)$, (^{0}C)), grain moisture $W(x,t)$ and air humidity $d(x,t)(\%)$:

$$\frac{\partial W}{\partial t} = K(W_p - W), \qquad t > 0, \quad x > 0 \qquad (1)$$

$$\frac{\partial d}{\partial t} + a_1 \frac{\partial d}{\partial x} = \frac{K}{a_2}(W - W_p), \qquad t > 0, \quad x > 0 \qquad (2)$$

$$\frac{\partial \Theta}{\partial t} = c_1(T - \Theta) + c_2(W_p - W) \qquad t > 0, \quad x > 0 \qquad (3)$$

$$\frac{\partial T}{\partial t} + a_1 \frac{\partial T}{\partial x} = c_0(\Theta - T) \qquad t > 0, \quad x > 0 \qquad (4)$$

where $x(m), t(h)$ are the variables of layer thickness and drying time.
Initial and boundary conditions for the system (1)–(4) can give following:

$$T_{t=0} = \Theta_{t=0} = \Psi_1(x) \quad , \quad W_{t=0} = \Psi_2(x) \quad , d_{t=0} = \Psi_3(x)$$

$$T_{x=0} = \Phi_1(t) \quad , \quad d_{x=0} = \Phi_2(t)$$

In our situation initial condition are given as follows:

$$T_{t=0} = \Theta_{t=0} = \Theta_s \quad , \quad W_{t=0} = W_s \quad , \quad d_{t=0} = d_s \quad , \tag{5}$$

where Θ_s (^0C) is grain and intergrain air temperature in layer, W_s, d_s (%) are grain moisture and intergrain air humidity. Weather conditions in Latvia don't make possibility to dry grains until $W = 14 - 15$ % / for grain keeping / with ambient air, i.e. air must be a litle bit heated [3]. Boundary conditions for drying are choosed constant :

$$T_{x=0} = T_r \quad , \quad d_{x=0} = d_r \tag{6}$$

There $\quad a_1 = 3600\nu \quad , \quad a_2 = \dfrac{\gamma_a \varepsilon}{10\gamma_g} \quad , c_0 = \dfrac{\alpha_q}{m\gamma_a c_a} \quad ,$

$$c_1 = \dfrac{\alpha_q}{(m-1)c_g\gamma_g} \quad , \quad c_2 = \dfrac{Kr}{100c_g} .$$

To determine the drying rate K we used Bovden's recommended expression :

$$K = \exp\left(20.95 - \dfrac{6942}{T + 273}\right).$$

The difficulties to solve the system (1) – (4) include the determination of α_q. Grain shape distinguish from sphere.These degrees of comparision characterize sphericity ψ, which indicate phere's , equal capacity of grain , surface ratio to real surface of grain. The wheat sphericity ψ=0.82-0.85, barley sphericity ψ=0.80 [5]. Using heat boundary layer theory / for plate and sphere / and grain sphericity ψ we obtained heat transfer rate average in time [3] :

$$\alpha_q = (12.6(1 - \psi) + 62.8\psi)\dfrac{\lambda}{R^2}$$

Notations are :

$\quad \nu$ – air velocity (m/s) ,
$\quad \gamma_a, \gamma_g$ – capacity of weight / air, grain respectively / (kg/m^3),

c_a, c_g – heat of the drying air and moist grain (kJ/kg),
r – latent heat for water vaporization (kJ/kg),
$\varepsilon = m/(1-m)$ (m – porosity of grain),
W_p – equilibrium moisture content, dry basis (%),
α_q – interphase heat exchange coefficient (kJ/mh°C),
K – drying coefficient (1/h),
λ – rate of grain heat transfer (kJ/m^3h°C),
R – grain radius (m).

Equilibrium moisture content W_p was obtaining using S.Henderson's modified equation in M.Fortes's interpretation [5]:

$$W_p = \left(-\frac{1}{5869}\ln\left(1-\frac{\varphi}{100}\right)(T+273)^{0.775}\right)^{\frac{(T+273)^{1.363}}{5203}}$$

φ – ambient air relative humidity (%).

3 Numerical solution of model

The system (1)–(4) was approximated by time and space weighted σ_k (k=1,2,3,4) finite differences [6].

Taking finite differences $\tau = \Delta t$, $h = \Delta x$ we obtain numerical equations of model.

$$\frac{W_i^{j+1} - W_i^j}{\tau} = K\left[\sigma_1(W_p - W_i^{j+1}) + (1-\sigma_1)(W_p - W_i^j)\right] \tag{7}$$

$$\frac{d_i^{j+1} - d_i^j}{\tau} + a_1\frac{d_i^j - d_{i-1}^j}{h} = \frac{K}{a_2}\left[\sigma_2(W_i^{j+1} - W_p) + (1-\sigma_2)(W_i^j - W_p)\right] \tag{8}$$

$$\frac{\Theta_i^{j+1} - \Theta_i^j}{\tau} = c_1\left[\sigma_3(T_i^{j+1} - \Theta_i^{j+1}) + (1-\sigma_3)(T_i^j - \Theta_i^j)\right]$$
$$+ c_2\left[\sigma_3(W_p - W_i^{j+1}) + (1-\sigma_3)(W_p - W_i^j)\right] \tag{9}$$

$$\frac{T_i^{j+1} - T_i^j}{\tau} + a_1\frac{T_i^j - T_{i-1}^j}{h} = c_0\left[\sigma_4(\Theta_i^{j+1} - T_i^{j+1}) + (1-\sigma_4)(\Theta_i^j - T_i^j)\right] \tag{10}$$

As result we receive following expressions :

$$W_i^{j+1} = W_i^j \left(1 - \frac{K\tau}{1 + K\tau\sigma_1}\right) + \frac{K\tau W_p}{1 + K\tau\sigma_1} \tag{11}$$

$$\begin{aligned}d_i^{j+1} &= d_i^j \left(1 - \frac{a_1\tau}{h}\right) + \frac{a_1\tau}{h}d_{i-1}^j + \\ &\quad + \frac{K\tau}{a_2}\left[\sigma_2(W_i^{j+1} - W_p) + (1-\sigma_2)(W_i^j - W_p)\right]\end{aligned} \tag{12}$$

From Eq. (12) we convey Θ_i^{j+1} and substituting in Eq. (13) we obtain T_i^{j+1}:

$$A \cdot T_i^{j+1} = B \cdot T_i^j + C \cdot T_{i-1}^j + D \cdot \Theta_i^j + E, \tag{13}$$

where $A = 1 + \tau(c_1\sigma_3 + c_o\sigma_4)$,

$B = (1 + \tau c_1\sigma_3)\left(1 - \frac{a_1\tau}{h}\right) + \tau c_o[\tau c_1(\sigma_4 - \sigma_3) - (1 - \sigma_4)]$,

$C = \frac{a_1\tau}{h}(1 + \tau c_1\sigma_3)$, $\quad D = \tau c_o[1 + c_1\tau(\sigma_3 - \sigma_4)]$,

$E = \tau^2 c_o c_2 \sigma_4 \left[\sigma_3(W_p - W_i^{j+1}) + (1 - \sigma_3)(W_p - W_i^j)\right]$.

Knowing T_i^{j+1}, we can determine Θ_i^{j+1}:

$$A \cdot \Theta_i^{j+1} = B_1 \cdot \Theta_i^j + C_1 \cdot T_i^j + D_1 \cdot T_{i-1}^j + E_1, \tag{14}$$

where $B_1 = 1 + \tau^2 c_o c_1(\sigma_3 - \sigma_4) - \tau c_1 + \tau(c_o\sigma_4 + c_1\sigma_3)$,

$C_1 = \tau^2 c_o c_1(\sigma_3 - \sigma_4) + \tau c_1 - \frac{a_1\tau^2}{h}c_1\sigma_1$, $\quad D_1 = \frac{a_1}{h}\tau c_1^2\sigma_3$,

$E_1 = \tau c_2(1 + \tau c_o\sigma_4)\left[\sigma_3(W_p - W_i^{j+1}) + (1 - \sigma_3)(W_p - W_i^j)\right]$.

Initial conditions (5) are expressed following:

$$T_i^o = \Theta_i^o = \Theta_s, \quad W_i^o = W_s, \quad d_i^o = d_s \tag{15}$$

Boundary conditions for deep-bed grain layer drying are given following:

$$T_0^j = T_r, \quad d_0^j = d_r. \tag{16}$$

Sufficient conditions for schemes (11) and (12) stability are $\sigma_1 > 0.5$; $\sigma_2 > 0.5$ by $\tau < \frac{h}{a_1}$ [6, 7]. For stability of (13) we need $A > 0, B > 0, C > 0$ and

$A \geq B + C$ / max principle [6, 7] / . $A > 0$ and $C > 0$ if $\sigma_3 > 0$, $\sigma_4 > 0$. By determining $B > 0$ we assume $\sigma_3 = \sigma_4$ and estimate τ ,

$$\tau \leq \frac{h}{a_1 + hc_0 - h\sigma_3(c_o + c_1)} .$$

Estimation for σ_3 is :

$$\sigma_3 \leq 1 - \frac{hc_1 - a_1}{h(c_1 + c_0)} .$$

For $A \geq B + C$ we obtain inequality:

$$\tau \leq \frac{1}{c_1(\sigma_4 - \sigma_3)}.$$

For stability of (14) we need $A > 0$, $B_1 > 0$. Estimation for τ (using $\sigma_3 = \sigma_4$ and $B_1 > 0$) is following:

$$\tau \leq \frac{1}{c_1 + \sigma_3(c_o + c_1)} ,$$

and σ_3 is

$$\sigma_3 \leq 1 - \frac{c_0}{c_1 + c_0} .$$

Finally the given schemes (11)–(14) with conditions (15)–(16) is stable (at $\sigma_3 = \sigma_4$) if

$$\tau \leq \min \left[\frac{h}{a_1} \; ; \; \frac{h}{a_1 + hc_0 - h\sigma_3(c_o + c_1)} \; ; \; \frac{1}{c_1 - \sigma_3(c_o + c_1)} \right] .$$

4 Practical results

Using nonlinear regression to simulate modeling results and assuming that the permissable storage time for wet grain of wheat is 6 hours an analytical expression has been obtained for maximal layer thickness H_1 depending on grain moisture W at optimal air velocity $v = 0.1(m/s)$ and equilibrium moisture content $W_p = 14 - 15(\%)$:

$$H_1 = \frac{83}{W} - 2.62$$

In practice an important problem is to determine the max. thickness of second and next strew upon grain layers for various grain moistures when the drying front reaches the top of the previous layer. Assuming that the second and the next layers of grain can be strewed when the moisture of previous grain layer has decreased by 0.1 (%) on top, an expression has been obtained for the permissable thickness of the second and the next layers H_1 depending on the moistures W_{k-1} and W_k of the previous and the new grain layers:

$$H_k = 7.89 - (0.03 \cdot W_{k-1} + 0.46 \cdot W_k) + 0.0076 \cdot (W_k)^2 \qquad \text{where}$$

k – grain layers number ($k = 2, 3, .., n$). The permissable thickness of layers is 1.5–2 times greater as that is used in production in Republic of Latvia [3].
This mathematical model and solving algorithm can be used in modelling of hot grain cooling in bins [8] and other situations.

References

[1] Berzins,E., Aboltins, A., Rajeckis, P.:Type of grain drying. Patent No- 2116504 (1994) (Russian Federation)

[2] Miketinas, M.J., Sokhansanj S., Tutek Z.: Determination of heat and mass transfer coefficients in thin layer drying of grain. Trans. ASAE, Vol.35,(1992) pp.1853–1858

[3] Aboltins, A.: The solving methods of the deep-bed grain layer drying process by ventilation . Ph.D. Thesis, Latvia University of Agriculture, Jelgava (1993) 15 pp.

[4] Egorov, G.A.: Heat and moisture influence on grain processing and storage. Kolos, Moscow (1973), 264 pp. (in Russian).

[5] Fortes, A., Okos, M.R., Barrett, J.R.: Heat and mass transfer analysis of intrakernel wheat drying and rewetting. J. Agr. Engng. Res. 26 (1981) pp. 109–125

[6] Samarskii, A.A.: Theory of difference scheme. Nauka, Moscow (1983), 653 pp. (in Russian)

[7] Samarskii, A.A., Goolin, A.V.: Numerical methods. Nauka, Moscow (1989), 430 pp. (in Russian)

[8] Aboltins, A.: Mathematical modelling of grain ventilation. In: F. Beitenecker, et.al.(eds): Proc. of the EUROSIM Simulation Congress EUROSIM 1995 , Elsevier Science B.V. (1995) pp.369-372

A model of oil burnout from glass fabric

A.Buikis[1], A.D.Fitt[2], N.Ulanova[1]

[1] Institute of Mathematics of Latvian Academy of Sciences
and University of Latvia,
1 Akademijas lauk., LV–1524 Riga, Latvia
Fax: (+371) 7227520, e-mail: lzalumi@cclu.lv

[2] University of Southampton
Highfield, Southampton, S017 1BJ, United Kingdom
Fax: (+44)01703 592225, e-mail: adf@maths.soton.ac.uk

Abstract

A mathematical model is proposed for the process of the removal (by burning) of oil contained in a glass fibre insulation fabric manufactured in Latvia. The small aspect ratio of the fabric allows simplifications to the modelling which reduce the problem to a single nonlinear ordinary differential equation. When the effects of reflected radiation are also included, the differential equation is supplemented by two integral equations. Predictions of the position of the 'burning zone' accord well with observations made at the factory. The effect of the inclusion of extra heating chambers is also examined, and it is found that the temperature gradient in the fabric may be greatly decreased in this way.

1 Introduction

During the industrial process of manufacturing glass fibre insulation fabric, oil is used to assist in the weaving process. The oil must be removed from the fabric, and to do this a special furnace is used. The fabric is passed into the furnace, where the oil is heated to burnout temperature by diesel fuel powered heaters. The complete burnout process takes place on the surface of the moving fabric. The resultant high fabric temperatures are known to influence the intrinsic structure of the material and may cause the tensile strength of the glass fabric to decrease.

For this reason it is necessary to model the processes of fabric heating and oil burnout. As well as the models reported below, simultaneous investigations were carried out to determine some of the unknown parameters in the problem. These help to justify the assumptions used in the modelling.

2 The formulation of the heat problem

To propose a simple model of the process we do not consider combustion of the oil as it contributes little to the overall heat balance. In any case, considering only diffusive, convective and radiative effects reveals important aspects of the technological process inside the furnace.

The temperature distribution T in the fabric is described by the basic two-dimensional equation:

$$\rho c_p \frac{\partial T}{\partial t} = \lambda \left(\frac{\partial^2 T}{\partial x^2} + \frac{\partial^2 T}{\partial y^2} \right) - v \rho c_p \frac{\partial T}{\partial x}, \qquad 0 < x < D,\ 0 < y < \delta, \qquad (1)$$

where D is the length of the furnace and δ is the thickness of the fabric. It is assumed that the fabric moves along the X-axis with the velocity v; ρ, c_p, λ are the density, heat capacity and heat conductivity of the fabric material.

The initial condition is:
$t = 0: \qquad T = T_0.$

The boundary conditions along X-axis are given by
for $x = 0: \qquad T = T_0,$
for $x = D: \qquad \frac{\partial T}{\partial x} = 0,$

For the fabric surface two kinds of boundary conditions are considered:
I. A simple Stefan-Boltzmann law
for $y = 0: \qquad -\lambda \frac{\partial T}{\partial y} = \epsilon_L \sigma (T_N^4 - T^4) - \alpha (T - T_g), \qquad (2)$

for $y = \delta: \qquad \lambda \frac{\partial T}{\partial y} = \epsilon_L \sigma (T_k^4 - T^4) - \alpha (T - T_g). \qquad (3)$

Here $T_N = 850^0\ C$, $T_k = 700^0\ C$ are the temperatures of the heaters at the bottom and the top of the furnace, respectively, $T_g = 720^0\ C$ is the temperature of the gas in the furnace, ϵ_L is the emissivity of the fabric material and α is the coefficient of the convective heat transfer between the fabric and the gas in the furnace, taken to be (see [1])

$$\alpha = Nu\ \lambda_g / D, \qquad Nu = 0.044 Re^{0.77} T/T_g.$$

The form of the boundary conditions (2) and (3) assumes that the distance between the fabric and the heater is large enough so that the effect of possible reflections of the heat flux from the material surface may be neglected. In the specific problem of interest, the distance from the fabric to the bottom heater was given by $a = 0.15$m, whilst the distance to the top heater was $b = 0.2$m. Since the length of the furnace is $D = 1.16$m this suggests that instead of the simple boundary condition (2) allowance should be made for the reflected heat fluxes arising from radiative heat exchanges between the bottom surface of the fabric and the bottom heater: such effects were not considered (and therefore (3) was used) for the top heater since it is further from the fabric and has a lower temperature.

II. With the reflected heat fluxes we therefore obtain

for $y = 0$:
$$-\lambda \frac{\partial T}{\partial y} = -\frac{\epsilon_L}{1-\epsilon_L}(\sigma T^4 - q_0) - \alpha(T - T_g), \quad (4)$$

for $y = \delta$: condition (3) is used,

where q_0 and q_L, the reflected heat fluxes, are given by (for details see [2])

$$\begin{cases} q_0(x,t) - (1-\epsilon_0)\int_0^D q_L(\xi_1,t)\frac{a^2}{2[(\xi_1-x)^2+a^2]^{3/2}}d\xi_1 = \epsilon_0\sigma T_N^4(x,t) \\ q_L(\xi,t) - (1-\epsilon_L)\int_0^D q_0(x_1,t)\frac{a^2}{2[(x_1-\xi)^2+a^2]^{3/2}}dx_1 = \epsilon_L\sigma T^4(\xi,t). \end{cases} \quad (5)$$

$0 < x < D$, $0 < \xi < D$, $0 < x_1 < D$, $0 < \xi_1 < D$.

Here ϵ_0 is emissivity of the heater material.

Since the thickness δ of the fabric (typically equal to 0.2 mm) is an order of magnitude less than its other characteristic sizes, we can define an averaging along the y-axis as

$$u(x,t) = \frac{1}{\delta}\int_0^\delta T(x,y,t)dy$$

$$\rho c_p \frac{\partial u}{\partial t} = \lambda \frac{\partial^2 u}{\partial x^2} + \frac{1}{\delta}\left(\lambda\frac{\partial T}{\partial y}\bigg|_{y=\delta} - \lambda\frac{\partial T}{\partial y}\bigg|_{y=0}\right) - v\rho c_p\frac{\partial u}{\partial x}, \quad 0 < x < D.$$

Assuming that the temperature does not vary in the y-direction and using the boundary conditions (4) and (3) we obtain for the fabric:

$$\rho c_p \frac{\partial u}{\partial t} = \lambda\frac{\partial^2 u}{\partial x^2} - v\rho c_p\frac{\partial u}{\partial x} +$$

$$\frac{1}{\delta}\left(\epsilon_L\sigma(T_k^4 - u^4) - \alpha(u - T_g) + \frac{\epsilon_L}{1-\epsilon_L}(q_0 - \sigma u^4) - \alpha(u - T_g)\right), \quad 0 < x < D. \quad (6)$$

where

$$\begin{cases} q_0(x,t) - (1-\epsilon_0)\int_0^D q_L(\xi_1,t)\frac{a^2}{2[(\xi_1-x)^2+a^2]^{3/2}}d\xi_1 = \epsilon_0\sigma T_N^4(x,t) \\ q_L(\xi,t) - (1-\epsilon_L)\int_0^D q_0(x_1,t)\frac{a^2}{2[(x_1-\xi)^2+a^2]^{3/2}}dx_1 = \epsilon_L\sigma T^4(\xi,t). \end{cases} \quad (7)$$

The simple model using conditions (2),(3) for $u(x,t)$ gives the following one-dimensional equation:

$$\rho c_p \frac{\partial u}{\partial t} = \lambda\frac{\partial^2 u}{\partial x^2} - v\rho c_p\frac{\partial u}{\partial x} +$$

$$\frac{1}{\delta}\left(\epsilon_L\sigma(T_k^4 - u^4) - 2\alpha(u - T_g) + \epsilon_L\sigma(T_N^4 - u^4)\right), \quad 0 < x < D. \quad (6')$$

This equation, together with the initial condition $u(x,0) = T_0$, and both the boundary conditions in the X−direction, is solved by a finite difference method. A scheme with a uniform mesh $x_i = ih$, $i = \overline{0,N}$ was used, and the calculation was performed until a quasistationary process was established, i.e. $u(x,t) \equiv u(x)$. The system (6),(7) for the reflected heat fluxes is solved at each time-step $t = t_n$ with the additional iterations $u^{(p)}(x_i, t_n)$, $q_0^{(p-1)}$, $q_L^{(p-1)}$, $p = \overline{1,P}$ until convergence. The integrals in (7) were approximated by Simpson's quadrature formula, the quadrature points being assumed coincident with those of the difference scheme. The initial estimates for the fluxes were the following:

$$\begin{cases} q_0^0(x_i, t_n) = 0. \\ q_L^0(x_i, t_n) = 0. \end{cases}$$

Once again, the calculation was performed until a quasistationary process was established, i.e. $u(x,t) \equiv u(x)$, $q_0(x,t) \equiv q_0(x)$, $q_L(x,t) \equiv q_L(x)$.

3 The results of the calculations

The numerical results for the models described above were compared with experimental observations to allow parameter ranges to be determined for the emissivities ϵ_0 and ϵ_L and the heat transfer coefficient α.

Comparison of results from the two different models also showed that it is essential to take into account the effects of reflected radiation by using (4).

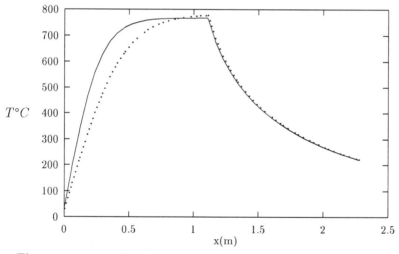

Fig.1 The temperature distribution of the fabric in and outside the furnace for $\epsilon_L = 0.6$.

....-no reflected radiation (2),(3)

— -reflected radiation included (3)-(5).

In Fig.1 the temperature distribution in the fabric is given as a function of the distance along the x-axis. In this and all other calculations described an initial temperature $T_0 = 30^0\ C$ was used. For distances up to 1.16m the fabric remains in the furnace; after this the fabric has left the furnace and is therefore cooling. The fabric temperature predicted using (3)-(5) (i.e. taking into account reflected radiation) is denoted by a continuous curve, whilst results for the simple Stefan-Boltzmann model ((2) and (3)) are denoted by asterisks. In Fig. 1 a fabric emissivity of $\epsilon_L = 0.6$ was used. As Fig.2 shows, the material temperature depends very strongly on the emissivity of the fabric. That is why a separate experiment was carried out, which showed a very high emissivity of the fabric $\epsilon_L = 0.92$. The point $x = 0.4\ m$ used in the calculations (see Fig.2) was chosen because it corresponded to the actual beginning of the oil burnout. Pronounced burnout in reality occurs within a narrow region, from $x = 0.4\ m$ to $x = 0.5\ m$.

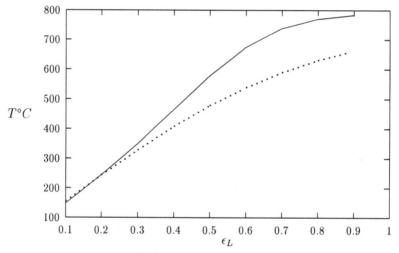

Fig.2 The temperature of the fabric at the point $x = 0.4\ m$ as a function of the emissivity coefficient ϵ_L of the material of the fabric.
.....-boundary conditions (2),(3)
— -boundary conditions (3)-(5).

After the real emissivity of the fabric had been determined the temperature distribution of the fabric in the furnace was checked again, as well as its cooling down after it emerged from there. The results are shown in Fig.3.

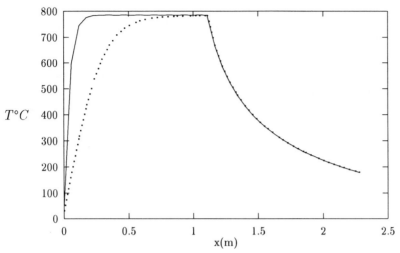

Fig.3 The temperature distribution of the fabric in and outside the furnace for $\epsilon_L = 0.92$.
..... -boundary conditions (2),(3)
— -boundary conditions (3)-(5).

As illustrated there, the difference between the results predicted using the boundary conditions (2),(3) and (3)-(5) became even more important. If we take into account that every point of the material is heated not only by the nearest point of the heater, but by all the points of the bottom heater, it means that there is a much faster heating of the material after entering the furnace. It is obvious that the technological process is time-independant: the fabric temperature inside the furnace is constant almost all the time and it is near to 800^0 C.

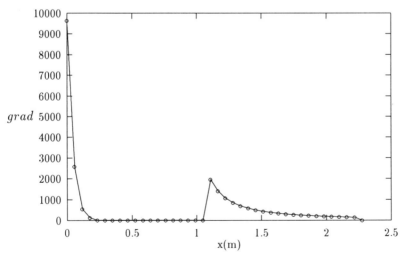

Fig.4 The temperature gradient on the fabric in and outside the furnace.

Another important observation is follows: when entering the furnace and after emerging from it the temperature gradients (^{0} C/m) along the fabric are very large (see Fig.4). This causes high thermal stresses in the fabric and may lead to adverse effects in its mechanical characteristics.

The results from Fig. 4 reveal that, even when combustion reactions are ignored and a simple heating process is considered, large thermal gradients are present in the fabric. It is thought that it may be possible to decrease these by using slow additional pre- and post- heating chambers.

Fig.5 shows the temperature distribution along the fabric when such chambers are present (the length of each chamber being 0.6 m). The two curves correspond to two different heat conditions in the chambers. The continuous curve shows the temperature distribution when the upper walls of the extra chambers are unheated and the temperature on the lower walls varies linearly between T_0 and T_N. The dotted curve shows results when the upper and lower walls are both heated; on the bottom walls the temperature varies linearly between 250^0 C and T_N, whilst on the top walls of the extra chambers it is held constant at 250^0 C.

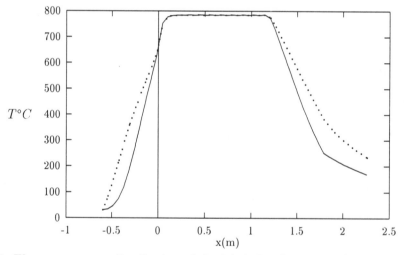

Fig.5 The temperature distribution of the fabric for the case with two additional chambers,
— - the chambers without heating,
... - the chambers with additional heating.

Fig.6 shows the temperature gradient along the X−axis when there are extra heating chambers. It is obvious that having an additional chamber (even if there is no special heater inside it) leads to a decrease in the temperature gradient of approximately five times. In the same manner a strong decrease in the temperature gradient is observed in the slow-cooling chamber. As has already been noted, so far the effects of the burnout process were not included here.

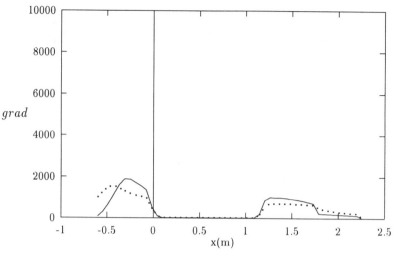

Fig.6 The temperature gradient on the fabric in the furnace.

This process can cause additional temperature gradients inside the furnace, but they seem to be not so large as at the chamber inlet, due to the simple reason that the fabric temperature (about $800^0\ C$) has already almost reached that of the hottest heater $T_N = 850^0\ C$ and is higher than the furnace gas temperature $T_g = 720^0\ C$.

4 Conclusion

It has been shown above how a simple mathematical model may be proposed for the process of oil burnout in glass fabric manufacture. By comparing the results to observations made in the factory, it has become evident that it is necessary to include the effects of reflected radiation if accurate predictions of the burning zone are to be obtained. It is also clear that the mathematical model may be used as a predictive tool to allow the investigation of alternative heating programmes and geometries within the furnace. It is worth pointing out that another key element in the process concerns the fate of the fabric *after* it exits from the furnace. In some circumstances the permanent stress field set up by the cooling of the fabric may adversely affect the quality of the final product. This aspect of the problem is actively being considered at present.

References

[1] M.A.Miheev: Heat transfer.Moscow, 1940.(Russian)

[2] R.Siegel/ J.B.Howell: Thermal radiation heat transfer.McGraw/Hill Book Company,1972, Publishing "Mir",Moscow,1975,part 8,pp.294-296.(Russian)

Computation of the Trajectories Generated by 2D Discrete Model Approximations of the Dynamics of Differential Linear Repetitive Processes

K. Galkowski[1], E. Rogers[2] and D. H. Owens[3]

[1] Institute of Robotics and Software Engineering,
Technical University of Zielona Gora, Poland.
[2] Image, Speech and Intelligent Systems (ISIS) Research Group,
Department of Electronics and Computer Science,
University of Southampton, UK.
[3] Centre for Systems and Control Engineering,
School of Engineering, University of Exeter, UK.

Abstract

Differential linear repetitive processes are a class of 2D linear systems which can be used, for example, to model industrial processes such as long-wall coal cutting . Also they can be used to study the properties of classes of iterative learning control schemes and the convergence properties of iterative algorithms for solving nonlinear dynamic optimal control problems based on the maximum principle. The key unique feature of interest in this paper is the fact that information propagation in one of the two separate directions evolves continuously over a fixed finite interval and in the other it is, in effect, discrete. This paper describes the development of discrete approximations for these processes, resulting in 2D linear systems state space models of the well known Fornasini Marchesini form on which to base further analysis. In this context, the remainder of this paper develops formulas for computing the trajectories generated by these 2D representations which, by analogy with the standard (1D) case, can be expected to play a key role in characterising basic systems theoretic properties such as controllability and observability. Some on-going work and areas for further development in these and related areas will also be briefly discussed.

1 Introduction

The essential unique characteristic of a repetitive, or multipass, process is a series of sweeps, termed passes, through a set of dynamics defined over a fixed finite duration known as the pass length. On each pass an output, the so-called pass length, is produced which acts as a forcing function on, and hence contributes to, the next pass profile. Industrial examples include long-wall coal cutting and metal rolling operations and algorithmic examples include image processing and certain classes of iterative learning control schemes.

A detailed treatment of a cross-section of the industrial examples can be found in [1] and a detailed treatment of the iterative learning control application can, for example, be found in [2]. The work of Smyth also details the essential unique control problem for these processes which is that the output sequence, generated in response to bounded (in some well defined sense) inputs, can produce an output sequence of pass profiles with unbounded dynamics in the pass to pass direction. Such behaviour is easily generated in simulation studies and in experiments on scaled models of industrial examples and cannot be removed by standard, ie 1D, control action. Rogers and Owens [3] have developed a basic stability/systems theory for linear constant pass length repetitive processes using an abstract model in a Banach space setting which contains all such processes as special cases.

Repetitive processes clearly have strong structural links with two-dimensional, or 2D, systems, ie systems which propagate information in two separate directions. At the most basic level this arises because it is necessary to use two co-ordinates to specify a variable in a repetitive process. These are the pass number, or index, $i \geq 0$ and the position $0 \leq t \leq \alpha$ along a pass of constant finite length α. A key difference, however, arises from the fact that the pass length of a repetitive process (which corresponds to one direction of information propagation in a 2D systems representation) is always finite. It is clear, therefore, that appropriately constructed linear 2D systems models (state-space or transfer function based) should be of major use in the analysis and control of the dynamics of at least some classes of linear repetitive processes.

The continual developments in digital techniques for the design/implementation of filters/controllers is, of course, of direct relevance to the control of continuous time linear (and nonlinear) repetitive processes. By analogy with other areas, one possible approach is to convert filters/controllers designed in the continuous domain to a discrete equivalent for implementation. This could be done in either the time domain or the frequency domain where the latter option would require the use of appropriate tools for mapping to the frequency domain. One possible means of mapping to the frequency domain is to use numerical integration techniques such as Euler, or trapezoidal, rule or, equivalently, the well known bilinear transformation in the frequency domain.

This paper describes the development of discrete approximations for the dynamics of the class of so-called differential linear repetitive processes which have many theoretical and practical applications. For example, they can be used to study the convergence properties of iterative algorithms for solving nonlinear dynamic optimal control problems using the maximum principle (see, for example, Roberts [5]). The results of this analysis are 2D linear systems state-space models of the Fornasini- Marchesini type [4]. Formulas for computing the trajectories generated by these 2D representations are also developed which, by analogy with the standard (1D) case, can be expected to play a key role in the characterisation of key systems theoretic properties such as controllability and observability. Finally, some on-going work and areas for further development of these basic results are briefly noted.

2 Basic Approximations

The major industrial example of a repetitive process is long-wall coal cutting which is the major method of extracting coal from deep cast mines in Great Britain. This application area is 'rich' in problems across a broad spectrum of 'industrially targeted' mathematics such as mathematical modelling and control/optimisation. For a detailed treatment of this general area see [1] and the relevant cited references.

Following Rogers and Owens [3], the state-space model of a differential linear repetitive process has the form

$$\frac{\partial x(i+1,t)}{\partial t} = \hat{A}x(i+1,t) + \hat{B}u(i+1,t) + \hat{B}_0 y(i,t) \quad (1)$$

$$y(i+1,t) = \hat{C}x(i+1,t) + \hat{D}u(i+1,t) + \hat{D}_0 y(i,t) \quad (2)$$

where $x(i,t)$ is the $n \times 1$ state vector, $y(i,t)$ is the $p \times 1$ vector pass profile, $u(i,t)$ is the $m \times 1$ input vector and the constant matrices $\hat{A}$, $\hat{B}$, $\hat{B}_0$, $\hat{C}$, $\hat{D}$, and $\hat{D}_0$ have appropriate dimensions. The integer $i \geq 0$ denotes the pass number or index and the continuous time variable $t \in [0, \alpha]$ where, by definition, α is a finite constant and is termed the pass length.

Consider first the case when the elements in the input and pass profile vectors have the so-called step-wise property, ie

$$u(i,k) \simeq u(i,k+1) \quad (3)$$

$$y(i,k) \simeq y(i,k+1) \quad (4)$$

where k denotes the sample index. Then under these assumptions, the dynamics of (1)-(2) can be approximated by a member of the class of so-called discrete linear repetitive processes with state space model of the form

$$\begin{aligned} x(i+1,k+1) &= Ax(i+1,k) + Bu(i+1,k) + B_0 y(i,k) \\ y(i+1,k) &= Cx(i+1,k) + Du(i+1,k) + D_0 y(i,k) \end{aligned} \quad (5)$$

where the matrices A, B, B_0, C, D, and D_0 are defined by the particular numerical approximation technique employed.

One possible choice here is to use numerical integration techniques such as the Euler, or trapezoidal, rule or, equivalently, the well known bilinear transform in the frequency domain. In particular, with step length T, apply the following formula to the state equation of a differential linear repetitive process described by (1)-(2) under the step wise assumptions of (3)-(4).

$$x(i,k+1) = x(i,k) + \frac{T}{2}[\dot{x}(i,k+1) + \dot{x}(i,k)] \quad (6)$$

Then the result is the state equation of (5) with

$$A = [I - \frac{T}{2}\hat{A}]^{-1}[I + \frac{T}{2}\hat{A}] \tag{7}$$

$$B_0 = T[I - \frac{T}{2}\hat{A}]^{-1}\hat{B}_0 \tag{8}$$

$$B = T[I - \frac{T}{2}\hat{A}]^{-1}\hat{B} \tag{9}$$

Applying the same rule to the output equation of this differential linear repetitive process defined by (2) gives the output equation of (5), ie the output equation of differential linear repetitive processes are invariant under this approximation. It is also possible to derive bounds on the errors induced by these approximation techniques which are omitted here since they are not central to the main theme of the paper. A full treatment of them can be found in [6].

The dynamics of discrete linear repetitive processes of the form (5) can be written as a 2D linear system described by a singular version of the well known Fornasini Marchesini state space model [4] for such systems. In particular, define the so-called extended vector for (5) as

$$z(i,k) = \begin{bmatrix} x(i,k) \\ y(i,k) \end{bmatrix} \tag{10}$$

Then the resulting 2D linear state space model is

$$Ez(i+1, k+1) = A_1 z(i+1, k) + A_0 z(i, k) + B_1 u(i+1, k) \tag{11}$$

where

$$E = \begin{bmatrix} I_n & O_{nm} \\ O_{mn} & O_{mm} \end{bmatrix}, A_1 = \begin{bmatrix} A & O_{nm} \\ C & -I_m \end{bmatrix}, A_0 = \begin{bmatrix} O_{nn} & B_0 \\ O_{mn} & D_0 \end{bmatrix}, B_1 = \begin{bmatrix} B \\ D \end{bmatrix} \tag{12}$$

where $O_{p,q}$ denotes the zero matrix of dimension $p \times q$. (For background on the structure and analysis of 2D linear systems described by singular state space models see, for example, [7] and the cited references.)

Consider now the computation of the transition matrix for 2D linear systems described by the singular version of the Fornasini Marchesini model defined by (10)-(12) and hence the system response to a given input sequence and initial conditions. Then it can be shown (for a proof see [6]) that the transition matrix, also known as the fundamental matrix sequence in the discrete case, and denoted by $T_{p,q}$ here is given by

$$ET_{p,q} = \begin{cases} A_0 T_{-1,-1} + A_1 T_{0,-1} + I, p = q = 0 \\ A_0 T_{p-1,q-1} + A_1 T_{p,q-1}, p \neq 0 \vee q \neq 0, p \geq -\mu_1, q \geq -\mu_2 \end{cases} \tag{13}$$

where (μ_1, μ_2) denotes the index of (10)-(12). To expand on this concept of the system index, first consider the Laurent expansion about infinity of the so-called characteristic matrix of (10)-(12), ie $(s_1 s_2 E - s_1 A_1 - A_0)^{-1}$ written (formally) in the following form where s_1 and s_2 are complex variables.

$$(s_1 s_2 E - s_1 A_1 - A_0)^{-1} = s_1^{-1} s_2^{-1} \sum_{p=-\mu_1}^{\infty} \sum_{q=-\mu_2}^{\infty} T_{pq} s_1^{-p} s_2^{-q} \qquad (14)$$

Then following Lewis [7] it is routine to show that the following condition (expressed in terms of an arbitrary two variable polynomial here) is a sufficient condition for the existence of (14) with finite lower limits.

$$deg(\Delta(s_1, s_2)) = deg_{s_1}(\Delta(s_1, s_2)) + deg_{s_2}(\Delta(s_1, s_2)) \qquad (15)$$

Here the degree of the two-variable polynomial $\Delta(s_1, s_2)$ is defined as the degree in s of $\Delta(s, s)$ and $deg_{s_i}(\Delta(s_1, s_2))$, denotes the degree in s_i.

To evaluate (13), first partition $T_{p,q}$ as follows

$$T_{p,q} = \begin{bmatrix} T_{p,q}^{11} & T_{p,q}^{12} \\ T_{p,q}^{21} & T_{p,q}^{22} \end{bmatrix} \qquad (16)$$

where $T_{p,q}^{11}$ is of dimension $n \times n$ and $T_{p,q}^{22}$ is of dimension $m \times m$. Then it follows immediately that the index (μ_1, μ_2) in this case is defined by $\mu_1 = 0$ and $\mu_2 = 1$ and hence $T_{0,-1}$, ie

$$T_{0,-1} = \begin{bmatrix} O & O \\ O & I_m \end{bmatrix} \qquad (17)$$

is the initial matrix in this case. To obtain $T_{p,q}$, $p \geq 0, q \geq -1$, it is necessary to solve the following set of equations for $T_{p,q}$

$$T_{p,q}^{1\alpha} = A T_{p,q-1}^{1\alpha} + B_0 T_{p-1,q-1}^{2\alpha}, \alpha = 1, 2 \qquad (18)$$

and the following set of equations for $T_{p-1,q-1}$

$$T_{p,q}^{2\alpha} = C T_{p,q}^{1\alpha} + D_0 T_{p-1,q}^{2\alpha}, \alpha = 1, 2 \qquad (19)$$

Using these formulas, gives, for example, the following expressions for the 'boundary' values of the transition matrix.

$$T_{0,0} = \begin{bmatrix} I_n & O \\ C & O \end{bmatrix}, T_{p,-1} = \begin{bmatrix} O & O \\ O & D_0^p \end{bmatrix}, T_{0,q} = \begin{bmatrix} A^q & O \\ C A^q & O \end{bmatrix}, p = 1, 2, ..., q = 1, 2, ... \qquad (20)$$

Given the transition matrix, it is now possible to give the main result of this paper.

Theorem 1 *Consider the discrete approximation of (5)-(9) to the dynamics of differential linear repetitive processes defined by the state space model (1)-(2). Then the general response formula for this discrete approximation is given by*

$$z(M,N) = \sum_{p=0}^{M} \sum_{q=-1}^{N-1} T_{p,q} B_1 u(M-p+1, N-q) +$$

$$\sum_{p=0}^{M} T_{p,N} \left[A_1 z(M-p+1, 0) + B_0 u(M-p+1, 0) \right] +$$

$$\sum_{p=0}^{M-1} T_{p,N} A_0 z(M-p, 0) + \sum_{q=-1}^{N-1} T_{M,q} A_0 z(0, N-q) \quad (21)$$

$$+ T_{M,N} A_0 z(0,0) \quad (22)$$

Proof: This involves a long, but essentially routine, argument based on induction. Hence the details are omitted here and can be found in [6].

3 Refinements

The discrete approximation results of the previous section are, in effect, based on the assumptions of (3) and (4). In a number of practically relevant cases, however, (see [6] for more details) the assumption of (4) on the pass profile dynamics is too restrictive. A more realistic assumption in such cases is that only the entries in the input vector are step wise. With this assumption, employing (6) again yields the following state space model describing the dynamics of the resulting discrete approximation to the dynamics of differential linear repetitive processes described by (1)-(2)

$$x(i+1, k+1) = Ax(i+1, k) + Bu(i+1, k) + B'_0 y(i, k) + B'_0 y(i, k+1)$$
$$y(i, k) = Cx(i+1, k) + Du(i+1, k) + D_0 y(i, k) \quad (23)$$

where

$$B'_0 = \frac{1}{2} B_0 \quad (24)$$

and the other matrices in this state space model are also defined by (9) and the output equation of (5) respectively. Note here that the state space model contains an additional 'time shifted' contribution from the previous pass profile, ie the term $B'_0 y(i, k+1)$ which (potentially) considerably enriches the dynamics of this model.

As before, the extended state vector of (10) can be used to develop a singular 2D Fornasini Marchesini state space model description for the dynamics of (24) whose structure is as follows.

$$Ez(i+1, k+1) = A_1 z(i+1, k) + A_2 z(i, k+1) + A_0 z(i, k) + B_1 u(i+1, k) \quad (25)$$

where the matrices A_1, E, and B_1 are again given as in (12) and

$$A_0 = \begin{bmatrix} O_{nn} & B'_0 \\ O_{mn} & D_0 \end{bmatrix}, A_2 = \begin{bmatrix} O_{nn} & B'_0 \\ O_{mn} & O_{mm} \end{bmatrix} \tag{26}$$

In 2D linear systems terms this is the full version of the Fornasini Marchesini model (see also [6]) for a full discussion of this point).

The results of the previous section generalise in a natural manner to (25) - (26). Hence the details are omitted here and can be found in [6].

4 Conclusions

This paper has considered the general problem of constructing discrete approximations to the dynamics of differential linear repetitive processes. It has been shown that if numerical integration routines are used (here attention has been restricted to Euler techniques) then the resulting approximations are in the form of so-called discrete linear repetitive processes with a well defined state space model. Also it is possible to derive bounds for the errors resulting from using such approximations, the details of this analysis have been omitted here for brevity and can be found in [6].

It has also been shown that these discrete linear approximation models can be written as singular versions of the well known Fornasini Marchesini state space model for 2D linear systems recursive in the positive quadrant. In this context, note that singularity is not an intrinsic feature of 2D linear systems state space interpretations of the dynamics of discrete linear repetitive processes and that there exist standard, or nonsingular, 2D state space descriptions of these processes. A detailed treatment of this general area can be found in [8]

This paper has also shown that the singular 2D Fornasini Marchesini state space model interpretations lead naturally to a transition matrix for discrete linear repetitive processes either intrinsically or, as here, when used to approximate the dynamics of differential linear repetitive processes. This then has led to an analytic expression for the trajectories generated by such processes in response to a given input control sequence and initial conditions. By analogy with the standard (1D) case it is to be expected that this 'closed form solution' will play a key role in characterising key systems theoretic properties such as controllability and observability for these processes. Work is currently in progress in this and a number of related general areas. To date this has shown that controllability and observability for linear repetitive processes is much more complex than in the 1D case with, for example, an number of quite distinct definitions required. Also the results of this paper can be used to provide necessary and sufficient conditions for a number of these cases but as yet computationally feasible tests are not available. Results from all of these areas will be reported in due course.

References

[1] K. J. Smyth Computer Aided Analysis for Linear Repetitive Processes, PhD Thesis, University of Strathclyde, Glasgow, UK, 1992.

[2] N. Amann, D. H. Owens, and E. Rogers, Iterative Learning Control using Optimal Feedback and Feedforward Actions, Int. J. Control, Vol 65, No. 2, pp 277-293, 1996.

[3] E. Rogers and D. H. Owens, Stability Analysis for Linear Repetitive Processes, Springer Verlag Lecture Notes in Control Information Sciences Series, Vol 175, 1992.

[4] E. Fornasini and G. Marchesini, Doubly Indexed Dynamical Systems: State-Space Models and Structural Properties, Math. Syst. Theory, Vol 12, pp 59-72, 1978.

[5] P. D. Roberts, Computing the Stability of Iterative Optimal Control Algorithms through the use of Two-Dimensional Systems Theory, Proc UKACC Int. Conf. Control 96, Exeter, UK, Vol 2, pp 981-986.

[6] K. Galkowski, E. Rogers, and D. H. Owens, Discrete Approximations to the Dynamics of Differential Linear Repetitive Processes, Research Report, ECS 96/5, University of Southampton, 1996.

[7] F. L. Lewis, A Review of 2-D Implicit Systems, Automatica, Vol 28,, No.2, pp 345-352, 1992

[8] K. Galkowski, E. Rogers, and D. H. Owens, Singularity in the Analysis and Control of Linear Repetitive Processes, Research Report, ECS96/7, University of Southampton, 1996.

A CERTAIN MATHEMATICAL MODEL OF THE GLASS FIBRE MATERIAL PRODUCTION

Jānis Cepītis, Harijs Kalis
Institute of Mathematics
Latvian Academy of Sciences and University of Latvia
Akadēmijas laukums 1, Rīga, Latvia
E-mail:cepitis@fmf.lu.lv, kalis@fmf.lu.lv

Abstract

There is considered the full mathematical model of chemical reactions on the surface of glass fibre material that was imbedded in the flow of acid solution and was pulled longitudionally. Self-similar forms of this model are obtained and their approximations by monotone schemes of differences are proposed. Some special cases which make possible to get the analytic solutions are underlined. The self-similar forms of the differential equations of the substances transport allow to calculate the emission of the alkaline oxide from the glass fibre material under the influence of the acid solution flow. Some conclusions with practical significance for the technological process is made up according to the provided computational experiments.

1 Introduction

In order to minimize substance of alkaline metal oxides in glass fibre material it was pulled through bathes filled with acid solution. In such a way raised substance of the silicon oxide in the glass fibre material has been reached and therefore thermal endurance of material has been attained.

The full mathematical model of chemical reactions on the surface of glass fibre material that was imbedded in the flow of acid solution and was pulled longitudinally consists of

a) differential equations of hydrodynamics (equation of flow continuity and equations of momentum conservation in the directions of coordinate axes),

b) differential equations of substances transport in the acid solution,

c) boundary conditions determined by chemical reactions on the surface of glass fibre material.

Of course, we can consider the steady-state equations since the industrial process is continued and established.

2 Mathematical Model

For the sake of simplicity let us consider that only one oxide of alkaline metal was involved in the chemical reaction on the surface of the glass fibre material and:
  x_1 is the spatial coordinate in the lengthwise direction of the glass fibre material,
  x_2 is the spatial coordinate in the normal direction of the glass fibre material,
  u_1, u_2 are velocity components of the acid solution flow in the directions corresponding to axes x_1, x_2,
  ρ is the density of the acid solution,
  m_1 is mass concentration of acid in the acid solution flow,
  D_1 is diffusion coefficient of acid,
  m_2 is mass concentration of alkaline metal salt formed by the chemical reaction,
  D_2 is diffusion coefficient of alkaline metal salt.

We assume for the sake of simplicity that the diffusion coefficients D_1, D_2 do not depend on the mass concentrations although it is easy to overcome this restriction.

Then the differential equations of substances transport taking into account the differential equations of hydrodynamics are following

$$\rho(u_1 \frac{\partial m_j}{\partial x_1} + u_2 \frac{\partial m_j}{\partial x_2}) = D_j \left(\frac{\partial}{\partial x_1}(\rho \frac{\partial m_j}{\partial x_1}) + \frac{\partial}{\partial x_2}(\rho \frac{\partial m_j}{\partial x_2}) \right), \qquad (1_j)$$

$j = 1, 2$.

If axis x_1 is located on the surface of the glass fibre material and R_i is the rate of the chemical reaction ($i = 0$ for water, $i = 1$ for acid, $i = 2$ for alkaline metal salt), then boundary conditions which are determined by the chemical reaction on the surface of the glass fibre material are following

$$-D_j \rho \frac{\partial m_j}{\partial x_2} = R_j - (R_0 + R_1 + R_2)m_j, j = 1, 2,$$

Let us further assume that

$$R_i = -\gamma_i M_i A \exp\left(-\frac{E}{RT}\right) \times \left(\frac{\rho m_1}{M_1}\right)^\alpha \left(\frac{\bar{\rho} m_3}{\lambda M_3}\right)^\gamma, i = 0, 1, 2,$$

where γ_i are stochiometric coefficients in the equation of chemical reaction,
  M_i are molecular weights of substances,
  E is the energy of the chemical reaction,
  T is the temperature of the acid solution flow,
  $R = 8314[\frac{J}{kmol.deg}]$,
  $\bar{\rho}$ is the area density of the glass fibre material,
  λ is the coefficient of unevenness of the surface of glass fibre material ($\lambda \geq 1$),

m_3, M_3 are mass concentration in the glass fibre material and molecular weight of the alkaline metal oxide,

A is a chosen coefficient of proportionality,

α, γ are chosen exponents and designate $A_0 = A\exp(-\frac{E}{RT})$.

Then the boundary conditions obtain the following form:

$$D_j \frac{\partial m_j}{\partial x_2} = A_0 M_1^{-\alpha} M_3^{-\gamma} \rho^{\alpha-1} (\bar{\rho} m_3)^\gamma \lambda^{-\gamma} m_1^\alpha (\gamma_3 M_3 m_j + \gamma_j M_j), \qquad (2_j)$$

$j = 1, 2.$

3 Self-similar Problems

We assume that the acid solution flows between two surfaces of the glass fibre material that was imbedded in parallel in the distance d one of other and was moved in the opposite directions with the velocity $v_0 = u_1|_{x_2=0}$ and the acid solution flow is a stationary laminar flow with velocity components $u_1 = v(x_2)$, $u_2 = 0$ and constant pressure gradient $\frac{\partial p}{\partial x_1} = c_p$. In this case from the differential equations of hydrodynamics we obtain $v(x_2) = \frac{1}{2\mu} c_p x_2^2 + A_2 x_2 + A_1$. The sticking conditions $v(0) = -v(d) = v_0$ and flowing of the current $Q = \int\limits_0^d v(\tau)d\tau$ give

$$A_1 = v_0, A_2 = -\frac{2v_0}{d} + \frac{6Q}{d^3}, \frac{1}{2\mu} c_p = -\frac{6Q}{d^3}.$$

If $Q = 0$, then we have the Quetta flow with the velocity profile $u_1 = v_0(1 - \kappa \frac{2x_2}{d}), \kappa = 1$.

The considered velocity profiles and the boundary conditions determined by the chemical reaction on the moving surfaces of the glass fibre material are symmetric. Therefore problems for the differential equations (1_j) we can solve merely in the boundary layer with the width $\delta < \frac{d}{2}$, which contained in the interval $0 < x_2 < \frac{d}{2}$, taking as the conditions of symmetry at $x_2 = \frac{d}{2}$ $m_1 = m_1^*, m_2 = 0$, where m_1^* is the mass concentration of the acid at the beginning ($x_1 = 0$). If in the formula of the Quetta flow velocity profile $\kappa = 0$, then we can to consider that the glass fibre material moves jointly with the boundary layer.

We assume that $\kappa = 0$ and $\frac{\partial^2 m_j}{\partial x_1^2} \ll \frac{\partial^2 m_j}{\partial x_2^2}$ in the boundary layer, then using variable $\eta = x_2 \sqrt{\frac{v_0}{x_1 D_1 D_2}}$ the partial differential equations (1_j) we can alter in the system of self-similar ordinary differential equations

$$\frac{1}{D_i} f_j'' + \frac{\eta}{2} f_j' - \frac{D_j \rho}{\rho_0 \rho_1 \rho_2} \left(\frac{1}{D_i D_j} + \frac{\eta^2}{4v_0} \right) \times$$

$$\times \left[\rho^2 (\rho_0 - \rho_1) f_1' f_j' + \rho_1 (\rho_0 - \rho_2) f_2' f_j' \right] = 0,$$

where $f_j(\eta) = m_j, j = 1,2, i = 1,2, i \neq j$, $\rho_k, k = 0,1,2$ is the density of the chemical substance with number k. The right side of this system does not satisfy the classical Bernstein condition and therefore the solvability conditions of boundary value problems for this system are not clear and causes mathematical interest.

In general case of Quetta flow we can put

$$\eta_j = x_2\sqrt{\frac{v_0}{x_1 D_j}}, m_j = f_j(\eta_j), j = 1,2,$$

and for any fixed value x_1 the system of differential equations (1_j) alters in the following form

$$(2 + \epsilon_j \eta_j^2) f_j'' + [(1 - \kappa \eta_j \tilde{\gamma}_j)\eta_j + 3\epsilon_j \eta_j - 2\rho B_2 - \rho \epsilon_j \eta_j B_1] f_j' = 0, \qquad (3)$$

where $\epsilon_j = \frac{D_j}{2v_0 x_1}$, $\tilde{\gamma}^j = \frac{2x_1}{d}\sqrt{2\epsilon_j}$, $B_1 = B_{11}\eta_1 + B_{22}\eta_2$, $B_2 = B_{11} + B_{22}$,

$$B_{jj} = \left(\frac{1}{\rho_2} - \frac{1}{\rho_0}\right) f_j'(\eta_j), j = 1,2,$$

$$\rho = \left(\left(\frac{1}{\rho_1} - \frac{1}{\rho_0}\right)f_1 + \left(\frac{1}{\rho_2} - \frac{1}{\rho_0}\right)f_2 + \frac{1}{\rho_0}\right)^{-1}$$

From the boundary conditions (2_j) we obtain

$$f_j'(0) = \tilde{A}_j f_j^\alpha(0)(\gamma_3 M_3 f_j(0) + \gamma_j M_j), \qquad (4)$$

where $\tilde{A}_j = A_0 \rho^{\alpha-1} M_1^{-\alpha} M_3^{-\gamma}(\bar{\rho}m_3)^\gamma \lambda^{-\gamma}\sqrt{\frac{x_1}{D_j v_0}}, j = 1,2$.

Finally, the conditions for the mass concentrations of the acid and the salt at the beginning ($x_1 = 0$) give the boundary conditions

$$f_1(\infty) = m_1^*, f_2(\infty) = 0. \qquad (5)$$

4 Difference Schemes

Let $L \in (0, +\infty)$ is such, that

$$f_1(\eta_1) \approx m_1^*, f_2(\eta_2) \approx 0.$$

for $\eta_j \geq , j = 1,2$. It means that the estimates

$$\delta_j(x_1) \approx L\sqrt{\frac{D_j x_1}{v_0}}. \qquad (6)$$

are true for the thickness δ_j of the boundary layer. Using the difference method for the rough solving of problem (3)-(5) we can introduce the uniform grid with values

$$\eta_i = ih, i = 0, ..., N, Nh = L(\eta_1 = \eta_2 = \eta)$$

and to approximate the derivatives $f'_j(0)$ like this

$$f'_j(0) = \frac{f_j(h) - f_j(0)}{h} + O(h). \tag{7}$$

Then for sharpening of $f_1(0)$ we can use the following process of iteration

$$f_1^{(k)}(0) = \frac{2f_1^{(k-1)}(h)}{B_k + \sqrt{B_k^2 + 4f_1^{(k-1)}(h)h\gamma_3 M_3 \tilde{A}_1 (f_1^{(k-1)}(0))^{\alpha-1}}}, \tag{8}$$

where $B_k = 1 + \gamma_1 M_1 \tilde{A}_1 h (f_1^{(k-1)}(0))^{\alpha-1}, k = 1, 2, f_1^{(0)}(0)$ is an initial approximation, but $f_1^{(0)}(h)$ we can find approximately solving the differential equations (3).

Expressing $f_2(0)$ from (4),(7), we can use such process of iteration

$$f_2^{(k)}(0) = \frac{f_2^{(k-1)}(h) + h\gamma_2 M_2 \tilde{A}_2 (f_1(k)(0))^{\alpha}}{1 + h\gamma_3 M_3 \tilde{A}_2 (f_1^{(k)}(0))^{\alpha}}, \tag{9}$$

where $f_2(0)(h)$ we can also find, solving the differential equation (3).

The approximation of derivatives we can improve using the scheme

$$f'_j(0) = \frac{4f_j(h) - 3f_j(0) - f_j(2h)}{2h} + O(h^2.)$$

Then in the (8) and (9) instead of $f_j^{(k-1)}(h)$ we have to take

$$\frac{4f_j^{(k-1)}(h) - f_j^{(k-1)}(2h)}{3}, j = 1, 2,$$

but instead of h we have to take $\frac{2h}{3}$.

Each of the differential equations (3) we can approximate (see [1]) by the difference equations using for the approximation monotone 3 point pattern in the following form

$$a_i f_{i-1} - (a_i + b_i) f_i + b_i f_i + 1 = 0, i = 1, ..., N-1,$$

where

$$f_i \approx f(\eta_i), f_0 = f(0), f_N = f(L),$$

$$a_i = \nu_i^- \gamma_i^- - \frac{h\alpha_i^-}{2}, b_i = \nu_i^+ \gamma_i^+ + \frac{h\alpha_i^+}{2},$$

$$\nu_i^\pm = \nu(\eta_{i\pm 1/2}), \alpha_i^\pm = \alpha(\eta_{i\pm 1/2}),$$

$$\gamma_i^\pm = \frac{\alpha_i^\pm h}{2\nu_i^\pm} \coth \frac{\alpha_i^\pm h}{2\nu_i^\pm}, \eta_{i\pm 1/2} = \frac{\eta_i + \eta_{i\pm 1}}{2},$$

$$\nu(\eta) = 2 + \epsilon\eta^2, \alpha(\eta) = (1 - \kappa\eta\tilde{\gamma})\eta + 3\epsilon\eta - 2\rho B_2 - \rho\epsilon\eta B_1.$$

This monotone scheme of differences ($a_i > 0, b_i > 0$) is precise in the case of piecewise constant coefficients $\nu(\eta), \alpha(\eta)$ and we can realize it for each functions $f_1(\eta), f_2(\eta)$ and for every fixed x by the method of factorization fixing in the expressions for B_{11}, B_{22} the values of the derivatives $f_1'(\eta), f_2'(\eta)$ on the previous layer of iteration with number $k-1$. For $k = 1$ we assume that $B_1 = B_2 = 0$. The perturbed coefficient $\gamma = z \coth z$ of the difference scheme we can calculate in the following way:

if $|z| > 10^{-4}$, then $\gamma = z\frac{1+exp(-2z)}{1-exp(-2z)}$, if $|z| \le 10^{-4}$, then $\gamma = \frac{1+exp(-2z)}{2-2z+\frac{4z^2}{3}}$.

5 Analytic Solutions

In the special case, when $\rho = const$, $\kappa = 0$, the differential equations (3) become linear and mutually independent. Their general solutions we can to write in the following form

$$f_j(\eta_j) = C_j \Phi_{\epsilon_j}(\eta_j) + K_j, j = 1, 2,$$

where C_j, K_j are arbitrary chosen constants,

$$\Phi_\epsilon(\eta) = \frac{1}{B_\epsilon} \int_0^\eta \left(1 + \frac{\epsilon t^2}{2}\right)^{-(3/2+1/2\epsilon)} dt, B_\epsilon = \frac{B(1/2, 1+1/2\epsilon)}{\sqrt{2\epsilon}},$$

$B(a, b)$— beta-function, $\Phi_\epsilon(0) = 0, \Phi_\epsilon(\infty) = 1$, and the conditions (5) implies

$$f_1(\eta_1) = m_1^* - f_1(0))\Phi_{\epsilon_1}(\eta_1) + f_1(0), f_2(\eta_2) = f_2(0)(1 - \Phi_{\epsilon_2}(\eta_2)).$$

Since $\Phi_\epsilon'(0) = (B_\epsilon)^{-1}$, the conditions (4) are realized, and from the connections

$$f_1'(0) = \frac{m_1^* - f_1(0)}{B_{\epsilon_1}}, f_2'(0) = -\frac{f_2(0)}{B_{\epsilon_2}}$$

we obtain the system of nonlinear algebraic equations for the receiving of values $f_1(0), f_2(0)$.

If $\alpha = 1$, these algebraic equations are quadratic and have solution

$$f_1(0) = \frac{2m_1^*}{B_0 + \sqrt{B_0^2 + 4m_1^* B_{\epsilon_1} \gamma_3 M_3 \tilde{A}_1}}, \tag{10}$$

where $B_0 = 1 + \gamma_1 M_1 B_{\epsilon_1} \tilde{A}_1$,

$$f_2(0) = -\frac{\gamma_2 M_2 B_{\epsilon_2} \tilde{A}_2 f_1(0)}{1 + \gamma_3 M_3 B_{\epsilon_2} \tilde{A}_2 f_1(0)}.$$

If $\alpha < 1$, then solving these algebraic equations we can use following iteration process which converge with high speed

$$f_1^{(k)}(0) = \frac{2m_1^*}{B_k + \sqrt{B_k^2 + 4m_1^* B_{\epsilon_1} \gamma_3 M_3 \tilde{A}_1 (f_1^{(k-1)}(0))^{\alpha-1}}},$$

where $k = 1, 2, ..., B_k = 1 + \gamma_1 M_1 \tilde{A}_1 B_{\epsilon_1}(f_1^{(k-1)}(0))^{\alpha-1}$.

If $\epsilon \ll 1$, for the calculation we can use the expansion

$$B_\epsilon = \frac{\sqrt{\pi}}{\epsilon + 1} \sqrt{\frac{\epsilon}{(\epsilon+1)^{\epsilon-1}}} \frac{1 + a_1\epsilon + a_2\epsilon^2 + a_3\epsilon^3 + a_4\epsilon^4}{1 + a_1\tilde{\epsilon} + a_2\tilde{\epsilon}^2 + a_3\tilde{\epsilon}^3 + a_4\tilde{\epsilon}^4},$$

where

$$a_1 = \frac{1}{6}, a_2 = \frac{1}{72}, a_3 = -\frac{139}{6480}, a_4 = -\frac{571}{155520},$$

$$\tilde{\epsilon} = \frac{\epsilon}{1+\epsilon}, (\epsilon+1)^{\epsilon-1} \approx exp(1 - \frac{\epsilon}{2} + \frac{\epsilon^2}{3} - \frac{\epsilon^3}{4} + \frac{\epsilon^4}{5}).$$

If $\kappa = 1$, then the general solution of the differential equation (3) we can also write in the form of analogous formula

$$f_j(\eta_j) = \tilde{C}_j \int_0^{\eta_j} \exp(-\frac{t^2}{4} + \tilde{\gamma}_j \frac{t^3}{6}) dt + \tilde{K}_j, j = 1, 2.$$

In this case analytic expansions still are too complicated therefore it is better to use the grid method.

Note, that $\Phi_0(\eta_j) = erf(\frac{\eta_j}{2})$ in the case $\epsilon_j = 0$, where

$$erf(z) = \frac{2}{\sqrt{\pi}} \int_0^{\eta_j} \exp(-t^2) dt$$

is the error integral, $\lim_{\epsilon \to 0} B_\epsilon = \sqrt{\pi}$ and we can obtain more simple expressions of the algebraic equations for finding values $f_1(0), f_2(0)$.

These expressions allow to obtain some useful connections between parameters of chemical process which must be identified. For example, if $\alpha = 1$ and $\tilde{A}_1 \gg 1$, then (10) implies

$$f_1(0) \approx \frac{m_1^*}{\gamma_1 M_1 \sqrt{\pi} \tilde{A}_1}, f_2(0) \approx \frac{-\gamma_2 M_2 \sqrt{\frac{D_1}{D_2}} m_1^*(\gamma_1 M_1)^{-1}}{1 + \gamma_3 M_3 \sqrt{\frac{D_1}{D_2}} m_1^*(\gamma_1 M_1)^{-1}}.$$

Since $f_2(0) \leq 1$, then the following connection must be true

$$-(\gamma_2 M_2 + \gamma_3 M_3) \sqrt{\frac{D_1}{D_2}} m_1^* \leq \gamma M_1.$$

6 Emission of Alkaline Metal Oxide

The self-similar forms of the differential equations of the substances transport allow to calculate the emission of the alkaline metal oxide from the glass fibre material under the influence of the acid solution flow.

Let m_3^* is the mass concentration of the alkaline metal oxide in the glass fibre material when it leaves the acid solution flow ($x_1 = 0$). The value $m_3(x_1 + \Delta x_1)$ is possible to obtain from the values $m_3^*, m_3(x_1)$. Let further ρ_3 is the density of the alkaline metal oxide, ρ_4 is the density of the silicon oxide, l is the thickness of the glass fibre material, $\bar{\rho}^i = l\rho_i, i = 3, 4$, then the density area expresses by the formula

$$\bar{\rho} = c \frac{\bar{\rho}_3 \bar{\rho}_4}{m_3 \bar{\rho}_4 + (1 - m_3)\bar{\rho}_3},$$

where c is the porosity coefficient of the glass fibre material. We can consider that

$$m_3 \bar{\rho}_3 \mid_{x_1 + \Delta x_1} = m_3 \bar{\rho} \mid_{x_1} - \frac{(R_0 + R_1 + R_2)}{v_0} \Delta x_1.$$

Denoting $u = m_3 \bar{\rho} \mid_{x_1 + \Delta x_1}$, we obtain

$$m_3(x_1 + \Delta x_1) = \frac{\bar{\rho}_3 u}{c(x_1)\bar{\rho}_3 \bar{\rho}_4 + (\bar{\rho}_3 - \bar{\rho}_4)u},$$

furthermore $\bar{\rho}(x_1 + \Delta x_1) = \frac{u}{m_3(x_1 + \Delta x_1)}$,

$$c(x_1 + \Delta x_1) = \frac{\bar{\rho}(x_1 + \Delta x_1)}{\bar{\rho}_3 \bar{\rho}_4}(m_3 \bar{\rho}_4 + (1 - m_3)\bar{\rho}_3).$$

7 Conclusions

The provided computational experiments allow to make up some conclusions with practical significance for technological process.

1. The alkaline metal salt formed in the chemical reaction is concentrating near the glass fibre material and there is considerably diminished mass concentration of the acid. Therefore in order to increase the effectiveness of the process it is necessary to rinse periodically the glass fibre material in clean water.

2. The exact estimate of the thickness of the boundary layer $\delta_1(x_1)$ provides the opportunity to establish the optimal distance between the surfaces of the glass fibre material imbedded in parallel in the bath and to establish the optimal length of the glass fibre material stages that are imbedded in the bath filled with the acid solution.

3. The chemical reaction courses actively in the bath stage with lower mass concentration of the acid in the solution and higher mass concentration of the alkaline metal oxide in the glass fibre material.

References

[1] H. Kalis: Numerical integration by finite difference methods for some nonlinear problems, Latv. Univ. Zinātn. Raksti, 592, 1994, pp. 73–90.

Electronics, circuits, and filters

Numerical Solutions for the Simulation of Monolithic Microwave Integrated Circuits

Georg Hebermehl, Rainer Schlundt
Weierstrass Institute for Applied Analysis and Stochastics
Mohrenstr. 39, D-10117 Berlin, Germany
hebermehl@wias-berlin.de, schlundt@wias-berlin.de

Horst Zscheile, Wolfgang Heinrich
Ferdinand-Braun-Institut für Höchstfrequenztechnik
Rudower Chaussee 5, D-12489 Berlin, Germany
zscheile@fbh-berlin.de, heinrich@fbh-berlin.de

Abstract

The electric properties of monolithic microwave integrated circuits can be described in terms of their scattering matrix using Maxwellian equations. The corresponding three-dimensional boundary value problem of Maxwell's equations can be solved by means of a finite-volume scheme in the frequency domain. This results in a two-step procedure: a time and memory consuming eigenvalue problem for nonsymmetric matrices and the solution of a large-scale system of linear equations with indefinite symmetric matrices. Improved numerical solutions for these two linear algebraic problems are treated.

1 Introduction

The design of monolithic microwave integrated circuits (MMIC) requires efficient CAD tools in order to avoid costly and time-consuming redesign cycles. Commonly, network-oriented methods are used for this purpose. With increasing frequency and growing packaging density, however, the coupling effects become critical and the simple low-frequency models fail. Thus, field-oriented simulation methods become an indispensable tool for circuit design.

Figure 1 illustrates the principal structure under investigation. Since the electric properties are described in terms of the scattering matrix, transmission-line sections have to be attached at the ports. This defines propagation constants and mode patterns required for scattering matrix calculation. Typical line structures are planar lines (microstrip, coplanar waveguide), coaxial lines, or rectangular waveguides. The scattering matrix describes the structure in terms of wave modes

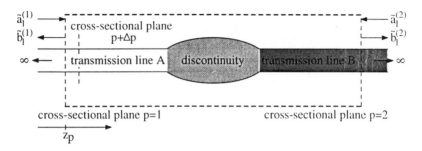

Figure 1: Structure under investigation

at the ports [2], [3], [1], [4], [5], which can be computed from the electromagnetic field. A three-dimensional boundary value problem can be formulated using Maxwell's equations in order to compute the electromagnetic field.

The application of the finite-volume method to the three-dimensional boundary value problem for the Maxwellian equations results in the so-called Finite-Difference method in the Frequency Domain (FDFD).

The program package F3D (Finite Differenzen dreidimensional) [3], [1] allows to simulate the electromagnetic field of nearly arbitrarily shaped structures.

2 Scattering Matrix

The transmission lines of the structure must be longitudinally homogeneous. The junction of the transmission lines, the so-called discontinuity, may have an arbitrary structure. A short part of the transmission lines is considered as a part of the discontinuity. The whole structure may be surrounded with an enclosure. Cross-sectional planes $p = 1$ and $p = 2$, so-called ports, are defined on the transmission lines. The incoming modes $\tilde{a}_l^{(p)}$ are changed in the discontinuity (see Figure 1). The complex generalized scattering matrix S describes the energy exchange and phase relation between all outgoing modes $\tilde{b}_l^{(p)}$ and all incoming modes $\tilde{a}_l^{(p)}$ [5].

$$S = \begin{pmatrix} S_{11} & S_{12} & \cdots & S_{1m_{\bar{s}}} \\ S_{21} & S_{22} & \cdots & S_{2m_{\bar{s}}} \\ \cdots & \cdots & \cdots & \cdots \\ S_{m_{\bar{s}}1} & S_{m_{\bar{s}}2} & \cdots & S_{m_{\bar{s}}m_{\bar{s}}} \end{pmatrix} = (S_{\rho,\sigma}), \tag{1}$$

$$\rho, \sigma = 1(1)m_{\bar{s}}, \quad m_{\bar{s}} = \sum_{p=1}^{\bar{p}} m^{(p)}.$$

$m^{(p)}$ is the number of modes which have to be taken into account on the cross-sectional plane p. $\bar{p}$ is the number of cross-sectional planes.

The scattering matrix can be extracted from the orthogonal decomposition of the electric field at a pair of two neighboring cross-sectional planes p and $p + \Delta p$ (see Figure 1) on each waveguide for a number of linear independent excitations of the transmission lines. Therefore, we need the electric field. The electric field is computed using a boundary value problem for the Maxwellian equations.

3 The Boundary Value Problem

Because the scattering matrix is defined in the frequency domain, it is convenient to restrict oneself to fields which vary with the time t according to the complex exponential function $e^{\jmath\omega t}$. Thus, we use the integral form of the Maxwellian equations in the frequency domain:

$$\oint_{\partial\Omega} \frac{1}{\tilde{\mu}\mu_0} \vec{B} \cdot d\vec{s} = \int_{\Omega} (\jmath\omega\tilde{\epsilon}\epsilon_0 \vec{E}) \cdot d\vec{\Omega}, \quad \oint_{\Omega} (\tilde{\epsilon}\epsilon_0 \vec{E}) \cdot d\vec{\Omega} = 0,$$

$$\oint_{\partial\Omega} \vec{E} \cdot d\vec{s} = \int_{\Omega} (-\jmath\omega\vec{B}) \cdot d\vec{\Omega}, \quad \oint_{\Omega} \vec{B} \cdot d\vec{\Omega} = 0, \tag{2}$$

taking into account the constitutive relations

$$\vec{B} = \mu \vec{H}, \quad \vec{D} = \underline{\epsilon} \vec{E}, \quad \text{with} \quad \underline{\epsilon} = \epsilon + \frac{\kappa}{\jmath\omega}, \quad \mu = \tilde{\mu}\mu_0, \quad \underline{\epsilon} = \tilde{\epsilon}\epsilon_0.$$

The two equations on the right-hand side of (2) correspond to Gauss' flux laws. The field vectors $\vec{E}$, $\vec{H}$, $\vec{D}$ and $\vec{B}$ (electric and magnetic field intensity, electric and magnetic flux density, respectively) are complex functions of the spatial coordinates only. ω is the circular frequency and $\jmath^2 = -1$. The permeability μ, the permittivity ϵ, and the conductivity κ are assumed to be scalar functions of the spatial coordinates. $\underline{\epsilon}$ is the complex permittivity.

Boundary conditions

At the enclosure, except at the cross-sectional planes, the tangential electric or the tangential magnetic field is known. At the cross-sectional planes p the transverse electric field $\vec{E}_t^{(p)} = \vec{E}_t(z_p)$ is given by superposing transmission line modes $\vec{E}_{t,l}^{(p)} = \vec{E}_{t,l}(z_p)$ with weighted mode-amplitude sums $w_l^{(p)} = w_l(z_p)$. The transverse electric mode fields $\vec{E}_{t,l}^{(p)}$ are computed using an eigenvalue problem for transmission lines (see section 6).

The transverse mode fields $\vec{E}_{t,l}^{(p)}$ satisfy an orthogonality relation ($\delta_{l,m}$ Kronecker symbol).

$$\vec{E}_t^{(p)} = \sum_{l=1}^{m^{(p)}} w_l^{(p)} \vec{E}_{t,l}^{(p)}, \quad \int_{\Omega} (\vec{E}_{t,l}^{(p)} \times \vec{H}_{t,m}^{(p)}) \cdot d\vec{\Omega} = \eta_m \delta_{l,m}, \quad \eta_m = 1[W]. \tag{3}$$

This means that a three-dimensional boundary value problem of the Maxwellian equations is formulated.

The orthogonality relation is applied at two neighboring cross-sectional planes z_p and $z_{p+\Delta p} = z_p + \Delta z_p$ (see Figure 1):

$$\frac{1}{\eta_m} \int_\Omega (\vec{E}_t^{(p)} \times \vec{H}_{t,m}^{(p)}) \cdot d\vec{\Omega} = \tilde{a}_m^{(p)} + \tilde{b}_m^{(p)} = w_m^{(p)},$$

$$\frac{1}{\eta_m} \int_\Omega (\vec{E}_t^{(p+\Delta p)} \times \vec{H}_{t,m}^{(p)}) \cdot d\vec{\Omega} = \tilde{a}_m^{(p+\Delta p)} + \tilde{b}_m^{(p+\Delta p)} = w_m^{(p+\Delta p)}$$

(4)

The weighted mode-amplitude sums $w_l^{(p)}$ are given. Because of

$$\tilde{a}_m^{(p+\Delta p)} = \tilde{a}_m^{(p)} e^{-jk_{z_l}^{(p)} \Delta z_p}, \qquad \tilde{b}_m^{(p+\Delta p)} = \tilde{b}_m^{(p)} e^{+jk_{z_l}^{(p)} \Delta z_p} \qquad (5)$$

we can eliminate $\tilde{a}_m^{(p+\Delta p)}$ and $\tilde{b}_m^{(p+\Delta p)}$ from (4). Obtaining the mode amplitudes $\tilde{a}_m^{(p)}$ and $\tilde{b}_m^{(p)}$ we can compute the scattering matrix (1) [5].

4 The Maxwellian Grid Equations

It is advantageous to solve the Maxwellian equations directly rather than solving a partial differential equation of second order derived therefrom, because the quantities $\tilde{\mu}$ and $\tilde{\epsilon}$ can be different from cell to cell when using Maxwellian equations. The region is divided into elementary rectangular parallelepipeds by using a three-dimensional nonequidistant orthonormal Cartesian grid. The components E_x, E_y, and E_z of the electric field $\vec{E}$ are located in the centers of the edges of the elementary cells. The components B_x, B_y, and B_z, on the other hand are normal to the face centers [9], [8]. Thus, the electric field components form a primary grid, and the magnetic flux density components a dual grid (see Figure 2). Using the lowest-order integration formulae

$$\oint_{\partial\Omega} \vec{f} \cdot d\vec{s} \approx \sum (\pm f_i s_i), \qquad \int_\Omega \vec{f} \cdot d\vec{\Omega} \approx f\Omega \qquad (6)$$

to approximate the integrals in (2) we get the matrix representation of Maxwell's equations:

$$A^T D_{s/\tilde{\mu}} \vec{b} = j\omega\epsilon_0\mu_0 D_{A\tilde{\epsilon}} \vec{e}, \qquad B D_{A\tilde{\epsilon}} \vec{e} = 0,$$

$$AD_s \vec{e} = -j\omega D_A \vec{b}, \qquad B^T D_A \vec{b} = 0$$

(7)

with

$$\vec{e} = \begin{pmatrix} \vec{e}_x \\ \vec{e}_y \\ \vec{e}_z \end{pmatrix}, \qquad \begin{aligned} \vec{e}_x &= (e_{x_1}, e_{x_2}, \ldots, e_{x_{n_{xyz}}})^T, \\ \vec{e}_y &= (e_{y_1}, e_{y_2}, \ldots, e_{y_{n_{xyz}}})^T, \\ \vec{e}_z &= (e_{z_1}, e_{z_2}, \ldots, e_{z_{n_{xyz}}})^T, \end{aligned} \qquad \begin{aligned} e_{x_l} &= E_{x_{i,j,k}}, \\ e_{y_l} &= E_{y_{i,j,k}}, \\ e_{z_l} &= E_{z_{i,j,k}}, \end{aligned} \qquad (8)$$

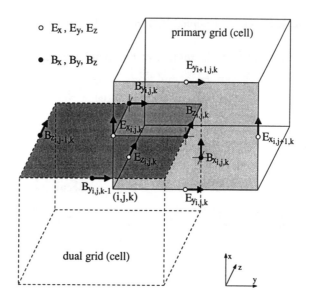

Figure 2: Primary and dual grid

$$\vec{b} = \begin{pmatrix} \vec{b}_x \\ \vec{b}_y \\ \vec{b}_z \end{pmatrix}, \quad \begin{aligned} \vec{b}_x &= (b_{x_1}, b_{x_2}, \ldots, b_{x_{n_{xyz}}})^T, \\ \vec{b}_y &= (b_{y_1}, b_{y_2}, \ldots, b_{y_{n_{xyz}}})^T, \\ \vec{b}_z &= (b_{z_1}, b_{z_2}, \ldots, b_{z_{n_{xyz}}})^T, \end{aligned} \quad \begin{aligned} b_{x_l} &= B_{x_{i,j,k}}, \\ b_{y_l} &= B_{y_{i,j,k}}, \\ b_{z_l} &= B_{z_{i,j,k}}, \end{aligned} \quad (9)$$

$$\ell = (k-1)n_{xy} + (j-1)n_x + i, \quad n_{xy} = n_x n_y, \quad n_{xyz} = n_x n_y n_z.$$

The vectors $\vec{e}$ and $\vec{b}$ contain the components of the electric field and the components of the magnetic flux density of the elementary cells, respectively. The diagonal matrices D_s and D_A contain the information on material and on dimension for the structure and the corresponding mesh. A represents the curl operator in the second Maxwellian equation of (2) using the primary grid. B represents the surface integral of the divergence. A and B are sparse and contain the values 1, -1, and 0 only.

5 The System of Linear Algebraic Equations

The two equations of the left-hand side of (7) form a system of linear algebraic equations of the electromagnetic field in the absence of any boundary conditions. Eliminating the components of the magnetic flux density the number of unknowns in this two equations can be reduced by a factor of two:

$$Q_1 \vec{e} = 0, \quad Q_1 = A^T D_{s/\tilde{\mu}} D_A^{-1} A D_s - k_0^2 D_{A\tilde{\epsilon}}, \quad k_0 = \omega\sqrt{\epsilon_0 \mu_0}. \tag{10}$$

Taking into account the boundary conditions we get a partitioning of the matrix Q_1 into a sum of two matrices

$$Q_1 \vec{e} = (Q_{1,A} + Q_{1,r})\vec{e} = 0,$$

and we have to solve

$$\tilde{Q}_{1,A}\vec{\tilde{e}} = \vec{\tilde{r}}, \quad \tilde{Q}_{1,A} = D_s^{\frac{1}{2}} Q_{1,A} D_s^{-\frac{1}{2}}, \quad \vec{\tilde{r}} = D_s^{\frac{1}{2}}\vec{r}, \quad \vec{r} = -Q_{1,r}\vec{e}. \tag{11}$$

Adding the matrix representation of the gradient of the electric-field divergence, which is equal to zero for fields without space charge,

$$\tilde{\epsilon}\epsilon_0 \nabla(\frac{1}{(\tilde{\epsilon}\epsilon_0)^2} \nabla \cdot \tilde{\epsilon}\epsilon_0 \vec{E}) = 0 \quad \Rightarrow \quad Q_2 \vec{e} = 0 \text{ with } Q_2 = D_s^{-1} D_{A\tilde{\epsilon}} B^T D_{V\tilde{\epsilon}\tilde{\epsilon}}^{-1} B D_{A\tilde{\epsilon}}$$

to (11) gives the system

$$(\tilde{Q}_{1,A} + \tilde{Q}_{2,A})\vec{\tilde{e}} = \vec{\tilde{r}} \tag{12}$$

with

$$\tilde{Q}_2 = D_s^{\frac{1}{2}} Q_2 D_s^{-\frac{1}{2}}, \quad \tilde{Q}_2 = \tilde{Q}_{2,A} + \tilde{Q}_{2,r}, \quad \vec{\tilde{e}} = D_s^{\frac{1}{2}}\vec{e}, \quad \tilde{Q}_{2,r}\vec{\tilde{e}} = 0, \quad \tilde{Q}_{2,A}\vec{\tilde{e}} = 0$$

which can be solved numerically faster [1]. The effect of this additional term can be interpretated as preconditioning. The system of linear algebraic equations is solved using multicoloring or independent set orderings to reduce the dimension of the system. The reduced systems are solved using iterative methods with preconditioning. The execution time was reduced by a factor of 10 in comparison to the original package F3D.

6 The Eigenvalue Problem

Before we can solve the system of linear equations we have to compute the transverse electric mode fields $\vec{E}_{t,l}^{(p)}$ at the port for the boundary condition using an eigenvalue problem. We consider a selected transmission line in the discussion to follow. $\tilde{\epsilon}$ and $\tilde{\mu}$ are functions of transverse position but are independent of the longitudinal direction. Thus, we assume that the fields vary exponentially in the longitudinal direction :

$$\vec{E}(x, y, z \pm 2h) = \vec{E}(x, y, z) e^{\mp j k_z 2h}. \tag{13}$$

k_z is the propagation constant. A substitution of the ansatz (13) into the system of linear equations (10) taking into account boundary conditions, and the elimination of the longitudinal electric field components by means of the electric-field divergence equation (see 7)) gives an eigenvalue problem [4] for the transverse electric field on the transmission line region:

$$M(h^2)\vec{e} = \gamma(h)\vec{e}, \quad \text{type}(M) = (2n_{xy} - n_b, 2n_{xy} - n_b). \quad (14)$$

The sparse matrix M is nonsymmetric. $\vec{e}$ consists of components $E_{x_{i,j,k}}$ and $E_{y_{i,j,k}}$, $k = const$, of the eigenfunctions. The size of n_b depends on the boundary conditions at the port. The propagation constants k_z (16) can be computed from γ (15) after the solution of the eigenvalue problem (14):

$$\gamma_\iota(h) = e^{-jk_{z_\iota}2h} + e^{+jk_{z_\iota}2h} - 2 = -4\sin^2(k_{z_\iota}h) = u_\iota + jv_\iota, \quad (15)$$

$$k_{z_\iota} = \frac{j}{2h}\ln\left(\frac{\gamma_\iota}{2} + 1 + \sqrt{\frac{\gamma_\iota}{2}\left(\frac{\gamma_\iota}{2} + 2\right)}\right) = \beta_\iota - j\alpha_\iota, \quad \iota = 1(1)2n_{xy} - n_b. \quad (16)$$

The energy of the complex and evanescent modes decreases exponentially with the distance from the discontinuity. Thus, most of the modes can be neglected within the limit of accuracy. Generally speaking, the larger the magnitude of the imaginary part of k_z the stronger the decay. Therefore, the propagation constants k_{z_ι}, $\iota = 1(1)2n_{xy} - n_b$, are sorted in ascending order of $|\alpha_\iota|$. In the case if some $|\alpha_\iota|$ have the same value the constants k_{z_ι} are sorted in descending order of $|\beta_\iota|$.

In the original version of the program package F3D all propagation constants are computed and sorted in order to select the wanted propagation constants. This way is very time-consuming. The full matrix is stored.

In order to avoid the time-consuming computation of all eigenvalues γ we use the implicitly restarted Arnoldi iteration [7], [6] now, which is carried out twice to find the first propagation constants of the sorted set.

Using the iterative method the computation of the needed propagation constants in a typical example is 40-fold faster than in the old version. The reduction of the memory consumption amounts the 20-fold, since the sparse storage technique is applied.

7 Conclusions

The Finite Difference method in Frequency Domain allows the calculation of the scattering matrix for a number of simultaneously excited modes. This is an advantage compared with computations in the Time Domain. The price to be paid is the high memory consumption and the time-consuming solution of an eigenmode problem of high dimension and of large systems of linear algebraic equations.

We avoid the time-consuming computation of all eigenvalues in order to calculate a selected set of propagation constants using an iterative method that is carried out twice. We find that the SSOR preconditioning combined with the multicoloring or independent set orderings is a very efficient method to solve the systems of linear equations. The numerical effort and the storage requirements can be reduced considerably.

References

[1] Beilenhoff, K. / Heinrich, W. / Hartnagel, H. L.: Improved Finite-Difference Formulation in Frequency Domain for Three-Dimensional Scattering Problems, IEEE Transactions on Microwave Theory and Techniques, Vol. 40, No. 3 (1992), pp. 540-546.

[2] Christ, A. / Hartnagel, H. L.: Three-Dimensional Finite-Difference Method for the Analysis of Microwave-Device Embedding, IEEE Transactions on Microwave Theory and Techniques, Vol. MTT-35, No. 8 (1987), pp. 688-696.

[3] Christ, A.: Streumatrixberechnung mit dreidimensionalen Finite-Differenzen für Mikrowellen-Chip-Verbindungen und deren CAD-Modelle, Fortschrittberichte VDI, Reihe 21: Elektrotechnik, Nr. 31 (1988), pp. 1-154.

[4] Hebermehl, G. / Schlundt, R. / Zscheile, H. / Heinrich, W.: Simulation of Monolithic Microwave Integrated Circuits, Weierstraß-Institut für Angewandte Analysis und Stochastik im Forschungsverbund Berlin e.V., Preprint No. 235 (1996), pp. 1-37.

[5] Hebermehl, G. / Schlundt, R. / Zscheile, H. / Heinrich, W.: Improved Numerical Solutions for the Simulation of Monolithic Microwave Integrated Circuits, Weierstraß-Institut für Angewandte Analysis und Stochastik im Forschungsverbund Berlin e.V., Preprint No. 236 (1996), pp. 1-43.

[6] Lehoucq, R. B.: Analysis and Implementation of an Implicitly Restarted Arnoldi Iteration, Rice University, Houston, Texas, Technical Report TR95-13 (1995), pp. 1-135.

[7] Sorensen, D. C.: Implicit Application of Polynomial Filters in a k-Step Arnoldi Method, SIAM Journal on Matrix Analysis and Applications., Vol. 13, No.1 (1992), pp. 357-385.

[8] Weiland, T.: Eine numerische Methode zur Lösung des Eigenwellenproblems längshomogener Wellenleiter, Archiv für Elektronik und Übertragungstechnik, Band 31, Heft 7/8 (1977), pp. 308-314.

[9] Yee, K. S.: Numerical Solution of Initial Boundary Value Problems Involving Maxwell's Equations in Isotropic Media, IEEE Transactions on Antennas and Propagation, Vol. AP-14, No. 3 (1966), pp. 302-307.

Symbolic Modelling in Circuit Simulation

W. Klein[*]

Abstract

Simulation models for semiconductor elements like MOS transistors are of great interest in the field of circuit simulation. Because of the nonlinearity of the terminal current and charge functions with respect to the input voltages, furthermore the derivatives of these transistor functions are needed.

Developing a new model is computing a large set of equations governing the behaviour of the MOSFET's and the influence of the various model parameters. Due to the complexity of expressions and to the large number of the involved parameters, the necessary differentiation of these expressions 'by hand' is almost impossible. Moreover an implementation of a new model in a numerical code is a tedious and risky task.

We describe two different ways to improve the process of model development. The use of Maple enables the generation of a transistor model from scratch including generation of derivatives and Fortran implementation. In contrast to this intermediate step of code generation, the coupling of circuit simulation to an operator based system Syperb leads directly to an evaluation of model function and their derivatives.

1 Introduction

The development of new transistor simulation models is an important task in the field of circuit simulation. Due to faster changing product cycles in the manufacturing of integrated circuits, new simulation models for larger and more complicated transistor models have to be developed in even shorter times. Besides the physical modelling, compact analytical models of MOS transistors lead to a non-linear dependence of the terminal current and charge functions at drain, source and bulk on the three input voltages V_{GS}, V_{DS} and V_{BS} between gate and source, drain and source as well as between bulk and source. The connection of these transistor models to the entire circuit simulation model leads to a non-linear, time-dependent set of algebraic differential equations. This sytem is mathematically treated by a discretization in time coupled to a Newtonian iteration scheme within an inner loop. The derivatives of the transistor model functions are needed with respect to the input voltages for this Newtonian process. Here, the use of tools for automatic differentiation can accelerate and improve essentially the process of developing new transistor models.

Many different computer algebra tools can be used for the automatic generation of simulation models. The application of these tools in the field of transistor modelling has to handle with very large simulation models whith dynamically piecewise defined transistor functions. The use of common subexpressions in model function and its derivative must be supplied in order to accelerate the model evaluation. The reason to apply Maple to this problem is the possibility to generate complete Fortran subroutines even for piecewise defined functions as well as to use the reverse mode of automatic differentiation (at least in Maple V release 4).

---

[*]SIEMENS AG, Corporate Research and Development, Otto-Hahn Ring 6, 81730 München, Germany. Phone: 49 89 636 41119, E-mail: wolfram.klein@zfe.siemens.de

In order to avoid the intermediate step of code generation for model functions and their derivatives as well as to reduce the memory requirements for the symbolic generation of derivatives, an operator based system Syperb was developed. This system enables reading, evaluation and differentiation of functions in different kinds of arithmetic using C++.

2 MOS Transistors

A rough description of the functionality and modelling of a MOS transistor will be given in this section. The four different terminals, drain, gate, source, and bulk of a MOS transistor are obtained by the construction of its different layers (MOS : metal oxide semiconductor, Fig. 1). The input voltage V_{GS} between gate and source controls the terminal current I_{DS} between drain and source, the charges of the transistor are furthermore depending on the voltage V_{GB} between gate and bulk. The voltage V_{GS} determines theconcentration of the carriers in the transistor 'channel', i.e. the surface region of the semiconductor material under the isolating oxide layer. The gate-source voltage for which the channel current I_{DS} becomes significant is called the threshold voltage V_{TH}, thus characterizing the 'on-off' behavior of the transistor (Fig. 2). Even one of the most simple MOS transistor simulation models (Level 1 model by Shichman and Hodges, [1]) contains a rather difficult equation for the threshold voltage

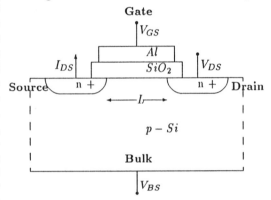

Fig. 1 : MOS Transistor

(1) $$V_{TH} = V_{FB} + 2\phi_p + \gamma\sqrt{2\phi_p - V_{BS}},$$

which requires the additional knowledge of the flat-band voltage V_{FB} for the condition of an identical carrier concentration between surface and substrate by

(2) $$V_{FB} = \phi_{MS} - \frac{Q_0}{C_{OX}}.$$

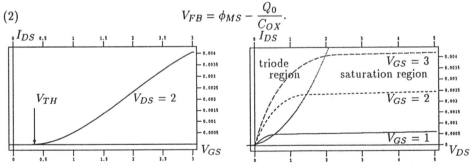

Fig. 2 : Drain Current (DC) over controlling V_{GS} Fig. 3 : MOSFET DC Characteristics

Here, the charge at the oxide semiconductor interface Q_0 and the capacitance per unit area of the oxide layer C_{OX} are needed, ϕ_{MS} denotes the contact potential between the gate and semiconductor material and is a constant depending on technology (ϕ_p : Fermi potential,

γ : body-effect parameter).

Even this very fundamental threshold voltage equation shows the problem of a deep physical understanding of semiconductor physics as well as the mathematical problem of quickly growing dependencies between a large number of different formulas. Neglecting some intermediate formulas, the terminal current between drain and source of the transistor of the Level 1 model is given in the case of $V_{GS} > V_{TH}$ and $V_{DS} < V_{GS} - V_{TH}$ ('triode region', Fig. 3) by

$$I_{DS} = KP \frac{W}{L - 2X_{jl}} \cdot (V_{GS} - V_{TH} - \frac{V_{DS}}{2}) \cdot V_{DS} \cdot (1 + \lambda V_{DS}).$$

Additional characteristics of the model are introduced for the channel length and width L and W, respectively, and for the channel length modulation parameter λ, as well as for the lateral diffusion X_{jl} and for the transconductance parameter KP.

In the saturation region (Fig. 3), i.e. in the case of $V_{GS} > V_{TH}$ and $V_{DS} > V_{GS} - V_{TH}$, the replacement of V_{DS} by $V_{GS} - V_{TH}$ leads to the following formula for the terminal current

$$I_{DS} = \frac{KP}{2} \frac{W}{L - 2X_{jl}} \cdot (V_{GS} - V_{TH})^2 \cdot (1 + \lambda V_{DS}).$$

Here, the threshold voltage V_{TH} mentioned above leads to a first branching of all successivly defined formulas into a linear ('triode'), and saturation region of the current vs. voltage characteristics (Fig. 3, Level 3 and all finer models require additional branches to further depletion and accumulation regions). This results in piecewise defined functions for all further formulas depending dynamically on the input voltages V_{GS}, V_{DS}, V_{BS} between gate and source, drain and source as well as between bulk and source. The charges Q_G and Q_B at gate and bulk are of further special interest in transistor modelling besides the terminal current I_{DS}.

3 Present state

New models for MOS transistors are currently essentially developed 'by hand'. Based on a given mathematical model description, the next step of the phase of model development is the implementation (by hand) of the functional behaviour of the drain and substrate current of the transistor as well as the implementation of the functions for three different charges at the drain, gate and bulk. This leads for a transistor model Level 3 of medium abstraction granularity to ~2,200 lines of FORTAN 77 code, e.g.. The subsequent step of model development is concerned with the determination of derivatives for the terminal current and charge functions with respect to certain input variables. Furthermore, a corresponding implementation of these derivatives has to be done in FORTRAN 77, e.g.. This leads to additional 600 lines of FORTRAN 77 code for the transistor model mentioned above. Even this second step is done currently 'by hand'. The aim of this paper to improve and accelerate both steps of model development.

The essential advantage of differentiation 'by hand' is a very effective code with respect to run time and memory costs. The knowledge of common subexpressions in the functional description simplifies the step of differentiation by hand as well as accelerates the subsequent model execution substantially. Furthermore, avoiding 'trivial' operations during differentiation such as multiplication with '1' or '0' or addition of '0' resulting from differentiation of constants or independent variables accelerates the code. A third advantage of this method is the additional knowledge of the range of arguments of standard functions such as sqrt, tan, ... Here, no time consuming branching is necessary.

Concerning the memory costs, the use of an 'EQUIVALENCE' statement in FORTRAN 77 in order to assign COMMON variables to internal model parameters or in order to return computed values to COMMON variables needs no additional memory for the gradients of these common variables: In spite of that fact, an automatically executed differentiation will require additional memory for these common variables.

But the main disadvantage of the method 'by hand' such as error-prone implementation of model functions as well as time consuming and error-prone determination of derivatives cannot be neglected.

4 Maple V

Maple V is a widely used computer algebra system (CASE-tool). Compared to other CASE-tools and to the old Maple V release 3, the new release 4 offers a better handling of piecewise defined functions as well as the reverse mode of automatic differentiation. Because of problems concerning the availability of release 4, the results of this section are concerned with Maple V release 3.

4.1 Generation of a transistor model in Maple

The application of Maple to a MOS transistor model consists essentially of four steps:

In a first preparing step, the internal kernel of Maple has to be enlarged. Here, essentially the functionality of piecewise defined functions must be generalized. The evaluation and differentiation of piecewise defined functions as well as the generation of corresponding Fortran code needs additionally defined user supplied functions.

The essential step during the model development is the definition of the four terminal functions of the transistor model in Maple. Besides the dynamically defined piecewise continous functions, the transistor functions can be implemented in a mathematical notation. The symbolic manipulation of these formulas in Maple combines all intermediate functions leading to only one functional expression for the corresponding transistor functions.

In a third step, the four terminal functions are automatically differentiated with respect to three independent variables. This is done using the differentiation statement of Maple (combined to user defined functions, see step 1).

The final step generates automatically an optimized Fortran code for functions and derivatives. Here, again the Maple functionality for generating program code can be used, which is able to detect common subexpressions of functions and derivatives.

4.2 Remarks to Maple

Using Maple V release 3 for the automatic generation of a transistor model leads to several problems.

- As mentioned above, the functionality of piecewise defined functions has to be generalized,
- instead of a complete Fortran code only the expression of the final formula is translated to Fortran. This leads to an additional user supplied postprocessing (header for subroutines, variable declaration),

- Maple is rather slow. Differentiation and code generation for both functions and derivatives needs about 1.5 hrs (SUN IPX, ca 350 lines Maple input). Larger models will run into memory and runtime problems,

- the runtime of the automatically generated model is about a factor 1,2 to 1,8 slower than the code developed 'by hand' (see Table 1).

In spite of these facts, there are some interesting advantages applying a computer algebra tool to a transistor model. Furthermore, using CASE tools leads to a new quality in developing simulation models.

- The time for the development of a new transistor model can be reduced essentially,
- the derivatives as well as the Fortran implementation generated by Maple for both functions and derivatives are correct (which is a new quality in transistor modelling),
- using Maple enables to generate easily 'test'-models besides the really wanted simulation model,
- the final transistor function can be plotted (Fig. 4).

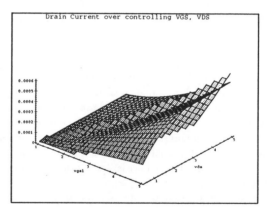

Fig. 4 : Piecwise defined Drain Current Characteristics, implemented in Maple, compare with Fig. 3.

4.3 Final Results of Maple

In order to get a feeling how Maple works, Maple was applied to the MOS transistor Level 3, which is a moderately sized model. The Maple input for this MOS3 model needs about 350 lines; Maple generates about 300 lines of Fortran code. An additional automatically done postprocessing and optimization of this Fortran code leads to about 1.500 lines of Fortran code ('hand coded': 700 lines). The postprocessing step is concerned with the generation of complete Fortran code. The optimization step ('Maple-Opt', see Table 1) improves essentially the generated Fortran code for the piecewiese defined functions.

The following Table 1 shows the runtime for the different realizations of the MOS3 model. Three different CPU-times are listed for each of the different input circuits. **Eval-Moseq** denotes the summarized CPU-time for evaluating functions and derivatives of the MOS transistor several times within a coupled Newton iteration scheme and time-discretization. **Load-Mosfet** means the repeated call to the just mentioned subroutine for evaluation of functions and derivatives followed by a correct placement and storing of the resulting values into a large coefficient matrix. Therefore, **Eval-Moseq** is included in the time for **Load-Mosfet**. The **Total** CPU-time denotes the time for one execution of the entire simulation process. These three CPU-times are tabulated for the MOS3 transistor model developed either 'by hand' or generated by the use of Maple. Furthermore, the CPU-times for an optimization of the Maple generated Fortran code, as mentioned above, are shown. The corresponding ratios of these times are also listed.

Model	input circuit	CPU-time	Hand	Maple	Maple-Opt	Ratio	Ratio-Opt
MOS3	Mos3	Eval-Moseq	2.2	6.5	3.85	2.9	1.74
		Load-Mosfet	4.6	9.2	6.6	1.95	1.43
		Total CPU	9.1	14.3	11.65	1.57	1.27
MOS3	Mosringo	Eval-Moseq	5.8	18.8	10.3	3.17	1.77
		Load-Mosfet	11.6	24.6	16.6	2.12	1.44
		Total CPU	19.2	36.6	26.4	1.9	1.37
MOS3	Mosringo.long	Eval-Moseq	12.5	40.9	23.5	3.26	1.88
		Load-Mosfet	25.2	53.6	37.1	2.13	1.46
		Total CPU	46.1	91.4	66.5	1.97	1.42

Table 1 : Handcoded Model vs. Maple coded Model (Comp.: SUN IPX)

5 Syperb

SYPERB (**SY**stem with o**PER**ator-**B**ased notation) is based on C++ and uses techniques for compiler development (LEX, YACC). The concept of operator overloading in Syperb enables reading, evaluation and differentiation of functions in different kinds of arithmetic using an identical operator notation. The essential difference to tools for automatic differentiation (AD-tools) is the additional feature of reading formulas in mathematical notation (AD-tools need a Fortran implementation of the formula). Compared to CASE-tools like Maple, no intermediate step of code generation is necessary. Furthermore, the derivatives are evaluated directly in arbitrary arithmetic instead of the symbolic generation of formulas.
The main features of the new operator-based system Syperb are highlighted below.

- *Reading.* A particular transistor is usually represented by up to 25 pages of model functions. Therefore, an efficient method for reading formulas is of great interest. An interpreter reads the formulas consisting of basic operations $+,-,\cdot,/$, and some elementary functions such as sqr, sqrt, exp, sin, cos, ... A function tree is generated automatically in order to simplify and accelerate further processing. In this framework the insertion of formulas into other ones, as well as piecewise defined formulas plays a dominant role.

- *Function evaluation.* Application of the Newton scheme leads to linear systems with model functions as coefficients. Hence, a careful analysis of the generated function tree to sum common, but independent subexpressions is indispensable to speed the simulation process. Employing different types of computer arithmetics (real, staggered, interval) enables a control of errors in the arithmetic process.

- *Operator overloading.* The features mentioned above are partially formulated in a computer program written in PASCAL-XSC using an operator concept. For example, an operator symbol '+' is used to add two formulas f and g by '$f + g$' as well as to generate the corresponding function tree. Furthermore, the operator symbol '+' can be used to evaluate the formula $f(x_0) + g(x_0)$ at a particular point x_0. In addition, the differentiation either in forward or reverse mode is done for the '+' sign by '$f+g$'. Analogous features are valid for other arithmetic operations.

- *Differentiation.* As already mentioned, the crucial point in circuit simulation is the computation of derivatives. Here, the technique of automatic differentiation [11] is used. Therefore, the evaluation of the derivative at a single point in an efficient manner is possible even when many functions must be differentiated.

Practical experiences

Table 2 lists execution times for 500 evaluations of function and derivatives for a function tree with 208, 1253 and 2089 nodes. The experiments were executed on a Pentium 75 with LINUX. The functions in Table 2 are depending on one variable, 5, 10 and 50 variables. In the first differention mode ('Forward, one Var.'), n-calls (n = 1, 5, 10, 50) of the forward differentiation with respect to one variable were executed. The second differention mode ('Forward, all Var.') uses a gradient arithmetic. All derivatives are stored in one dynamical array (of actual length 1, 5, 10, 50). Besides the overhead concerning the array handling with a small number of variables (1 and 5 variables), a larger number of independent variables (10 and 50 variables) leads to shorter execution times. Concerning the CPU times, the reverse mode of differentiation shows essential differences compared to the results of the forward mode. The execution time of the reverse mode is independent on the number of independent variables. Furthermore, even for 5 independent variables the reverse mode yields much faster execution times.

Table 2 : Results of SYPERB (time in sec)

Diff Mode	# Indep. Var.	208 nodes	1253 nodes	2089 nodes
Forward, one Var.	1 Var.	0.43	2.14	3.63
Forward, one Var.	5 Var.	1.74	10.92	17.70
Forward, one Var.	10 Var.		21.18	35.90
Forward, one Var.	50 Var.			184.18
Forward, all Var.	1 Var.	2.20	12.97	21.75
Forward, all Var.	5 Var.	2.61	15.96	26.93
Forward, all Var.	10 Var.		18.40	30.90
Forward, all Var.	50 Var.			62.39
Reverse, all Var.	1 Var.	1.01	5.95	10.17
Reverse, all Var.	5 Var.	1.01	6.08	10.19
Reverse, all Var.	10 Var.		6.09	10.31
Reverse, all Var.	50 Var.			10.53

6 Conclusion

The development of new transistor models for circuit simulation is a difficult and time consuming task. The implementation and differentiation of the terminal current and charge functions of the transistor are those parts of the model development, that can be accelerated by the use of software tools. Furthermore, the correctness of automatically generated model implementation and differentiation leads to a new quality in model development. The disadvantages of these systems are a slightly slower execution time as well as the additionally needed memory for the generation of the function tree.

References

[1] *P. Antognetti, G. Massobrio* : Semiconductor device modelling with SPICE; McGraw Hill Company, 1987.

[2] *C. Bischof, A. Carle, G. Corliss, A. Griewank, and P. Hovland* : ADIFOR: Generating derivative code from FORTRAN programs, Scientific Programming, 1 (1992), pp. 11-29.

[3] *C. Bischof, A. Carle, P. Khademi, A. Maurer and P. Hovland* : ADIFOR 2.0 User's Guide; ANL, 1994, technical memorandum, ANL/MCS-TM-192.

[4] *H.C. Fischer* : Automatic Differentiation and Applications in Scientific Computing with Automatic Result Verification (editors E. Adams, U. Kulisch), Academic Press 1993, 105 – 142.

[5] *H. Fischer* : Fast method to compute the scalar product of gradient and scalar vector, Computing 41, pp. 261-265, 1989.

[6] *A. Griewank, D. Jüdes, and J. Utke* : ADOL-C: A Package for the Automatic Differentiation of Algorithms written in C/C++. University of Dresden, 1994.

[7] *R. Klatte, U. Kulisch et al.* : PASCAL-XSC, Language Reference with Examples; Springer Verlag, 1992.

[8] *W. Klein* : Report on Symbolic Manipulators for Automatic Code Generation. EG Project Jessi AC 12 : 'Analog Expert Design System', Milestone Report, Dec 1995.

[9] *B. Lemaitre* : JESSI 0.8 μm CMOS Transistor model for analogue and digital Circuit Simulation. JESSI Report AC 41 94-2.

[10] *R. Lohner* : Verified Computing and Programs in PASCAL-XSC, Habilitationsschrift University of Karlsruhe, Inst. for Applied Math. with Examples; Springer Verlag, 1992.

[11] *L.B. Rall* : Automatic Differentiation, Springer Lecture Notes in Computer Science 120, 1981.

[12] *H. Spiro* : Simulation integrierter Schaltungen, Oldenbourg, 1985.

An extended hydrodynamic model for silicon semiconductor devices

Orazio Muscato

Dipartimento di Matematica, Viale Andrea Doria 6

95125 Catania (Italy), e-mail : muscato@dipmat.unict.it

Abstract

A set of closed hydrodynamic-like equations is derived from Boltzmann's Transport Equation describing charge transport in semiconductors. The moment hierarchy is closed by using the entropy principle and the model is compared with homogeneous Monte Carlo simulations for silicon.

1 Introduction

In modern-day submicron Si devices, high electric field (E $\geq$ 100 kV/cm) and high field-gradient ($\mid$ E $\mid$/$\mid \nabla E \mid \leq$ 200 $\mathring{A}$) conditions are encountered during operation and carrier dynamics are far from thermal equilibrium.

In this regime undesirable phenomena can be produced such as substrate currents and hot-carrier injection into the gate oxide in MOSFET's and impact ionization at the base-collector junction in Bipolar Junction Transistor's. Therefore it is of paramount importance to describe these phenomena with a CAD tool.

Carrier transport in CAD codes is described with the standard drift-diffusion equations,but these equations are not adequate because the carriers are supposed to be in thermal equilibrium with the lattice [1].

The semiclassical Boltzmann Transport Equation with the Poisson equation for the electric field is the starting point for a rigorous theory. The analytical solution of such a system is impossible from the practical point of view and for this reason the Monte Carlo (MC) numerical solution has been developed [2]. The MC method is able to describe the behavior of small semiconductor devices even far from thermal equilibrium, because the full band structure of the semiconductor and the scattering rates are taken into account. Microscopic and macroscopic quantities (as the distribution function, mean velocity, temperature etc.) are evaluated. However,the extensive computation required makes it impractical for a device design on a regular basis.

For this reason hydrodynamic models, derived from the moment equations of the Boltzmann Transport Equation,have been used. The aim of these models is to incorporate higher-order effects than those included in the drift diffusion equations. The use of these models requires solving the following important problems:

- how many moment equations are needed;
- the closure of the hierarchy;
- the modeling of the production terms.

The closure problem consists in finding an appropriate expression for the (N+1)-th moment which appears under the divergence operator of the N-th moment equation. Likewise, modeling the production terms consists in expressing them as suitable functions of the first N moments.

Several solutions have been proposed in order to answer these questions: these theories must be checked with experiments or with simulations. The MC method plays a very important role,infact it is possible

- to determine the trasport coefficients (which in general are unknowns) in the material;
- to test the closure equations and the number of moments needed;
- to obtain the characteristics (e.g. velocity, current, energy) of real devices which can be compared with those obtained by the hydrodynamic model.

In this paper we shall introduce a hydrodynamic model based on the principles of Extended Thermodynamics [3]-[4] and we check this model with our MC simulator for silicon in the homogeneous case. In sec.2 we set up the basic formalism, we tackle the problem of the number of equations and of the production terms. In sec.3 we summarize the closure problem according to several authors and a check on the validity of Extended Thermodynamics approach is shown. Finally in sec. 4 new trends and conclusions are drown.

2 Moment equations

Let $f(\mathbf{x}, t, \mathbf{k})$ be the one-particle distribution function appearing in Boltzmann Transport Equation (**k** is the electron momentum). In the following we shall consider parabolic bands, i.e. the electron energy and the microscopic velocity read:

$$\mathcal{E}(\mathbf{k}) = \frac{\mathbf{k}^2}{2m^\star}, \quad \mathbf{v}(\mathbf{k}) = \frac{\mathbf{k}}{m^\star}$$

with $m^\star$ being the effective electron mass. It is convenient to introduce the random component **c** of **v** ($\mathbf{v} = \mathbf{u} + \mathbf{c}$) and to take the moments in the Boltzmann transport equation by multipling with $(1, c_i, c_i c_j, ..)$ [3]. For the sake of simplicity we write the first five equations of the the infinite hierarchy:

$$\frac{\partial n}{\partial t} + \frac{\partial (nu_i)}{\partial x^i} = 0, \qquad (1)$$

$$\frac{\partial (nu_i)}{\partial t} + \frac{\partial (nu_i u_j + \frac{1}{m^*}\hat{\theta}_{ij})}{\partial x^j} + \frac{nqE_i}{m^*} = Q_i := -\frac{nu_i}{\tau_p}, \qquad (2)$$

$$\frac{\partial \hat{\theta}_{ii}}{\partial t} + \frac{\partial \hat{\theta}_{ijj}}{\partial x^i} + ... = \hat{Q}_{ii} := -\frac{\hat{Q}_{ii} - \hat{Q}_{ii}|_0}{\tau_w}, \qquad (3)$$

$$\frac{\partial \hat{\theta}_{<ij>}}{\partial t} + \frac{\partial \hat{\theta}_{<ij>k}}{\partial x^k} + ... = \hat{Q}_{<ij>} := -\frac{\hat{Q}_{<ij>}}{\tau_\sigma}, \qquad (4)$$

$$\frac{\partial \hat{\theta}_{ill}}{\partial t} + \frac{\partial \hat{\theta}_{ijll}}{\partial x^j} + ... = \hat{Q}_{ill} := -\frac{\hat{Q}_{ill}}{\tau_h}, \qquad (5)$$

where $n(\mathbf{x}, t) = \int d\mathbf{k} f(\mathbf{x}, t, \mathbf{k})$ is the particle density, $\mathbf{u}(\mathbf{x}, t) = \frac{1}{n} \int d\mathbf{k} \mathbf{v}(\mathbf{k}) f(\mathbf{x}, t, \mathbf{k})$ the mean velocity, $\hat{\theta}_{ij} = m^* \int d\mathbf{k} f c_i c_j$ the random part of the stress tensor, $h_i := 2\hat{\theta}_{ill} = m^* \int d\mathbf{k} f c_i c^2$ is the heat-flow vector; the symbol $\hat{\theta}_{<ij>}$ denotes the trace-free symmetric part.

The previous system (1)-(5) represents balance equations for the number of particles, momentum, energy, stress and heat flux. The Q_i, $\hat{Q}_{ii}$, $\hat{Q}_{<ij>}$, $\hat{Q}_{ill}$ represent the production terms (due to the collisions), which are supposed to be of relaxation type. The τ's are the relaxation times (which in this theory are unknown) and the symbol $(.)|_0$ means evaluation at termodynamic equilibrium.

This system of 13 equations in the 13 unknowns $(n, u_i, \hat{\theta}_{ii}, \hat{\theta}_{<ij>}, \hat{\theta}_{ill})$, is not closed because it comprises the higher moments $\hat{\theta}_{ijk}, \hat{\theta}_{ijll}$ and the production terms. In principle we can write more moment equations (e.g. up to 35), so the first problem is : *how many equations are needed ?*

With our MC simulator (with parabolic bands, for silicon, at room temperature, in the homogeneous case [4]) we simulated the production terms and consequently we evaluated the relaxation times. In fig. 1 we show the relaxation times $\tau_w, \tau_p, \tau_h, \tau_\sigma, \tau_3$ (corresponding to $\hat{\theta}_{<ijk>}$), τ_4 ($\hat{\theta}_{<ij>ll}$), τ_{41} ($\hat{\theta}_{<ijkl>}$), τ_{42} ($\hat{\theta}_{ill}$). The energy relaxation time τ_w (which is the characteristic time in which the electron energy relaxes to the lattice energy) is substantially larger than the other times. In fact, in silicon, most collision are of elastic type and therefore a large number of collisions is necessary in order to relax the carrier's energy to its equilibrium value ($k_B T_L$, where T_L is the lattice temperature), while the other relaxation times relax to the equilibrium values (zero) within a shorter time. During the relaxation to *global thermal equilibrium* an intermediate state (called *partial thermal equilibrium*) arises where the electron fluid is in its own thermal equilibrium. We observe also that higher order moments decay to their partial thermal equilibrium values with decreasing (with the order) relaxation times. Hence few moments, in this homogeneous case, are enough to describe the system.

3 The closure problem.

The first attempt to the closure problem was due to Blotekjaer [5]. He assumed that some higher-order moments can be calculated by utilizing a displaced Maxwellian distribution, which could be justified if the scattering among the carriers were sufficiently strong for the carrier system to be in thermal equilibrium at temperature T and drift velocity v_d. This question has been debated frequently in the literature and there is no definite consensus. However, at low electron density it is difficult to imagine that the Coulomb interaction would be sufficient to maintain equilibrium among carriers. MC simulations performed in the homogeneous case [6] and in a real device [7], show that the distribution function is not a drifted Maxwellian.

Baccarani and Wordeman [8] introduced an hydrodynamic model in which the closure problem is considered at level of the first five equations (1)-(3). The closure is obtained by assuming that the stress tensor is isotropic (i.e. $\hat{\theta}_{<ij>} = 0$) and that the heat-flow vector obeys the Fourier law:

$$\mathbf{h} = -\kappa \nabla T \tag{6}$$

where the heat conductivity κ is given by the Wiedemann-Franz law (holding for electrons near thermal equilibrium) which contains an adjustable constant. A tuning is made on this constant in order to fit MC data of a $n^+ - n - n^+$ diode [9]: Stettler [10] has shown that, for the same device, the mean velocity and the average energy are well fitted *but not the heat-flux*. In fact the heat flux is not zero when the temperature gradient vanishes and there is a small region near the channel-drain junction where $\mathbf{h}$ actually flows against the temperature gradient. The assumption of a Fourier law leads to serious difficulties.

The closure problem has been studied within the framework of Extended Thermodynamics by Anile and Muscato [4]. According to this theory the balance equations must be compatible with a supplementary conservation law to be interpreted as entropy production law; $\hat{\theta}_{ijk}$ and $\hat{\theta}_{ijll}$ can be considered as functions of the lower-order moments; the constitutive functions for the previous moments and the flux entropy can be expanded around the state of *partial thermal equilibrium*. With this machinery the following closure is obtained up to the first order around partial thermal equilibrium:

$$\hat{\theta}_{<ijk>} = 0, \quad \hat{\theta}_{ijll} = \frac{5n(k_B T)^2}{2m^\star}\delta_{ij}. \tag{7}$$

These equations are exactly those obtained with the drifted Maxwellian assumption: therefore the above closure are completely independent of the drifted Maxwellian assumption, and hence the question of the importance of the carrier-carrier scattering is completely irrelevant in this context.

With our MC simulator we tested the previous closure equations successfully [11]: eqs.$(7)_1$ and $(7)_2$ (for i=j) are satisfied with an error $\leq 7\%$. In fig. 2 we plot eq.$(7)_2$ (for i$\neq$j) in non dimensional units. This quantity is almost zero also for high electric fields.

More sophisticated closure equations (up to the second order) are tested with MC simulations in [6].

4 Conclusions

In this paper we have discussed the problems arising in modeling semiconductor transport with hydrodynamic models. We introduced an improved model in the framework of Extended Thermodynamics. The model was successfully checked with MC simulations in the homogeneous case for silicon.

What about the more realistic inhomogeneous case? Since the system (1)-(5) is very complicated, a computationally simpler system was obtained by the well known Maxwellian iteration procedure. The usual Navier-Stokes equations for shear viscosity and a modified Fourier law (the difference with the usual Fourier law being due to a convective term) are obtained with the first iteration.

With these equations a one dimensional $n^+ - n - n^+$ submicron diode was simulated by using finite difference and 1-D box method [12], [13]. The relaxation times were determined by MC simulations of the diode, with parabolic bands. These simulations show that, at the drain junction, the heat flow evaluated with our reduced model overestimates the MC data and that, as consequence, the hydrodynamic energy is lower than the corresponding quantity evaluated with the MC model and the velocity obtained by the hydro model exhibits the well-known 'spurious' peak. We ascribe this discrepancy to:

- the limitations of the Maxwellian iterative approximation, because in principle the complete system should be used;

- the production terms. In fact in the paper [4] we suggest that the production terms should be modeled as relaxation terms in such a way, near the thermal equilibrium, that they satisfy the Onsager Reciprocity Principle of linear irreversible thermodynamics. Hence another analytic form should be used.

We are currently investigating these routes, and the accuracy of the closure relations (at a given order of the moments hierarchy) in the strongly inhomogeneous situations occurring in realistic devices.

Acknowledgments

This work has been supported by the *G.N.F.M.* of CNR , the CNR *Progetto Speciale* (1996) "Mathematical models for semiconductors" , the MPI project (40%) "Problemi non lineari nell'analisi e nelle applicazioni fisiche, chimiche e biologiche: aspetti analitici modellistici e computazionali" and the MPI project (60%) (1995).

References

[1] W. Hänsch, *The drift-diffusion equation and its applications in MOSFET modeling*, (Springer-Verlag, Wien, 1991).

[2] M.V. Fischetti and S.E. Laux, Phys. Rev. B **48**, 2244 (1993).

[3] I. Muller and T. Ruggeri, *Extended Thermodynamics* ,(Springer-Verlag, Berlin, 1993).

[4] A.M. Anile, O. Muscato, Phys. Rev. B,**51**,16728,(1995)

[5] K. Blotekjaer, IEEE Trans.on Electron Devices,**ED-17**,38,(1970)

[6] A.M. Anile, O. Muscato, Cont. Mech. Therm.,**8**,131,(1996)

[7] O. Muscato, *Exponential representation of the distribution fuction in silicon semiconductors devices*,preprint (1996)

[8] G. Baccarani, M.R. Wordeman, Solid-state Electronics, **29**,970,(1982)

[9] A. Gnudi,F. Odeh and M. Rudan, Europ. Trans. on Telecom. and related Tecnologies,**1**,307,(1990)

[10] M.A. Stettler,M.A. Alam and M. Lundstrom,IEEE trans. elec. dev.,**40**, 733,(1993)

[11] A.M. Anile and O. Muscato,Transp. Theory and Stat. Phys.,**25**,20,(1996)

[12] A.M. Anile, C. Maccora and R.M. Pidatella,COMPEL,**14**,1,(1995)

[13] O. Muscato,R.M. Pidatella and M.V. Fischetti,*Monte Carlo and hydrodynamic simulation of a one dimensional $n^+ - n - n^+$ silicon diode*, to appear in the Proc. *4th International Workshop on Computational Electronics*, ed. C. Gardner, Phoenix (1996)

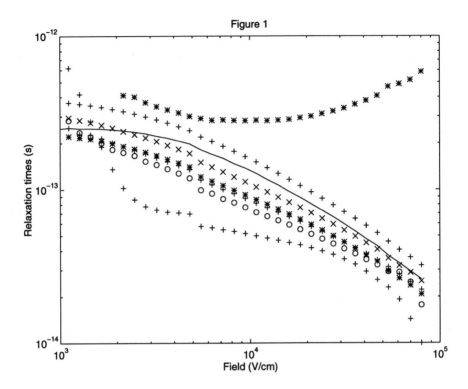

Figure 1: Relaxation times versus electric field, evaluated by Monte Carlo simulations for silicon in the homogeneous case, with parabolic bands. From top to bottom τ_w (energy), τ_p (momentum), τ_h (heat flux), τ_σ (stress), τ_3 ($\hat{\theta}_{<ijk>}$), τ_4 ($\hat{\theta}_{<ij>ll}$), τ_{41} ($\hat{\theta}_{<ijkl>}$), τ_{42} ($\hat{\theta}_{iill}$).

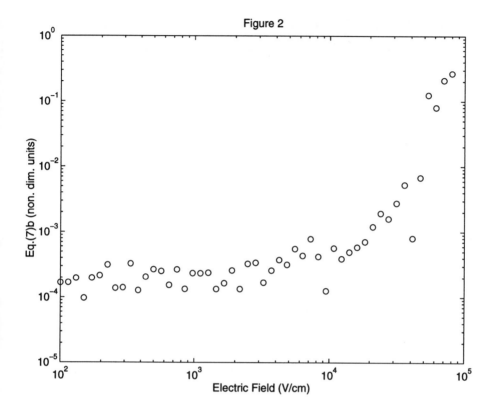

Figure 2: equation (7)$_2$ for i≠j in non dimensional units, obtained by Monte Carlo simulations for silicon , versus electric field.

OSCILLATIONS OF A MINIATURE SINGLE STATOR SYNCHRONOUS MOTOR

W.D.Collins

School of Mathematics and Statistics, The University of Sheffield, Sheffield S3 7RH, England, U.K.

ABSTRACT A mathematical model is established for the rotary motion of a single stator permanent magnet synchronous motor when both current- and voltage-fed. Pure synchronous running is shown to be possible only when there is a specific relation between the electrical and mechanical parameters of the motor. It is shown analytically and confirmed by numerical simulation that, when the design parameters approximately satisfy this relation, the motor goes into synchronous running but with small superimposed oscillations. The more the relation is deviated from, however, the less likely is synchronous running and, even when it is established, the oscillations on it become increasingly large.

A synchronous motor is a rotational motor used to convert electrical to mechanical energy which under ideal conditions operates at a constant angular speed ω_m determined by the supply frequency ω_s in that $\omega_m = \omega_s/p$ radians s^{-1} for p pole pairs. Miniature permanent magnet motors are manufactured in millions each year and widely used in both domestic and industrial appliances such as control or timing devices in, for example, motor driven valves, switches, central heating timers, hair dryers, coffee grinders, lemon squeezers, electric carving knives and electric fire flame effects. They operate at a variety of speeds, typically in the range 250 to 750 rpm. Generally the length and diameter do not exceed 3 cm. In operation these motors are however prone to relatively large amplitude oscillations and thus there is scope for improving their design. This work, which is in collaboration with Dr. Russell Brown of the Department of Electronic and Electrical Engineering at the University of Sheffield, gives a model for a single stator motor and investigates oscillations in this model, both analytically and numerically, with a view to controlling the design choice of electrical and mechanical parameters to reduce the occurrence of these oscillations and thus save energy.

A miniature synchronous motor is made up of an outer stationary element, the stator, and an inner element, the rotor, mounted on bearings fixed to the stator. The stator comprises two interleaved crowns of steel pole fingers issuing from two end plates with a single coil positioned around these and connected to the supply. The rotor is a permanent magnet, a ferrite cylinder, with a regular pattern of north and south poles magnetically imprinted around its circumference with as many poles on the rotor as on the stator. The magnetic flux in the air gap between the rotor and the stator can be represented by two rotating fields with opposite senses of rotation, the torque due to the coil supply arising from the interaction of these rotating fields with the field of the magnet. When synchronised, the magnetic field follows one of the rotating fields, the other field generating an alternating torque which may cause substantial oscillations on the synchronous speed at low values of damping and particular values of inertia.

To investigate these oscillations the rotor is treated as a rigid cylinder and d-q (direct-quadrature) theory used to model the motor by two coils on the d-axis, a magnet equivalent d-coil and the stator coil, and one coil on the q-axis, the magnet equivalent q-coil, the d-axis being taken to coincide with the stator coil axis [1] . The electromagnetic torque T_{elec} on the rotor is found as

$$T_{elec} = -pMI_m i_c \sin p\theta - 1/2p(L_d - L_q)I_m^2 \sin(2p\theta + \phi),$$

where θ is the angular displacement of the rotor, ϕ the starting offset angle and i_c the current in the stator coil. The parameters L_d, L_q, M and I_m represent the equivalent (fictitious) magnet d- and q- coil self inductances, the mutual inductance between the d-coil and the stator coil and the maximum current in the magnet coils respectively. The first term in T_{elec} is the contribution due to the coil current and the second the saliency torque.

The equation of motion of the rotor is then

$$Jd^2\theta/dt^2 = T_{elec} - Dd\theta/dt - (T_f + T_L)sgn(d\theta/dt),$$

where t is time, J is the total moment of inertia of the rotor and its shaft, $Dd\theta/dt$ a damping torque due to viscous friction, and T_f a friction torque from the bearings (assumed constant) whilst T_L is an external load torque applied at the shaft of the motor and arising from the mechanical system being driven and so opposing the motion, here taken as constant. The equation of motion of the rotor is then

$$Jd^2\theta/dt^2 + Dd\theta/dt + pMI_m i_c \sin p\theta$$

$$+ 1/2p(L_d - L_q)I_m^2 \sin(2p\theta + \phi) + (T_f + T_L)sgn(d\theta/dt) = 0.$$

When the stator is fed by an alternating current supply, the current-fed case, then $i_c = I_c \cos\omega t$, ω the supply frequency, but, when it is fed by an alternating voltage source, the practically important voltage-fed case, the equation of motion is coupled with the circuit equation

$$L_c di_c/dt + R_c i_c + Md(I_m \cos p\theta)/dt = V\cos(\omega t + \sigma),$$

the third term on the left being the voltage induced in the stator coil by the rotating magnet.

The offset starting angle ϕ arises because the magnet wants to pull itself into a non-starting position but, by slitting the pole fingers and correctly proportioning the width of the slots to the total width of the pole fingers, the rotor can be made to align itself in a good starting position midway between the poles. When the current i_c is switched off, the rotor rest position is stable and yet, when the current is present, must give rise to an initial torque to start the motor from the rest position. This means that $\sin p\theta$ is non-zero for the rest position. These conditions are met if $\phi = \pi$ and the rest position is $p\theta = \pi/2$. Neither is met, for instance, if $\phi = 0$ and the rest position is $p\theta = 0$. In practice some magnetic asymmetry is introduced and this can be satisfactorily modelled by the introduction of ϕ into the saliency term. Ideally $\phi = \pi$ and $p\theta = \pi/2$ to give the initial torque its maximum value but in practice ϕ is likely to be somewhat less than π.

These equations involve several parameters, so a first step in their investigation is to put them in scaled form by setting

$$y = p\theta, \quad \Omega\tau = \omega t, \quad i_c = I_c u,$$

where I_c is defined subsequently, and Ω, μ, α and γ are defined by

$$\Omega^2 p^2 M I_m I_c = J\omega^2,$$

$$2\mu = \Omega D/J\omega, \quad \alpha = (L_d - L_q)I_m/2MI_c, \quad \gamma = (T_f + T_L)/pMI_m I_c,$$

so that μ, α and γ are essentially damping, saliency and external torque parameters. The equation of motion then becomes

$$d^2y/d\tau^2 + 2\mu dy/d\tau + u\sin y + \alpha\sin(2y + \phi)$$

$$+ \gamma sgn(dyd\tau) = 0.$$

In the current-fed case I_c is the peak value of the current i_c and $u = \cos\Omega\tau$. In the voltage-fed case

$$du/d\tau + \Omega\rho u + \eta d(\cos y)/d\tau = \Omega\cos(\Omega\tau + \sigma),$$

where $\rho = R_c/L_c\omega$, $\eta = MI_m\omega/V$, $I_c = V/L_c\omega$.

A single stator motor is required to run at synchronous speed ω/p, but in its basic form this can be either forwards or backwards. In practice however synchronous running is accompanied by oscillations which can have periods including π/Ω or $2\pi/\Omega$. Further the motor only goes into synchronous running if sufficiently large oscillations are set up and, if this is not achieved, just oscillates about standstill.

These types of behaviour are displayed in numerical simulations for both the current- and voltage- fed cases. Figure 1 shows graphs of y and $dy/d\tau$ against τ for different frequencies Ω for the current-fed case. For each of the graphs, $\gamma = 0$, $\alpha = 0.42$ and $\mu = 0.09$, whilst $\phi = \pi$ in Figures 1(c) and 1(d) and $\phi = 7\pi/8$ in Figures 1(a) and 1(b). The initial conditions are in each case $y = \phi/2$, $dy/d\tau = 0$. In Figures 1(a), (b) and (c) there is synchronous running, both forwards and backwards, with oscillations at π/Ω and in (a) $2\pi/\Omega$ also, whilst in Figure 1(d) the motor oscillates about standstill.

Synchronous running, ie, $y = \pm\Omega\tau$ is not an exact solution of the current-fed equations except in the special case $\alpha = 1/2$, $\phi = \pi$, $\mu = 0$, $\gamma = 0$. This suggests that oscillations on synchronous running should be small in the neighbourhood of $\alpha = 1/2$, $\phi = \pi$. Approximate analytic solutions can thus be obtained by setting

$$y = \Omega\tau + \psi$$

and assuming $|d\psi/d\tau| < \Omega$, so that ψ is a perturbation on forward running which satisfies the equation

$$d^2\psi/d\tau^2 + 2\mu d\psi/d\tau + 1/2(\sin(2\Omega\tau + \psi) + \sin\psi)$$
$$+ \alpha\sin(2\Omega\tau + 2\psi + \phi) = -2\Omega\mu - \gamma.$$

A similar equation holds for backward running.

An approximate solution in the neighbourhood of $\alpha = 1/2$ can be obtained using the Poincaré-Linstedt method. A new time variable $s = \Omega\tau$ is introduced and the parameters α, μ and Ω expanded as

$$\alpha = 1/2 + \epsilon\beta, \quad \mu = \epsilon\mu_0, \quad \Omega = \Omega_0 + \epsilon\Omega_1 + \epsilon^2\Omega_2 + ...,$$

where $\epsilon > 0$ is a small parameter introduced for "bookkeeping" convenience. Only the case $\gamma = 0$, $\phi = \pi$, is considered here. Similar solutions however are obtained for γ and ϕ in the neighbourhoods of these values. The expansion

$$\psi = \epsilon\psi_1 + \epsilon^2\psi_2 + ...,$$

is made and ψ_1 found to satisfy

$$d^2\psi_1/ds^2 + (a - 2q\cos 2s)\psi_1 = \Omega_0^{-2}(\beta\sin 2s - 2\Omega_0\mu_0), \quad (1)$$

where $a = 1/(2\Omega_0^2)$, $q = 1/(4\Omega_0^2)$, an inhomogeneous Mathieu equation. For a given

FIGURE 1

(a) $\Omega = 0.64$, $\phi = 2.749$.

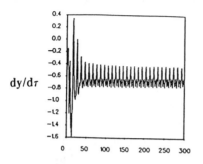

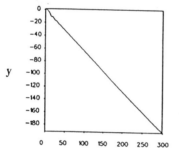

(b) $\Omega = 0.83$, $\phi = 2.749$.

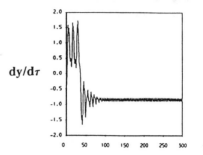

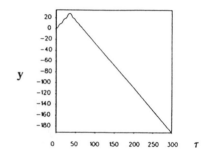

(c) $\Omega = 1.05$, $\phi = 3.142$.

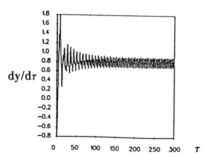

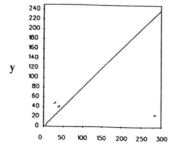

(d) $\Omega = 1.54$, $\phi = 3.142$.

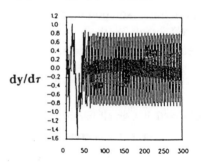

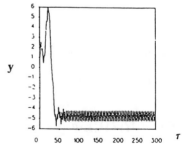

non-zero q one and only one basically periodic solution of the homogeneous equation, a solution with period π or 2π, exists when a has one of a set of characteristic values, which for small q are found as power series perturbations on $a = n^2$, n an integer [2]. Thus for $n = 1$

$$a = 1 + q - (1/8)q^2 + ..., \quad ce_1(s,q) = \cos s - (q/8)\cos 3s + ...,$$

and

$$a = 1 - q - (1/8)q^2 + ..., \quad se_1(s,q) = \sin s - (q/8)\sin 3s + ...$$

Between these two values of a however the solution of the homogeneous equation is unstable [2]. This leads to approximate stability bounds on Ω_0^2, between which the solution of the inhomogeneous equation is unstable, as

$$1/4 < \Omega_0^2 < 3/4.$$

When the homogeneous equation has stable solutions, the inhomogeneous equation for ψ_1 has a unique periodic solution, which is attained provided the right-hand side is orthogonal to both ce_1 and se_1 in the sense that

$$\int_0^{2\pi} f(s)\, ce_1(s)\, ds = 0, \quad \int_0^{2\pi} f(s)\, se_1(s)\, ds = 0.$$

Since these conditions are satisfied when $f(s)$ is the right-hand side of equation (1), an approximate solution of this equation is assumed of the form

$$\psi_1 = c + p\cos(s + \chi) + r\cos(2s + \delta) + ...$$

Substituting, equating coefficients and solving the resulting equations gives

$$r \sin \delta = 2\beta/(8\Omega_0^2 - 1), \quad r \cos \delta = 8\Omega_0\mu_0/(16\Omega_0^2 - 1),$$

$$c = -8\Omega_0\mu_0(8\Omega_0^2 - 1)/(16\Omega_0^2 - 1),$$

and either

$$\chi = 0, \pi, \text{ with } (1/4 - \Omega_0^2)p = 0$$

or

$$\chi = \pi/2,\ 3\pi/2,\ \text{with}\ (3/4 - \Omega_0^2)p = 0.$$

In the first case $p = 0$ unless $\Omega_0^2 = 1/4$ and in the second $p = 0$ unless $\Omega_0^2 = 3/4$. To a first approximation the first harmonic is not then present for $\Omega_0^2 > 3/4$, $\Omega_0^2 < 1/4$.

The solutions in the neighbourhoods of $\Omega_0^2 = 1/4,\ 3/4$, can be investigated further by considering the equations for ψ_2 and ψ_3. Thus for $\Omega_0^2 = 3/4$ it is found that $\Omega_1 = 7\beta/10\sqrt{3}$, whilst from the equation for ψ_3 the amplitude p is found from

$$p^2 = (120/29)[-2\Omega_0\Omega_2 - (983/1500)\beta^2 - (1050/121)\mu_0^2\ ].$$

The requirement that $p^2 > 0$ gives

$$2\Omega_0\Omega_2 < -(983/1500)\beta^2 - (1050/121)\mu_0^2,$$

but otherwise Ω_2 is arbitary. When $p = 0$, the critical value of Ω_2 above which the first harmonic does not arise is obtained, the corresponding value of Ω, Ω_+, being $\sqrt{3}/2 + 7\beta\epsilon/10\sqrt{3} + \Omega_2\epsilon^2$.

For $\alpha = 0.42$, $\mu = 0.09$, this gives $\Omega_+ = 0.79$, in agreement with simulations. However, the critical frequency corresponding to $\Omega_0 = 1/2$ is $\Omega_- = 0.46$, the value obtained from simulations being 0.62. Improved agreement between an approximate value of Ω_- and the value from simulations can be obtained by setting

$$\psi = c + r\sin(2\Omega\tau + \delta),$$

and using the Bessel function expansions of $\cos(r\sin(2\Omega\tau + \delta))$ and $\sin(r\sin(2\Omega\tau + \delta))$ to obtain approximate equations satisfied by c, r and δ, which need to be solved numerically. The stability of this solution can however be investigated by considering small perturbations of the form $p\sin(\Omega\tau + \chi)$ on it. This gives Ω_- as 0.62. Generally there is good agreement of the Poincaré-Linstedt value for Ω_+ with that from simulations for other values of α in the neighbourhood of 0.5 and a range of ϕ from $5\pi/6$ to π, but less satisfactory agreement for Ω_-, though usually better than for the example given and better the closer α is to 1/2. In practice frequencies in the neighbourhood of Ω_+ are the ones of interest.

Simulations show that, for Ω greater than Ω_+ and less than Ω_-, after transients have decayed, periodic solutions are obtained in which the first harmonic is not present and the oscillation is dominated by the second harmonic with period π/Ω. When Ω is decreased sufficiently below Ω_-, to less than 0.37 for the example given, however synchronous running cannot be achieved at least within reasonable time spans and the motor oscillates about standstill. For Ω between Ω_+ and Ω_- periodic solutions are attained after the transients decay but with both the first and second harmonics present. A Fourier analysis of one or two simulations shows that harmonics of higher order are present but their amplitudes are small compared to those of the first two harmonics. Period doubling at the critical frequencies suggests a transition to chaos, but this does not happen at least for the parameters in the examples so far considered.

For Ω greater than Ω_+ and less than Ω_- there is good agreement between the numerical and Poincaré-Linstedt solutions for the amplitude of the second harmonic for both forward and backward running. Amplitudes are smaller for backward than for forward running when ϕ is not equal to π, but are equal for $\phi = \pi$.

The voltage-fed case can be investigated in the same way by first determining conditions on the parameters for which perfect synchronous running satisfies the equations with $\mu = 0$, $\gamma = 0$. For

$$y = \pm\Omega\tau + \lambda, \quad u = 2\alpha\cos y,$$

the voltage-fed equations are satisfied if

$$\phi = \pi, \quad 4\alpha^2\rho^2 + (2\alpha + \eta)^2 = 1, \quad \lambda = \pm(\sigma - \chi),$$

where $\tan\chi = (2\alpha + \eta)/2\alpha\rho$, $0 \leq \chi \leq \pi/2$.

These conditions correspond to the conditions $\alpha = 1/2$, $\phi = \pi$, in the current-fed case. On setting $y = \pm\Omega\tau + \lambda + \psi$, with a corresponding perturbation of u the equations for perturbation on synchronous running are obtained. The techniques used to investigate the corresponding current-fed equations can be applied to these equations, though the analysis is more complicated. Results however are qualitatively similar to those for the current-fed case as are the numerical simulations and again there is good agreement between analytic and numerical solutions for oscillations at period π/Ω.

What emerges from this work are some pointers as to desirable parameter ranges for which oscillations should be small in amplitude, though more work still needs to be done. It is hoped to test the predictions of the model on real motors.

References

[1] Collins, W.D: Analytic solution of resonant oscillations in miniature synchronous motors, Math. Engng. Ind., 4 (1993), 179-210.

[2] McLachlan, N.W: Theory and application of Mathieu Functions, 1947, Oxford.

WHITE NOISE AND NONLINEAR DAMPING IN THE TWO-DIMENSIONAL MODEL OF ENERGY TRANSFER IN MOLECULAR SYSTEMS

Peter L. Christiansen, Magnus Johansson, K. Ø. Rasmussen
Department of Mathematical Modelling, The Technical University of Denmark, DK-2800 Lyngby, Denmark

Yuri B. Gaididei and Irina I. Yakimenko
Bogolyubov Institute for Theoretical Physics, Metrologicheskaya Str. 14b, 252 143, Kiev 143, Ukraine

Abstract

In the present paper a model of energy transfer in Langmuir-Blodgett Scheibe aggregates which takes into account the non-vanishing thermal fluctuations in molecular systems is derived. This is done using the two-dimensional NLS equation with multiplicative noise, which describes the exciton coherence destruction caused by phonons and providing energy input to the exciton system, and with nonlinear damping providing energy dissipation. Using a collective coordinate approach, we find that for initial conditions where total collapse occurs in the unperturbed NLS, the presence of the damping term will instead result in an exponentially decreasing width of the solution in the long-time limit. We also find that a sufficiently large noise variance may cause an initially localized distribution to spread instead of contracting, and that the critical variance necessary to cause dispersion will for small damping be the same as for the undamped system.

1 Introduction

Scheibe aggregates (or J-aggregates) are found in nature, where they function as energy funnels for sunlight to be used in photochemical processes, but they may also be produced artificially in the laboratory by the so-called Langmuir-Blodgett (LB) technique. In this way LB Scheibe aggregates may be used by the photo industry to identify the combination of molecules resulting in the most efficient photon energy transfer. The aggregates may also be used to study the important process of photosynthesis.

The Scheibe aggregate formation of monolayer LB films may be obtained by mixing dye molecules (fx oxacyanine) with inert molecules (fx octadecane) in the molar ratio 1:1, which yields a very compact structure resulting in a large intermolecular binding energy. This gives the Scheibe aggregates their strong and narrow coinciding absorption and fluorescence bands which are absent in less compact molecular structures, and leads to unique energy transfer properties. The confirmation of highly efficient energy transfer over unusually large distances in LB Scheibe aggregates was first made by Möbius and Kuhn in 1979 [1]. They found the efficiency of energy transfer to be proportional to the temperature in the range 20K-300K [1].

Theoretically, Möbius and Kuhn found that a coherent exciton model could explain their experimental results [1, 2]. In this model it is assumed that on excitation of the aggregate a coherent exciton, extending over a certain number of molecules, is produced. Since the decay rate associated with a single molecule is constant, this results in a radiative decay rate of the exciton which is proportional to its domain size. Experimentally the radiative decay rate of the exciton was found to be inversely proportional to the temperature. This implies that the domain size of the coherent exciton is also inversely proportional to the temperature [1, 2]. The exciton then moves across the aggregate with its domain size remaining constant.

It is important to note that the coherent exciton model suggested by Möbius and Kuhn displays mean behavior. It does not state anything about the dynamics. It seems likely that, e.g., the domain size will be a decreasing function of time, due to destruction of coherence through thermal fluctuations.

LB films and molecular architecture have many industrial applications in electronics (dielectric layers in semiconductors, capacitors, etc.), and especially optoelectronics (photoresistors, photodiodes, photomemory, gratings, 2nd harmonic generation, etc.). They may also serve as model systems for fundamental investigations of exciton processes as well as structure and function in 2D and 3D molecular systems in general.

Clearly these molecular aggregates are good candidates for novel technological and industrial advances. The first important step in advancing a technology based on such molecular aggregates is to build a mathematical model which is able to describe the observed phenomena. In [3] a model based on the 2D nonlinear Schrödinger (NLS) equation with energy transfer occuring through solitary waves was employed. It was demonstrated that the efficient mechanism of energy transfer from the host matrix to acceptor molecules can be provided by means of collapse of such solitary waves. Here, this model is generalized to take into account thermal fluctuations and energy dissipation in a way that allows the exciton system to reach thermal equilibrium.

2 Nonlinear model of energy transfer in two-dimensional molecular systems

We consider a two-dimensional molecular system in which thermal fluctuations are taken into account. The energy operator is given by

$$\hat{H} = \{\sum_{n=1}^{N} E_0 B_n^+ B_n - \sum_{n\neq p}^{N}\sum_{p=1}^{N} J_{np} B_n^+ B_p\} + \{\frac{1}{2} M \sum_{n=1}^{N} [\dot{u}_n^2 + \omega_0^2 u_n^2]\} + \{\chi \sum_{n=1}^{N} u_n B_n^+ B_n\} \tag{1}$$

$(n, p = 1, N)$ where B_n^+, B_n are the operators of creation and annihilation of the excitation on the site n, E_0 is the energy of the site, J_{np} is the dipole-dipole interaction energy, $u_n(t)$ represents the elastic degree of freedom at site n, M is the molecular mass, ω_0 is the frequency of the oscillation of the molecules, χ is the constant of exciton-phonon interaction which measures the amount of energy required to displace a molecule in the aggregate, N is the number of molecules.

Using the non-equilibrium density matrix approach the set of coupled equations describing the dynamics of excitons and phonons has been found as [3]

$$i\hbar\dot{\psi}_n + \sum_{p\neq n} J_{pn}\psi_p + \chi u_n \psi_n = 0, \tag{2}$$

$$M\ddot{u}_n + M\lambda \dot{u}_n + M\omega_0^2 u_n = \chi \mid \psi_n \mid^2 + \eta_n(t). \tag{3}$$

To describe the interaction of the phonon system with a thermal reservoir at temperature T, damping λ and noise $\eta_n(t)$ have been included in Eq. (3) after its quantum-mechanical derivation. $\eta_n(t)$ is assumed to be Gaussian white noise with the statistical properties:

$$<\eta_n(t)>= 0, \quad <\eta_n(t)\eta_{n'}(t')>= 2M\lambda kT\delta(t-t')\delta_{nn'}, \tag{4}$$

where the strength of the noise is chosen according to the classical fluctuation-dissipation theorem which assures thermal equilibrium. The damping coefficient λ is the linewidth of the infra-red absorption band, k is Boltzmann's constant, $<...>$ denotes an ensemble average.

The noise in Eq. (3) is additive, implying that phonons are being created and destroyed thermally. In contrast, in Eq. (2) the noise is multiplicative, implying that only the coherence of the exciton is disturbed by the thermal fluctuations. As a result, the exciton number $N\{\psi_n\}$ defined as the total probability for finding the exciton in the system is a conserved quantity:

$$N\{\psi_n\} = \sum_n \mid \psi_n(t) \mid^2 = 1. \tag{5}$$

For $t \gg 2/\lambda$ we obtain from Eq. (3) the approximate expression for the molecular displacements

$$u_n(t) \approx \frac{\chi}{M\omega_0^2}\left(\mid \psi_n(t) \mid^2 - \frac{\lambda}{\omega_0^2}\frac{d}{dt}(\mid \psi_n(t) \mid^2)\right) + \sigma_n(t) \tag{6}$$

where

$$\sigma_n = \frac{1}{M(S_+ - S_-)} \int_0^t dt'(e^{S_+(t-t')} - e^{S_-(t-t')})\eta_n(t'), \tag{7}$$

$$S_\pm = -\frac{\lambda}{2} \pm \sqrt{(\frac{\lambda}{2})^2 - \omega_0^2}. \tag{8}$$

Introducing the expression (6) into Eq. (2) we get the equation for the exciton variables only:

$$i\hbar\dot{\psi}_n + \sum_{n'} J_{nn'}\psi_{n'} + V \mid \psi_n \mid^2 \psi_n - V\frac{\lambda}{\omega_0^2}\psi_n \frac{d}{dt}(\mid \psi_n(t) \mid)^2 + \chi\sigma_n(t)\psi_n = 0. \tag{9}$$

Here, V is the nonlinearity parameter, $V = \chi^2/M\omega_0^2$. Making the additional assumption that ψ_n varies slowly in space and that only nearest-neighbour coupling J is of importance, we obtain the continuum approximation for the continuous exciton field $\psi(x,y,t) = e^{-i4Jt/\hbar}\psi_n(t)/l$:

$$i\hbar\psi_t + Jl^2 \nabla^2 \psi + Vl^2\psi \mid \psi \mid^2 - V\frac{\lambda}{\omega_0^2}l^2\psi(\mid \psi \mid^2)_t + \chi l^2\sigma\psi = 0 \qquad (10)$$

where l is the distance between nearest-neighbour molecules and $\sigma(x,y,t) = \sigma_n(t)/l^2$ is the noise density.

Introduction of the dimensionless variables

$$\frac{x}{l} \to x, \frac{y}{l} \to y, \frac{Jt}{\hbar} \to t, \sqrt{\frac{Vl^2}{J}}\psi \to \psi, \frac{\chi l^2}{J}\sigma \to \sigma \qquad (11)$$

leads to the equation:

$$i\psi_t + \nabla^2\psi + \mid \psi \mid^2 \psi - \Lambda\psi(\mid \psi \mid^2)_t + \sigma\psi = 0, \qquad (12)$$

where the nonlinear damping parameter Λ is given by

$$\Lambda = \frac{\lambda J}{\hbar\omega_0^2}. \qquad (13)$$

It may be shown easily that in spite of the presence of the nonlinear damping and multiplicative noise terms in Eq. (12), the norm, defined as

$$N = \int\int \mid \psi(x,y,z) \mid^2 dxdy \qquad (14)$$

will still be a conserved quantity, having the value $N = V/J$ if the exciton wave function is assumed to be normalized in the physical coordinates. Writing $\psi = \sqrt{n}e^{i\theta}$, the following equations for the amplitude and phase of the solution can be obtained from (12):

$$\frac{1}{2}n_t + \nabla(n\nabla\theta) = 0, \qquad (15)$$

$$-\theta_t - \Lambda n_t - (\nabla\theta)^2 + n + \frac{1}{\sqrt{n}}\nabla^2(\sqrt{n}) + \sigma(x,y,t) = 0. \qquad (16)$$

The norm conservation is immediately seen from the first of these equations, while the second equation shows that the role of the damping term is to destroy the phase coherence of the solution and cause a diffusion-like behavior for the phase. The ordinary NLS Hamiltonian, defined as

$$H = \int\int (\mid \nabla\psi(x,y,t) \mid^2 - \frac{1}{2}\mid \psi(x,y,t) \mid^4)dxdy, \qquad (17)$$

will no longer be conserved. Instead, we find that

$$\frac{dH}{dt} = \int\int \sigma(x,y,t)(\mid \psi \mid^2)_t dxdy - \Lambda\int\int [(\mid \psi \mid^2)_t]^2 dxdy. \qquad (18)$$

The two terms provide energy input and energy dissipation to the exciton system, making an energy balance possible. Consequently, there is a possibility for the system to reach thermal equilibrium.

3 A collective coordinate approach

In two (and more) dimensions the NLS equation is not integrable, and possesses unstable solutions that may collapse in finite time, the amplitude becoming infinite at some place in space [4]. In describing collapse in the NLS equation, its ground state solitary wave (GS) solution plays a vital role.

Using a variational approach [5] Anderson et al. [6] have derived a good approximation of the GS solution to the 2D NLS equation (Eq.(10) with $\lambda = \sigma = 0$):

$$\psi_s(r,t) = A_0 \operatorname{sech}(r/B_0) \exp(iJt/\hbar) \tag{19}$$

here given with zero initial velocity, and center at $r = \sqrt{x^2 + y^2} = 0$. The initial amplitude A_0, and the initial width B_0, are given by

$$A_0 = \frac{1}{l}\sqrt{\frac{J}{V}\frac{12\ln(2)}{4\ln(2) - 1}}, \quad B_0 = l\sqrt{\frac{2\ln(2) + 1}{6\ln(2)}}. \tag{20}$$

It is reasonable to model the initial excitation of the Scheibe aggregate by $\psi_s(r,0)$, since it is the most stable initial condition known for the 2D NLS equation. Furthermore, it represents a localized domain of coherence, as suggested by the experiments of Möbius and Kuhn [1]. ψ_s must be normalized, $N_s = 1$, and the following condition must be satisfied by the parameter values

$$\frac{V}{J} = 11.7. \tag{21}$$

If $\frac{V}{J} < 11.7$ and $N < N_s$ the exciton will quickly disperse with a coherence time less than can be achieved with $V/J = 11.7$ and $\psi_s(r,0)$ as initial condition.

If $\frac{V}{J} > 11.7$ and $N > N_s$ the exciton wave-function will collapse in finite time. For this the exciton-phonon coupling must be strong. The collapse regime is interesting since it may lead to an increased nonlinear coherence time and allow for an estimation of the exciton life time which is in agreement with experimental results.

We will assume isotropy, which effectively reduces the problem to one space dimension with the radial coordinate $r = \sqrt{x^2 + y^2}$. We also assume that the noise σ can be approximated by radially isotropic Gaussian white noise with autocorrelation function

$$<\sigma(r,t)\sigma(r',t')> = \frac{D_r}{r}\delta(r-r')\delta(t-t'), \tag{22}$$

where D_r is the dimensionless noise variance, and that the collapse process can be described in terms of collective coordinates by using a localized self-similar trial function for the exciton wave function $\psi(r,t)$ of the form:

$$\psi(r,t) = A(t)f(\frac{r}{B(t)})\exp\left[i\alpha(t)r^2\right] \tag{23}$$

where $f(x)$ is an arbitrary well-behaved, real function, which decreases sufficiently fast (e.g., exponentially) as $x \to \infty$. The three time-dependent parameters A, B and α determine amplitude, width and phase of the wave function, respectively. In the case $\Lambda = 0$ the particular choice $f(x)=\text{sech}(x)$ was considered [7]. This choice was motivated by regarding it as a generalization of the approximate ground-state solution (19) to the ordinary 2D NLS found in [6]. As long as Λ is not too large, one may expect this trial function to give a good description of the collapse process also in the presence of damping, but since the qualitative features of the results obtained below do not depend on the explicit choice of trial function, the function f will be left unspecified. From the definition (14) of the norm, with ψ given by (23), we obtain the relation between the amplitude and width:

$$A(t) = \frac{\sqrt{\frac{N}{s_{1,2,0}}}}{B(t)} \qquad (24)$$

where the coefficient $s_{1,2,0}$ is obtained from the definition of the integrals $s_{m,n,p}$

$$s_{m,n,p} = 2\pi \int_0^\infty r^m [f(r)]^n [f'(r)]^p dr. \qquad (25)$$

When $\Lambda = 0$, the relation

$$\alpha(t) = \frac{\dot{B}(t)}{4B(t)} \qquad (26)$$

can be derived from the Euler-Lagrange equations together with an ordinary differential equation for $B(t)$ [8]. However, in the presence of damping this technique is not applicable. Instead, we shall use the trial function (23) with $\alpha(t)$ given by (26) and derive an ordinary differential equation for $B(t)$ using the virial theorem. Defining the virial coefficient W as

$$W(t) = 2\pi \int_0^\infty r^3 |\psi(r,t)|^2 dr \qquad (27)$$

we obtain from (12) that it satisfies the equation

$$\frac{1}{4}\frac{d^2W}{dt^2} = 2H - 2\pi\Lambda \int_0^\infty r^2 |\psi|^2 \partial_r[\partial_t(|\psi|^2)]dr + 2\pi \int_0^\infty r^2 |\psi|^2 \partial_r(\sigma)dr \qquad (28)$$

where H is the Hamiltonian (17).

Using (23)-(26), we arrive at the following differential equation for the width B of the exciton wave function:

$$B^3\ddot{B} = \Delta - \Gamma\frac{\dot{B}}{B} - \frac{8\pi}{s_{3,2,0}} \int_0^\infty \left(1 + \frac{r}{B}\frac{f'(r/B)}{f(r/B)}\right) [f(\frac{r}{B})]^2 \sigma(r,t) r dr \qquad (29)$$

where the constants Δ and Γ are defined as

$$\Delta = \frac{4}{s_{3,2,0}}(s_{1,0,2} - \frac{Ns_{1,4,0}}{2s_{1,2,0}}), \text{ and } \Gamma = \frac{8N\Lambda s_{3,2,2}}{s_{1,2,0}s_{3,2,0}}, \qquad (30)$$

respectively. Γ is always positive, Δ can be either positive or negative. The collapse will occur if $\Delta < 0$.

It is possible to find a stochastic differential equation which is simpler than Eq. (29) but leads to the same Fokker-Planck equation for the system. This equation, which gives an equivalent description of the process, is

$$\ddot{B} = \frac{\Delta}{B^3} - \frac{\Gamma \dot{B}}{B^4} + \frac{h(t)}{B^2}, \tag{31}$$

where $h(t)$ is white noise with the autocorrelation

$$< h(t)h(t') > = 2D\delta(t-t'), \quad D = \frac{32\pi^2 D_r s_{3,2,2}}{s_{3,2,0}^2}. \tag{32}$$

4 Numerical solution of the collective coordinate equation

1. When there is neither damping nor noise in the system: $\Gamma = D = 0$ and $\Delta < 0$ there is an exact solution

$$B(t) = B_0 \sqrt{1 - \frac{t^2}{t_c^2}}, \quad t_c = \frac{B_0^2}{\sqrt{|\Delta|}} \tag{33}$$

for the initial conditions

$$B(0) = B_0, \quad \dot{B}(0) = 0. \tag{34}$$

The solution collapses and the collapse time is t_c. As it has been evaluated in [9] using a realistic estimate of the parameter values for the Scheibe aggregates, it is reasonable to assume that $\frac{V}{J} > 11.7$, so that the initial pulse will collapse.

2. When the damping is present the solution will be well-defined for all t and $B(t)$ will approach zero exponentially for large t,

$$B(t) \sim e^{-\frac{|\Delta|}{\Gamma}t}, t \to \infty, \tag{35}$$

so that the process can be considered as a collapse process with an infinite collapse time. This type of behaviour has been called "pseudo-collapse" [10] (Fig. 1). The initial stages of the pseudo-collapse process will resemble a pure collapse as long as the damping is small.

3. In the presence of noise but no damping the collapse time will increase with increasing noise variance (or temperature) and the collapse can be stopped if the variance is large enough [7]. As it was discussed in [9] the collapse process may play an important role in the energy transfer in Scheibe aggregates.

4. To illustrate how the noise affects the pseudo-collapse process, we show the behaviour of the average of the width $< B(t) >$ in Fig. 2 for the parameters values $\Gamma = 0.1, B_0 = |\Delta| = 1$ and different values of the noise variance D. For $D < D_{crit} \simeq 0.15$ the effect of the noise is to delay the pseudo-collapse in terms

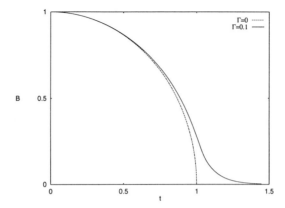

Figure 1: Width, B, as a function of time t in the absence of noise. Dashed line shows the analytical solution (33) for $\Gamma = 0$ with $B_0 = |\Delta| = 1$; solid line the numerical solution of (31) for $\Gamma = 0.1$.

of the ensemble average of the width in analogy with the result of [7] for the undamped case. For $D > D_{crit}$ initially the average width will decrease in a similar way as when $D < D_{crit}$, but after some time $< B(t) >$ will reach a minimum value and diverge as $t \to \infty$. This is due to the fact that for $D > D_{crit}$, the noise is strong enough to destroy the pseudo-collapse and cause dispersion for some of the systems in the ensemble. As $t \to \infty$ the dominating contribution to the mean value of $B(t)$ will come from the dispersing systems for which $B \to \infty$, and consequently $< B(t) >$ will diverge for $D > D_{crit}$. As can be seen in Fig. 2 the minimum value of $< B(t) >$ will increase towards B_0 as D increases. A more detailed description can be found in [10].

5 Concluding remarks

We have used the method of collective coordinates to study the influence of noise and nonlinear damping on the collapse process in the two-dimensional nonlinear Schrödinger equation. This model has been shown to result under certain approximations in the description of a coupled exciton-phonon system where thermal fluctuations are taken into account. We find that the main effect of the damping term is to replace the abrupt collapse process where the solution ceases to exist after a certain time with a physically more reasonable exponential decrease of the width as $t \to \infty$. We have found that if the variance of the noise is large enough the wave-packet may disperse instead of contract. The critical variance necessary to cause the dispersion will not change significantly compared to the undamped system unless the damping is large (in this case the critical variance is increased).

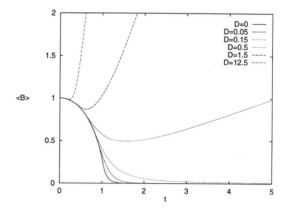

Figure 2: Ensemble average of the width, $<B>$, as a function of time t for different noise variances, $D = 0$, 0.05, 0.15, 0.5, 1.5, and 12.5, respectively, from bottom to top. $B_0 = |\Delta| = 1$ and $\Gamma = 0.1$ in all cases.

References

[1] D. Möbius and H. Kuhn, Isr. J. Chem. **18**, 375 (1979).

[2] D. Möbius and H. Kuhn, J. Appl. Phys. **64**, 5138 (1988).

[3] O. Bang, P. L. Christiansen, F. If, K. Ø. Rasmussen and Yu. B. Gaididei, Phys. Rev. E **49**, 4627 (1994).

[4] J. J. Rasmussen and K. Rypdal, Phys. Scr. **33**, 481 (1986).

[5] M. Desaix, D. Anderson, and M. Lisak, J. Opt. Soc. Am. B **8**, 2082 (1991).

[6] D. Anderson, M. Bonnedal, and M. Lisak, Phys. Fluids **22**, 1838 (1979).

[7] K. Ø. Rasmussen, Yu. B. Gaididei, O. Bang, and P. L. Christiansen, Phys. Lett. A **204**, 121 (1995).

[8] O. Bang, P. L. Christiansen, F. If, K. Ø. Rasmussen and Yu. B. Gaididei, Appl. Anal. **57**, 3 (1995).

[9] P. L. Christiansen, K. Ø. Rasmussen, O. Bang, and Yu. B. Gaididei, Physica D **87**, 321 (1995).

[10] P. L. Christiansen, Yu. B. Gaididei, M. Johansson, K. Ø. Rasmussen and I. I. Yakimenko, Phys. Rev. E **54**, 924 (1996).

Mathematics of Denominator-separable Multidimensional Digital Filters with Application to Image Processing

Xiaoning Nie and Rolf Unbehauen
Lehrstuhl für Allgemeine und Theoretische Elektrotechnik,
Universität Erlangen-Nürnberg,
Cauerstrasse 7, D-91058 Erlangen, Germany,
E-mail: unbehauen@late.e-technik.uni-erlangen.de

Abstract

In digital image processing, multi-dimensional digital filters play a central role. A special class of two-dimensional digital filters is considered based on their characterization by transfer functions. It will be shown that a number of analytical results can be employed to design and implement two-dimensional digital filters with a separable denominator of the transfer function. The results are not applicable to the design of general two-dimensional filters. In order to simplify the design and to make the realisation of two-dimensional filters efficient, filters with separable denominator transfer functions might be preferable in practice.

1 Introduction

Digital image processing is today involved in many fields of applications. Three basic areas of digital image processing are image preprocessing, image enhancement and image restoration. Other areas such as image reconstruction are also becoming more and more important. Digital image processing finds widely ranged applications in telecommunications, digital TV, biomedicine, remote sensing, and industrial automation [1]. In digital image processing, the multi-dimensional (M-D) digital filtering plays a key role. One of the fundamentals of the M-D digital filtering is considered to employ so-called discrete linear shift-invariant (LSI) systems [2]. A 2-D discrete signal is described as a function of two space variables: $x(n_1, n_2)$ where n_1 and n_2 are the space variables. A so-called causal LSI discrete system has in general the form [2]

$$y(n_1, n_2) = \sum_{l_1=0}^{N} \sum_{l_2=0}^{M} a(l_1, l_2) x(n_1 - l_1, n_2 - l_2)$$

$$-\sum_{\substack{k_1=0 \\ k_1+k_2\neq 0}}^{P}\sum_{k_2=0}^{Q} b(k_1, k_2)y(n_1 - k_1, n_2 - k_2). \tag{1}$$

A causal LSI discrete system is particularly useful for real-time processing of pixel-by-pixel scanned images. A signal is called causal if

$$x(n_1, n_2) = 0 \quad \text{for} \quad n_1 < 0 \quad \text{or} \quad n_2 < 0. \tag{2}$$

The 2-D z-transform of a causal signal $x(n_1, n_2)$ is defined as

$$X(z_1, z_2) = \sum_{n_1=0}^{\infty}\sum_{n_2=0}^{\infty} x(n_1, n_2)z_1^{-n_1}z_2^{-n_2}. \tag{3}$$

Provided $\sum_{n_1=0}^{\infty}\sum_{n_2=0}^{\infty} |x(n_1, n_2)| < \varepsilon$ for some positive number ε, the function $X(z_1, z_2)$ exists for every $(z_1, z_2) \in \{(z_1, z_2) \mid |z_1| \geq 1 \text{ and } |z_2| \geq 1\}$. The transfer function of the system defined in (1) can be written as [1, 2]

$$H(z_1, z_2) = \frac{\sum_{l_1=0}^{N}\sum_{l_2=0}^{M} a(l_1, l_2)z_1^{-l_1}z_2^{-l_2}}{1 + \sum_{\substack{k_1=0 \\ k_1+k_2\neq 0}}^{P}\sum_{k_2=0}^{Q} b(k_1, k_2)z_1^{-k_1}z_2^{-k_2}}. \tag{4}$$

The output and input are related by

$$Y(z_1, z_2) = H(z_1, z_2)X(z_1, z_2). \tag{5}$$

The system of (1) is defined to be BIBO-stable (Bounded Input-Bounded Output) if always

$$\sum_{n_1=0}^{\infty}\sum_{n_2=0}^{\infty} |y(n_1, n_2)| < \varepsilon_1$$

$$\text{for} \quad \sum_{n_1=0}^{\infty}\sum_{n_2=0}^{\infty} |x(n_1, n_2)| < \varepsilon_2$$

where ε_1 and ε_2 are some positive real numbers. The frequency response of $H(z_1, z_2)$ is defined as $H(e^{j\omega_1}, e^{j\omega_2})$ for $\omega_1, \omega_2 \in [-\pi, \pi]$.

2 Frequency response characteristics

In many applications, the frequency response of a system is required to possess a certain symmetry property. A typical example is the circular symmetry, which is itself a special case of the so-called quadrantal symmetry. For the magnitude $|H(e^{j\omega_1}, e^{j\omega_2})|$, the quadrantal symmetry is defined as

$$|H(e^{j\omega_1}, e^{j\omega_2})| = |H(e^{j|\omega_1|}, e^{j|\omega_2|})|$$

$$\forall \quad \omega_1, \omega_2 \in [-\pi, \pi]. \tag{6}$$

Theorem 1. The transfer function of a causal and stable LSI system with real coefficients has necessarily a **separable** denominator if its magnitude frequency response possesses quadrantal symmetry [4].

The proof of Theorem 1 was given by Rajan and Swamy in [4]. An extension to higher dimensions was made by Pitas and Venetsanopoulos [5]. Theorem 1 implies that the denominator-separable transfer functions are the natural choice for the design of digital filters with quadrantally symmetrical magnitude frequency responses. The advantage of having a separable denominator is particularly in testing and guaranteeing the BIBO-stability [1, 2]. A further advantage lies in the reduced number of coefficients because the number of coefficients of a two-variate polynomial increases approximately quadratically with the degree in each variable while the number of coefficients of a single-variate polynomial increases only linearly with the degree.

Design examples by incorporating the symmetry considerations can be found for example in [5].

3 Rational expansions

Giving a rational function in two variables, there is in general no partial fraction expansion comparable with the rational functions in one variable. However, if the two-variate rational function has a separable denominator, it does have a partial-fraction expansion. This fact is used for the derivation of several implementations of a 2-D denominator-separable transfer function [6, 7]. A 2-D denominator-separable transfer function can be written as

$$H(z_1, z_2) = \frac{P(z_1, z_2)}{D_1(z_1) D_2(z_2)}. \tag{7}$$

Theorem 2 [6]: The transfer function $H(z_1, z_2)$ of (7) has a partial-fraction expansion given by

$$H(z_1, z_2) = \sum_{i=1}^{P} \sum_{j=1}^{Q} \frac{c_{ij}}{(z_1 - \gamma_i)(z_2 - \lambda_j)}$$

$$+ \sum_{i=1}^{P} \frac{c_{iQ}}{z_1 - \gamma_i} + \sum_{j=1}^{Q} \frac{c_{Pj}}{z_2 - \lambda_j} + c_{00} \tag{8}$$

where γ_i, λ_j are the distinct roots of the denominator polynomials in each variable z_1 and z_2, respectively.

Theorem 2 can easily be generalized to the case of multiple roots of the denominator polynomials in each variable z_1 and z_2. A generalization of (8) is used in [3] to derive further interesting implementations. Eq. (8) can be generalized to

$$H(z_1, z_2) = \sum_{i=0}^{P}\sum_{j=0}^{Q} c_{ij} G_{1i}(z_1) G_{2j}(z_2) \tag{9}$$

or

$$H(z_1, z_2) = \boldsymbol{g}_1^t \boldsymbol{C} \boldsymbol{g}_2 \tag{10}$$

where $\boldsymbol{g}_1 = [G_{10}(z_1)\ G_{11}(z_1) \cdots G_{1P}(z_1)]^t$,
$\boldsymbol{g}_2 = [G_{20}(z_2)\ G_{21}(z_2) \cdots G_{2Q}(z_2)]^t$
and $\boldsymbol{C} = [c_{ij}]_{(P+1)\times(Q+1)}$.

In fact, we can choose two sets of functions $G_{1i}(z_1)$ and $G_{2j}(z_2)$ such that $G_{1i}(z_1)$ are orthogonal to each other and $G_{2j}(z_2)$ either. The realisations derived from the respective set of orthogonal functions $G_{1i}(z_1)$ and $G_{2j}(z_2)$ possess further advantages over those derived from (8), as it is shown in the work of [3]. An example of orthogonal functions is according to Walsh [8, 3]:

$$G_{10}(z_1) = 1, \quad G_{11}(z_1) = \frac{1}{z_1 - \gamma_1},$$

$$G_{12}(z_1) = \frac{1}{z_1 - \gamma_2} \frac{1 - \gamma_1^* z_1}{z_1 - \gamma_1}, \ldots$$

$$G_{1P}(z_1) = \frac{1}{z_1 - \gamma_P} \prod_{i=1}^{P-1} \frac{1 - \gamma_i^* z_1}{z_1 - \gamma_i},$$

and $\quad G_{20}(z_2) = 1, \quad G_{21}(z_2) = \dfrac{1}{z_2 - \lambda_1},$

$$G_{22}(z_2) = \frac{1}{z_2 - \lambda_2} \frac{1 - \lambda_1^* z_2}{z_2 - \lambda_1}, \ldots$$

$$G_{2Q}(z_2) = \frac{1}{z_2 - \lambda_Q} \prod_{j=1}^{Q-1} \frac{1 - \lambda_j^* z_2}{z_2 - \lambda_j}. \tag{11}$$

Another set of orthogonal functions can be derived using the well-known Schur-Recursion [9, 3].

The realizations derived from the orthogonal expansions are in general robust against the finite-wordlength effect of a digital realization. The experiments show [3] that e.g. the roundoff noise in a realization derived from (11) is of about the factor 10^3 below a direct realization using (4).

4 Some Remarks on the Implementation

In the following, only the realisation drived from (10) and (11) is treated. A further realisation based on the Schur-Recursion can be found in [3]. The implementation of a denominator-separable transfer function according to (10) and (11) involves the following two steps:

1. Calculation and realisation of the weighting matrix C,

2. The realisation of the functions $G_{1i}(z_1)$ and $G_{2j}(z_2)$.

An efficient way for the calculation of the weighting matrix is to consider an interpolation of $H(z_1, z_2)$ at the points $\{(\xi_1, \xi_2) \mid \xi_1 \in \{\infty, \frac{1}{\gamma_1^*}, \frac{1}{\gamma_2^*}, \ldots, \frac{1}{\gamma_P^*}\}$ and $\xi_2 \in \{\infty, \frac{1}{\lambda_1^*}, \frac{1}{\lambda_2^*}, \ldots, \frac{1}{\lambda_Q^*}\}$. The explicit formula for the interpolation reads

$$C = X_1^{-1} H X_2^{-1} \qquad (12)$$

where

$$H = \begin{bmatrix} H(\infty, \infty) & H(\infty, \frac{1}{\lambda_1^*}) & \cdots & H(\infty, \frac{1}{\lambda_Q^*}) \\ H(\frac{1}{\gamma_1^*}, \infty) & H(\frac{1}{\gamma_1^*}, \frac{1}{\lambda_1^*}) & \cdots & H(\frac{1}{\gamma_1^*}, \frac{1}{\lambda_Q^*}) \\ \vdots & \vdots & & \vdots \\ H(\frac{1}{\gamma_P^*}, \infty) & H(\frac{1}{\gamma_P^*}, \frac{1}{\lambda_1^*}) & \cdots & H(\frac{1}{\gamma_P^*}, \frac{1}{\lambda_Q^*}) \end{bmatrix},$$

$$X_1 = \begin{bmatrix} G_{10}(\infty) & G_{11}(\infty) & \cdots & G_{1P}(\infty) \\ G_{10}(\frac{1}{\gamma_1^*}) & G_{11}(\frac{1}{\gamma_1^*}) & \cdots & G_{1P}(\frac{1}{\gamma_1^*}) \\ \vdots & \vdots & & \vdots \\ G_{10}(\frac{1}{\gamma_P^*}) & G_{11}(\frac{1}{\gamma_P^*}) & \cdots & G_{1P}(\frac{1}{\gamma_P^*}) \end{bmatrix},$$

$$X_2 = \begin{bmatrix} G_{20}(\infty) & G_{20}(\frac{1}{\lambda_1^*}) & \cdots & G_{20}(\frac{1}{\lambda_Q^*}) \\ G_{21}(\infty) & G_{21}(\frac{1}{\lambda_1^*}) & \cdots & G_{21}(\frac{1}{\lambda_Q^*}) \\ \vdots & \vdots & & \vdots \\ G_{2Q}(\infty) & G_{2Q}(\frac{1}{\lambda_1^*}) & \cdots & G_{2Q}(\frac{1}{\lambda_Q^*}) \end{bmatrix},$$

As it can easily be verified, the matrices X_1 and X_2 are each lower- and upper trigonal matrices.

The realisation of the functions $G_{1i}(z_1)$ can be derived from the realisation of a cascaded first-order allpass with the transfer function

$$A_1(z_1) = \frac{(1 - \gamma_1^* z_1)(1 - \gamma_2^* z_1) \cdots (1 - \gamma_P^* z_1)}{(z_1 - \gamma_1)(z_1 - \gamma_2) \cdots (z_1 - \gamma_P)}. \qquad (13)$$

For $G_{2j}(z_2)$ the implementation can be derived from a cascaded realisation of

$$A_2(z_2) = \frac{(1 - \lambda_1^* z_2)(1 - \lambda_2^* z_2) \cdots (1 - \lambda_Q^* z_2)}{(z_2 - \lambda_1)(z_2 - \lambda_2) \cdots (z_2 - \lambda_Q)}. \qquad (14)$$

An efficient scheme of the realisation of the first-order allpasses and its analysis can be found in detail in [3].

5 Conclusion

A special class of 2-D digital filters is discussed in this paper. It has been shown that a number of analytical results are very useful for the design and implementation of 2-D digital filters with separable denominators. These results are very likely not existing for general 2-D digital filters. Thus 2-D digital filters with separable denominators might be preferable in practice if the simplicity of the design and the efficiency of the realisations are of primary interest.

References

[1] T.S. Huang, ed., Two-dimensional digital signal processing I and II, Springer Verlag, 1981.

[2] D.F. Dudgeon and R.M. Mersereau, Multidimensional digital signal processing, Prentice-Hall, 1984.

[3] X. Nie, On the synthesis of two-dimensional digital filters, Doctoral Dissertation at the Lehrstuhl für Allgemeine und Theoretische Elektrotechnik der Universität Erlangen-Nürnberg, 1992.

[4] P.K. Rajan and M.N.S. Swamy, "Quadrantal symmetry associated with two-dimensional digital transfer functions", IEEE Trans. CAS, vol. CAS-25 (1978), pp. 340–343.

[5] J.K. Pitas and A.N. Venetsanopoulos, "The use of symmetries in the design of multidimensional digital filters", IEEE Trans. CAS, vol. CAS-33 (1986), pp. 863–873.

[6] M.Y. Dabbagh and W.E. Alexander, "Multiprocessor implementation of 2-D denominator-separable digital filters for real-time processing", IEEE Trans. ASSP. vol. ASSP-37 (1989), pp. 872–881.

[7] D. Raghuramireddy and R. Unbehauen, "A new structure for multiprocessor implementation of 2-D denominator-separable filters", Proceedings of IEEE ISCAS-91, Singapore, (1991), pp. 472–475.

[8] J.L. Walsh, Interpolation and approximation by rational functions in the complex domain. Published by the American Mathematical Society, Rhode Island, 1965.

[9] J. Schur, "Über Potenzreihen, die im Inneren des Einheitskreises beschränkt sind", J. für Math., vol. 148 (1917), S. 122–195.

Comparison and Assessment of Various Wavelet and Wavelet Packet based Denoising Algorithms for Noisy Data

F. Hess*, M. Kraft**[1], M. Richter*, H. Bockhorn**
*Fachbereich Informatik, Universität Kaiserslautern,
**Fachbereich Chemie, Universität Kaiserslautern,
Erwin Schrödinger Straße, D-67663 Kaiserslautern, FRG
hess@informatik.uni-kl.de, mkraft@rhrk.uni-kl.de

Keywords *Wavelets, Wavelet Packets, Thresholding of Wavelet Coefficients, Adaptive Denoising*

1 Introduction

Denoising of measured data is an important method in data analysis and of great significance in many industrial applications. For example pattern recognition often needs signal denoising as a kind of preprocessing, followed by the actual classification. Denoising of measured data can be seen as a problem in nonparametric regression where an unknown function f has to be revealed from a signal $\tilde{f}$ containing undesired overlaying noise ξ. In the following we assume that noisy empirical data $\tilde{f}_i$ are measured at equidistant points $t_i \in [a, b]$ and $\tilde{f}_i$ results from a superposition of f at point t_i with white noise ξ_i of variance σ^2.

$$\tilde{f}_i = f(t_i) + \xi_i \qquad i = 0, ..., 2^{jmax} - 1 \qquad (1)$$

There exists a variety of different approaches to solve this regression problem, such as kernel estimators with global bandwidth or spline estimators [4]. A major disadvantage of these standard smoothing techniques is the fact, that they do not resolve local structures well enough. This, however, is necessary when dealing with signals that contain structures of different scales and amplitudes such as neurophysiological signals (see Figure 1). In recent years smoothing techniques based on the wavelet and wavelet packet transformation have been developed by Donoho et al. [2] and Majid et al.[6] which outperform standard denoising techniques for noisy data with local structures. It is the aim of this paper to describe a variety of wavelet and wavelet packet based denoising methods and compare them with each other by applying them to a simulated, noised signal. As the original signal f is known and the amount of added noise is a free parameter, the performance of each method can be rated by means of a Monte-Carlo simulation. The presented numerical investigations should serve as a guide for choosing the best wavelet based denoising method and a suitable denoising threshold in practical applications.

---
[1]corresponding author

2 Wavelets and Wavelet Packets

A sampled signal $\tilde{f}_i$ $i = 0,..,2^{jmax} - 1$ living in the sampling space can be represented by the discrete wavelet transformation

$$\tilde{f} = \tilde{c}_0^0 \cdot \phi_0^0 + \sum_{j=0}^{jmax-1} \sum_{i=0}^{2^j-1} \tilde{d}_i^j \cdot \psi_i^j. \qquad (2)$$

ϕ_0^0 and ψ_i^j are basis functions of a system of subspaces constructed by the successive application of convolution and subsampling filter operations. Here the ψ_i^j are *Daubechies Wavelets* with compact support as described for example in [1]. The empirical coefficients $\tilde{c}_0^0$ and $\tilde{d}_i^j$ are the projections of the empirical signal $\tilde{f}$ onto the basis functions ϕ_0^0 and ψ_i^j respectively and carry information about scale j and position i. So, the empirical wavelet coefficients describe which basis functions of scale j are needed to represent $\tilde{f}$ at position i in scale j which corresponds to position $[t_{i2^{jmax-j}}, t_{(i+1)2^{jmax-j}}[$ in the sampling space. The wavelet packet transformation is a generalization of the wavelet transformation. Using the same filter operations as in the case of the wavelet transformation one can construct a dyadic tree of subspaces of the sampling space. The algorithm and mathematical proofs are given in [8]. From this tree of subspaces a *best basis* of the sampling space is chosen in such a way that the coefficients according to this selection minimize a certain cost function and reconstruct $\tilde{f}$ completely. The wavelet packet coefficients carry besides information on scale j and position i in scale j a local frequency information f. The empirical signal $\tilde{f}$ can be expressed through:

$$\tilde{f} = \sum_{\mathcal{B}_{jfi}} \tilde{w}c_{jfi} \, w_{jfi} \qquad (3)$$

The index set $\mathcal{B}_{jfi}$ denotes the indices of the best basis according to the chosen cost function, the coefficients $\tilde{w}c_{jfi}$ are the projections of the empirical signal $\tilde{f}$ onto the basis functions w_{jfi}. In order to avoid errors at the boundary due to the finite length of the signal we assume the signal to be periodic. The high number of basis functions and the effective search for a best basis using a cost function lead to an efficient representation, e.g., of an image or a segment of data. Therefore, the comparison of data like audio recordings or 2d images are important application fields especially for the wavelet packet transformation.

3 Denoising Methods

The denoising algorithms considered here consist of three parts. First, the transformation of the empirical signal $\tilde{f}$, generating a set of signal coefficients, second, the application of a denoising technique onto this set, and last the back transformation. The reconstructed signal $\hat{f}$ is then the estimate for the original signal f. So, the presented denoising methods mainly differ in the transformation method used and the way coefficients are treated. We, however, investigate those methods based on the wavelet and wavelet packet transformation. The following denoising methods can then be separated into linear (L) and non-linear (NL) denoising methods. Linear denoising is independent of the size of empirical signal coefficients, and therefore not the coefficient size itself is taken into account, but the coefficient parameters scale (for wavelet coefficients) or scale and frequency (for wavelet packet coefficients). Linear denoising methods are based on the assumption, that signal noise can be found mainly in fine scale coefficients and not in coarse scale ones. Therefore, linear denoising for wavelet-transformed signals (WL) cuts off all coefficients with a scale finer than a certain scale threshold S_0. With this, $\tilde{d}_i^j$ denotes empirical coefficients, $\hat{d}_i^j$ denotes denoised coefficients:

$$\hat{d}_i^j = \begin{cases} 0 & , j \geq S_0 \\ \tilde{d}_i^j & , j < S_0 \end{cases} \quad (4)$$

A refinement of linear denoising for the wavelet transformation is used for wavelet packet based methods (PL). Here, fine scaled signal structures can be represented not only by fine scale coefficients, but also by coarse scale coefficients with high frequency (a high amount of zero-crossings of the underlying basis function). So, it is necessary to eliminate not only fine scale coefficients through linear denoising, but also coefficients of a scale and frequency combination which refer to a certain fine scale structure. Here, $\hat{w}c_{jfi}$ denotes denoised wavelet packet coefficients:

$$\hat{w}c_{jfi} = \begin{cases} 0 & , j \geq S_0 \\ 0 & , (j < S_0) \wedge (f > 2^{S_0 - j}) \\ \tilde{w}c_{jfi} & , else \end{cases} \quad (5)$$

Non-linear denoising (NL) on the other hand is based on the idea, that white noise can be found in every coefficient and is distributed over all scales. So, basis of a denoising is here the quantity of the coefficient itself, underlying a certain signal-to-noise ratio. Non-linear denoising of a coefficient can then be applied in two ways. Hard-thresholding (NLH) cuts off coefficients below a certain threshold λ, soft-thresholding (NLS) reduces all coefficients by this threshold. For wavelet packet based denoising, we get:

$$\hat{w}c_{jfl} = \begin{cases} 0 & , |\tilde{w}c_{jfl}| < \lambda \\ \tilde{w}c_{jfl} & , else \end{cases} \quad hard(PNLH) \quad (6)$$

$$\hat{w}c_{jfl} = \begin{cases} 0 & , |\tilde{w}c_{jfl}| < \lambda \\ sgn(\tilde{w}c_{jfl}) \cdot (|\tilde{w}c_{jfl}| - \lambda) & , else \end{cases} \quad soft(PNLS) \quad (7)$$

Donoho and Johnstone [2] introduced a denoising method (DJ) based on the wavelet packet transformation which uses a special cost function for generating a best basis. This cost function chooses a best basis with respect to best denoising results. The Donoho Johnston cost function is an entropy functional which includes a threshold which has to be specified. Linear and non-linear denoising methods can be applied in the same way as with standard cost functions (DJL, DJNLH, DJNLS). The so called *Coherence denoising* (COHERENCE), introduced

by Majid, Coifman and Wickerhauser [6] and applied in [3] for denoising acoustic
signals, considers denoising as a problem of signal compression. The idea is, that
a signal without noise can be compressed well, whereas noise itself has a poor
compression rate. We distinguish in the following between a coherent and an incoherent part of the signal. First the empirical signal is expanded in wavelet packet
coefficients. After applying a hard threshold the remaining N_0 coefficients are decreasingly ordered. The first M large coefficients represent the coherent part and
the others the incoherent part. M is chosen in such a way that the compression
rate of the incoherent part is below a certain threshold.

$$\tilde{f} = coher_M + incoher_M = \sum_1^M \tilde{w}cw + \sum_{M+1}^{N_0} \tilde{w}cw \qquad (8)$$

Then in an iterative process the incoherent part is reconstructed and the above
procedure is applied again. It is expanded in wavelet packet coefficients, denoised,
ordered, and split into a coherrent and incoherent part. This procedure is repeated
until no new coherent part is obtained or a certain number of iterations have been
completed. The resulting coherrent part is the estimated signal. All mentioned
denoising methods which are tested in section 4 are presented in Table 1, their
abbreviations refer to the above described methods.

4 Results

In this chapter we compare the described denoising methods by means of an simulated, noisy signal. For this, the necessary parameter investigations, especially the
right choice of basis functions, cost function and denoising threshold, is in the center of interest. For the following analysis a simulated signal f from neurobiology
(Figure 1, left side) was used.

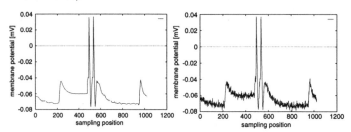

Figure 1: Neurophysiological signal f (left), and a noised signal $\tilde{f}$ with normally distributed white noise and signal-to-noise-ratio of 30:1 (right).

This signal is well suited for our purpose, because it contains fine scale and coarse
scale structures of different sizes as well, and therefore covers different "denoising
situations". From that, a noised signal $\tilde{f}$ can be generated, as described in equation
1. To determine the impact of the type of white noise distribution on the denoising

results, two types of noise are being examined: uniformly distributed white noise and normally distributed white noise, both with an underlying signal-to-noise ratio of 30 : 1 and a mean of zero. The applied signal-to-noise ratio is characteristic for a certain type of neurophysiological signals, investigated by the authors [5]. The analyzed denoising method creates an estimated denoised signal $\hat{f}$ out of f, which is then compared with the original signal f through distance measures. Here, three distance measures are used: l_1, l_2, and *entropy* norm. Small distances, or, a small *denoising error*, denote a good correspondence between f and $\hat{f}$, and therefore a good denoising performance.

$$l_p - norm: \quad \|f - \hat{f}\|_p = (\sum_i |f(t_i) - \hat{f}(t_i)|^p)^{\frac{1}{p}}, \quad p = 1, 2 \qquad (9)$$

$$entropy - norm: \quad -\sum_i \frac{|f(t_i) - \hat{f}(t_i)|^2}{\|f - \hat{f}\|_2^2} \cdot \log \frac{|f(t_i) - \hat{f}(t_i)|^2}{\|f - \hat{f}\|_2^2} \qquad (10)$$

In a first study it was investigated which wavelet and which cost function we should use for the subsequent numerical study. The right choice of the basis functions has an influence on the denoising performance. Figure 2 on the right side shows the relation between basis function and denoising error for one sample signal. A certain group of basis functions, the Daubechies wavelets introduced by I. Daubechies [1], has been investigated. Through Monte-Carlo simulation and variation of the threshold (see below), it can be shown that Daubechies basis functions with parameter $N = 3$ lead to the best denoising results in case of the PNLS method, expressed through a minimum denoising error which can be determined by applying any of the above distance measures. Therefore, Daubechies $N = 3$ basis functions are employed in the following test runs.

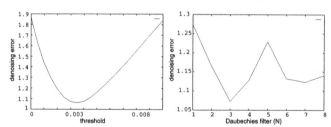

Figure 2: relation between denoising threshold (here for PNLS) and denoising error (left), relation between Daubechies base functions and denoising error (right).

Furthermore, the best cost function with respect to the denoising performance can be determined. A variety of standard cost functions such as *eps*, *lp*, *entropy* or *log-energy* are examined for a sample signal in order to find the smallest denoising error between f and $\hat{f}$. The best basis for denoising the here presented signal is given by the log-energy cost function. This basis is identical with the wavelet basis. Therefore we choose the second best result, the entropy cost function, for the parameter study. In the following the relation between denoising threshold (and denoising scale S_0 respectively) and the denoising performance is of interest. As described in Nason [7], an often used, *universal threshold* λ_u can be determined, only depending on the signal length and the noise standard deviation σ. In our case, $\lambda_u = \sqrt{2log(2^{jmax})}\hat{\sigma} = 0.0839$ for normally distributed noise. However, investigations reveal that λ_u lies above the optimal threshold, suppresses too much of the original signal therefore should not be utilized. Figure 2 on the left side shows the relation between denoising error and threshold, calculated for a set of thresholds. We approximate the optimal threshold by the best results of a Monte-Carlo simulation. With 30 independent trials, a mean denoising error and its standard deviation is generated for every denoising method and a given threshold/scale. Through variation of the threshold and scale respectively, the optimal threshold/scale and its mean minimum denoising error can then be determined. Mean minimum denoising error and standard deviation of every denoising method is then subject of a comparison. The Coherence denoising method needs special treatment. Here, the number of iterations is limited to one, and instead of a denoising threshold, the *compression* threshold τ is optimized in the course of the Monte-Carlo simulation. Table 1 summarizes the results of the denoising performance analysis. It presents the denoising performance of the introduced linear and non-linear denoising methods with optimal threshold/scale, cost-function and basis functions as a combination of mean minimum denoising error and standard deviation, dependent on the noise type. The comparison reveals that non-linear denoising with hard thresholding, using the wavelet transformation (WNLH) has the minimum mean denoising error and therefore the best denoising performance. Fig. 3 shows exemplary denoising results for this method, on the left side with uniformly distributed white noise, on the right side with normally distributed white noise.

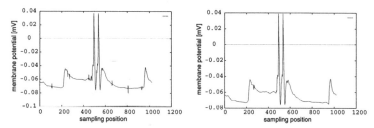

Figure 3: exemplary best denoising results for WNLH-method, using white uniformly (left) and normally (right) distributed noise. Note the fine-scale artifacts.

Obviously, denoising with the wavelet packet transformation and cost function does not lead to a better denoising performance, compared with the simpler wavelet denoising. It seems, for this particular signal, that the examined cost functions and the chosen wavelet packet basis are not optimal for denoising purposes, even with the Donoho-Johnstone cost function. The comparison also shows that denoising with the Coherence denoising method is a competitor for the WNLH method, and the authors believe that with an optimized parameter choice, its denoising performance can be improved.

Type of Noise	N(0,0.00225)			U[-0.004,0.004]		
Method	threshold	threshold	threshold	threshold	threshold	threshold
Daub. N=3	$\hat{E}\|\|_1$ std.dev.	$\hat{E}\|\|_2$ std.dev.	$\hat{E}(Entr.)$ std.dev.	$\hat{E}\|\|_1$ std.dev.	$\hat{E}\|\|_2$ std.dev.	$\hat{E}(Entr.)$ std.dev.
WL	8	8	8	8	8	8
	1.02776 0.04472	0.05027 0.00135	0.39545 0.01378	1.07019 0.04008	0.05115 0.00104	0.40694 0.01171
WNLH	0.007	0.007	0.007	0.006	0.006	0.006
	0.73024 0.04978	0.03369 0.00194	0.30076 0.01597	0.71280 0.05099	0.03229 0.00214	0.29159 0.01798
WNLS	0.003	0.003	0.003	0.003	0.003	0.003
	0.87034 0.04943	0.03839 0.00187	0.34352 0.01547	0.89008 0.04419	0.03857 0.00095	0.34697 0.01138
PL	8	8	8	8	8	8
	1.14461 0.08135	0.05406 0.00287	0.45877 0.04322	1.19165 0.07182	0.05626 0.00271	0.48823 0.03880
PNLH	0.00419	0.00390	0.00396	0.00416	0.00387	0.00409
	0.88898 0.06849	0.04138 0.00329	0.35604 0.02257	1.04809 0.56825	0.04740 0.01513	0.40225 0.13982
PNLS	0.00311	0.00300	0.00300	0.00362	0.00300	0.00312
	0.98594 0.07954	0.04260 0.00371	0.37870 0.02737	0.99727 0.05407	0.04364 0.00248	0.38481 0.01855
DJL	8	8	8	7	8	8
	1.61946 0.23498	0.06612 0.00637	0.57474 0.06918	2.02705 0.37177	0.07428 0.01042	0.67049 0.09041
DJNLH	0.00154	0.00142	0.00150	0.00160	0.00148	0.00156
	0.87264 0.07421	0.03957 0.00343	0.34804 0.02504	0.89884 0.07024	0.04166 0.00296	0.36053 0.02181
DJNLS	0.00300	0.00290	0.00296	0.00346	0.00296	0.00300
	0.98796 0.06553	0.04269 0.00258	0.37982 0.02064	1.13178 0.46297	0.04730 0.01403	0.42033 0.12289
COHER.	0.79	0.79	0.79	0.80	0.80	0.80
	0.83015 —	0.03944 —	0.33904 —	0.88573 —	0.04075 —	0.35541 —

Table 1: Comparison of different denoising methods with respect to their denoising performance. Basis functions: Daubechies Wavelets, filter length 6 ($N = 3$). Noised signals: white noise with uniform distribution (U) and normally distribution (N) (see figure 1). Denoising methods: P: wavelet packet transformation. W: wavelet transformation. L: linear denoising. NLH: non-linear hard thresholding. NLS: non-linear soft thresholding. DJ: Donoho-Johnstone wavelet packet denoising. Coherence: Majid-Wickerhauser Coherence denoising.

Due to undesired signal artifacts, the achieved denoising results are not optimal (see Figure 3). They originate from noisy coefficients, which are not suppressed by the global threshold - in particular hard thresholding methods leave coefficients above the threshold unchanged. Since the artifacts live only on fine-scales it is advisable to introduce an adaptive, scale-variant threshold λ_j. This threshold is computed through $\lambda_j = \lambda \cdot \mu_j$, with λ as the optimal threshold for the mean minimum denoising error and μ_j as a scale factor, presented in Figure 4 on the left side. Through this, the threshold λ_j for fine-scale structures is higher than for coarse scale ones.

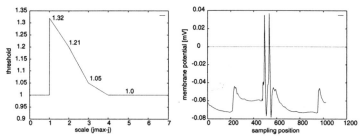

Figure 4: Adaptive, scale-variant threshold factor μ_j (left), and improved denoising result (right), achieved through WNLH-denoising.

As shown in Figure 4 on the right side, denoising with the adaptive threshold suppresses undesired artifacts and improves the denoising results. The minimal error is 10% less in the scale dependent threshold case than in the global threshold case.

References

[1] Daubechies I., *Orthonormal Bases of Compactly Supported Wavelets*, Comm. Pure Appl. Math., Vol. 41, 1988, pp. 909-996

[2] Donoho D.L., Johnstone I.M., *Ideal Denoising in an orthonormal basis chosen from a library of bases*, Dept. of Statistics, Stanford Univ., 1994

[3] Goldberg M.J., Coifman R.R., Berger J., *Removing Noise from Music using Local Trigonometric Bases and Wavelet Packets*, Tech. Report, 1994

[4] Härdle W., *Applied nonparametric regression*, Econometric Society Monographs, Cambridge University Press, 1990

[5] Hess F., Kraft M., *Mustererkennung mit Wavelet Packets am Beispiel neurophysiologischer Signale*, DAGM "Mustererkennung 96", Springer Verlag, 1996

[6] Majid F., Coifman R.R., Wickerhauser M.V., *The Xwpl system Reference Manual*, Yale Univ., New Haven, 1993

[7] Nason G.P., *Wavelet regression by cross-validation*, Dept. of Mathematics, Univ. of Bristol, 1994

[8] Wickerhauser M.V., *INRIA Lectures on Wavelet Packet Algorithms*, Numerical Algorithms Research Group, Dept. of Mathematics, Yale Univ., New Haven, 1991

Ship industry

Hydrodynamic Design of Ship Hull Shapes by Methods of Computational Fluid Dynamics

Prof. Horst Nowacki
Ship Design Division, Techn. Univ. of Berlin, Salzufer 17-19,
D-10587 Berlin, Germany, E-mail: nowacki@cadlab.tu-berlin.de

Abstract

In this article the problem of ship hull shape improvement for reduced drag is described from the viewpoint of mathematical optimization. The approach concentrates on methods of shape variation and flow analysis by Computational Fluid Dynamics. This still evolving methodology will be illustrated by a few examples of hydrodynamic design.

1 Objectives

The desire to design ship hull shapes with favourable speed performance is ancient. History knows many examples of ship types whose shape qualities were regarded as superior in their era, e.g., the Athenian trireme, the Viking ship, clippers, windjammers and many others. Their design in the past was generally based on intuition, trial and error.

The scientific era of hydrodynamic ship design began a little more than a century ago with ship model testing, flow analysis and systematic performance evaluation. But only recently have the analytical models and computational capabilities of ship hydrodynamics reached a stage of maturity that permits a detailed and accurate analysis of the cause and effect relationships between ship shape and flow. This is in large measure owed to the progress in Computational Fluid Dynamics in the last decade.

It has thus become possible to think about the mathematical optimization of hull shape for some given requirements. This paper will give an introduction to this problem, review some classical work, and formulate a mathematical optimization problem with constraints, which is of Nonlinear Programming type. This still evolving approach to hull form optimization based on Computational Fluid Dynamics will then be illustrated by a few characteristic examples.

2 Brief History

Table I gives a condensed overview with some of the important milestones in the development of ship flow analysis for resistance prediction and shape design. William Froude, the pioneer of ship model testing, strictly by observation already had an intuitive, but physically very clear understanding of ship wave patterns as one of the primary causes of the ship resistance (Figure 1). The objective to reduce the energy losses which are associated with the formation of diverging and transverse wave trains generated by the ship became one of the prime targets of hull shape design. Froude also developed an approach for dealing with frictional drag. The classical assumption that the total resistance is composed of wavemaking and viscous parts, known as Froude's hypothesis, remains an important foundation in our work. Kelvin's famous analysis of the wave pattern created by a travelling pressure point [3] and Michell's theory of ship wave resistance [4] based on linearized boundary conditions on the ship surface ('thin ship') and on the free surface ('small waves') are other pioneering contributions from the last century.

Table I: Brief History

- William Froude (1877): Ship Wave Patterns
- Lord Kelvin (1886-87): Waves of a Travelling Pressure Point
- John H. Michell (1898): Ship Wave Resistance, Linear Theory
- David W. Taylor (1915): Systematic Hull Form Variation
- Georg Weinblum (1930 ff.), C. Wigley (1931 ff.): Hull Form Variation and Wave Resistance Minimization
- Thomas H. Havelock (1913-1951): Ship Wave Resistance Theory
- Takao Inui (1960 ff.): Ship Wave Patterns and Bulbous Bows
- John Hess, A. M. O. Smith (1962): Panel Methods
- Lin, Webster, Wehausen (1963): Total Resistance Minimization
- Nagamatsu, Sakamoto, Baba (1983): Minimum Ship Viscous Drag
- S. Y. Ni (1987), G. Jensen (1988): Nonlinear Rankine Source Methods
- Many authors (since 1990): Ship RANSE Solvers
- Lars Larsson et al. (1992): Resistance Minimization Based on CFD

The idea of systematic hull form variation using a set of form parameters and a mathematical ship lines representation dates back to D. W. Taylor [5] in the beginning of this century. Weinblum [6] and Wigley [7] combined the principles of systematic hull form variation with wave resistance analysis and thus laid the ground for the wave resistance minimization of hull shapes. This work was far extended by Havelock's theories [8]. Many successful design concepts, including the application of bulbous bows, were rationally explained on this basis. However the agreement between experiments and linearized wave resistance theory was only qualitatively satisfactory.

After the advent of computers, beginning in the early sixties, this line of work was resumed and much intensified. Initially the minimization of wave resistance (Inui [9]) by linear theory remained a main target, though simple estimates of viscous drag were sometimes included (Lin [10]). The progress made in the calculation of potential flows for arbitrary shapes by panel methods (Hess and Smith [11]) was also soon exploited for ships.

Figure 1: William Froude's sketch of a ship bow wave train with diverging and transverse waves, (Froude, 1877) from [1], [2]

For slow ships with full forms the viscous drag component is predominant. The minimization of viscous drag was initially studied mainly by three-dimensioned boundary layer flow analysis (Nagamatsu et al. [12]). This flow analysis proved somewhat inaccurate for ships, especially in the afterbody [13]. Therefore the interest turned to Reynolds Averaged Navier-Stokes Equation (RANSE) solvers in the early nineties. About simultaneously an important breakthrough was the introduction of numerical solvers for the ship wave resistance with nonlinear free surface boundary conditions (Ni [14], Jensen [15]). Thereby in recent years ship flow predictions for wave making and viscous flows have become sufficiently reliable in a quantitative sense to warrant a fresh and promising look at the classical hull shape optimization problem (e.g. Larsson et al. [16]).

3 Problem Statement

3.1 Optimization Format

The hydrodynamic design problem for ship hull shapes can be stated in terms of a technical optimization problem as follows:

"Minimize the total resistance (R_T) per unit of displacement (Δ) for some given speed - or any equivalent measure of merit function - by suitable choice of the free variables (X) of hull shape definition subject to any given geometric or functional constraints !"

In formal, mathematical terms the equivalent problem statement is:
"Find an optimal state of the design variable vector $X = \{x_1, ..., x_n\}$, n positive integer, so as to minimize the measure of merit function $M(X)$ subject to a set of inequality constraints $C(X) = \{g_1(X), ...g_K(X)\}$, with $g_j(X) \geq 0$, $1 \leq j \leq K$."

Note that any existing equality constraints are accounted for separately and implicitly in this approach. The optimization problem in view of the usually nonlinear functional relationships for $M(X)$ and $g_j(X)$ is of Nonlinear Programming type.

The measure of merit function in the spirit of a cost/benefit ratio is chosen as
$$M = R_T / \Delta$$
where the total resistance is composed of
$$R_T = R_W + R_V + R_{WV},$$
with
R_W = wave resistance
R_V = viscous resistance
R_{WV} = wave-viscous interaction component
Δ = displacement

The ship speed V or Froude number $F_n = V/\sqrt{gL}$, L = ship length, is fixed in the optimization process.

3.2 Optimization Process

The sequence of steps in the optimization process is shown in Figure 2. The process can be organized in modular form with functional components for shape variation, CFD analysis, design evaluation and optimization strategy. The strategy can be some Nonlinear Programming algorithm which responds to the measure of merit and constraints evaluated by the analysis modules for any given state of hull shape. The numerical optimization process is automatically driven by the strategy. The algorithm must be properly tuned in terms of step widths, tolerances, starting points etc. to ensure convergence and accuracy. Multimodalities may exist.

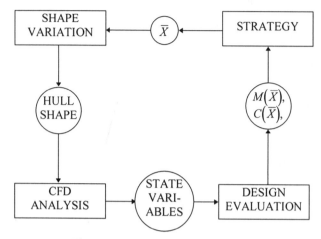

Figure 2: Hull Shape Optimization Process

3.3 Shape Variation

The objective of the shape variation process is to modify the hull shape by altering flow relevant form parameters while retaining a fair ship surface and usually conserving a given displacement.

The hull may be represented by discrete point sets (offset data) or curve meshes (wireframe models) or by surface representations.

The desired types of changes are in the following categories:

- Global changes:
  Principal dimensions (length, breadth, draught)
- Regional changes:
  Shift of volume centroids, shoulders etc.
- Local changes:
  Variation in streamlines, waterlines, sections

It is desired to represent each of these shape elements by form parameters so that a geometric modelling process can be driven by parametric shape design variation.

3.4 Constraints

The requirements of the design task and the computational assumptions yield the following categories of constraints:

a) Design constraints:
   Limits on shape variation for reasons of operational, technical, and legal requirements (e.g. draught limits, stability rules etc.)
b) Geometric constraints:
   Shape preservation, fairness (Convexity and monotonicity preservation etc.).
c) Validity constraints:
   Related to validity ranges of flow analysis (e.g. avoidance of separation).

4 Current Research

4.1 Shape Variation

Hull shape changes can in principle be achieved by any available geometric modeller for ship surfaces. However, a systematic approach is recommended here by which those form parameters and shape features are directly controlled which are of particular relevance for the flow and hydrodynamic performance of the underwater shape. These form parameters must be varied in a concerted way avoiding any undesired geometric degeneracies or physical drawbacks. This leads to a modelling requirement for parametric shape design with constraints.

A possible approach for parametric ship form variation is illustrated in Figure 3. This figure shows a set of four basic planar curves from which all other geometric properties of the underwater shape can be derived as dependent variables. The basic curves are:

- Sectional Area Curve (SAC),
  i.e., the distribution of underwater cross sectional area against ship length,
- Design Waterline (DWL),
  the floating condition waterline,
- Main Deck (top view)
- Lateral Plan (side of deck, stem, keel, stern profile)

Each of these basic curves is characterized by a set of form parameters, denoted by corresponding symbols in Figure 3, related to subareas, regional centroids, positional and derivative properties at certain points. This set of form parameters, varying between 20 and 50 potential variables, can serve as a set of free parameters for systematic hull shape control. It is often sufficient to control only a limited subset of these variables. Each curve segment is automatically generated from these form parameter requirements. Thus the basic curves can be uniquely defined from the free variables and hence also the dependent properties of hull shape. This approach must be modified for complex shape features like bulbous bows and appendages.

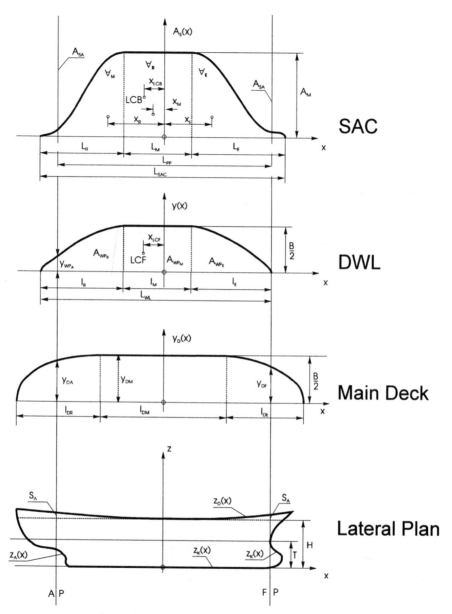

Figure 3: Basic Curves of Hull Form

Once an initial shape definition is generated, it is obviously possible, too, to modify the hull form via other free variables, especially by free B-spline control points for local changes.

The operations which may be performed to control hull form variation by systematic measures include the following:

- Scaling of principal dimensions
- Longitudinal or vertical stretching
- Centroid shifts
- Section shape control
- Streamline shape control
- Appendage design, filleting, blending
- Fairing

4.2 Flow Analysis

4.2.1 Potential Flow with Free Surface

This flow problem is a boundary value problem in an ideal fluid with a body and free surface boundary condition. The potential flow is governed by the Laplace equation

$$\Delta \phi = 0$$

where $\phi = \phi_0 + \varphi$ = transport + body flow potential.

The body boundary condition is

$$\frac{\partial \phi}{\partial n} = \frac{\partial \phi_0}{\partial n} + \frac{\partial \varphi}{\partial n} = 0 \text{ (on body)}$$

where $\varphi = \iint_S \frac{\sigma(q)}{r(P,q)} dS$ = body flow potential

$s(q)$ = surface singularity strength

The free surface boundary condition in its exact, nonlinear form, taken on the free fluid surface, is:

$$\frac{1}{2}\left(\nabla \phi \cdot \nabla (\nabla \phi)^2\right) - g\phi_z = 0$$

Further a radiation condition must be satisfied requiring that waves propagate only downstream.

This analytical problem formulation is usually discretized and then solved numerically by treating nonlinear algebraic systems of equations. Some of the relevant discretization schemes and numerical models are:

a) <u>Body surface panels:</u>
- Planar panels of constant source density, placed in the body surface (Hess and Smith) [11])
- Tringular panels with linear source strength function, submerged below body surface (Webster [17])
- Higher-order panel methods with curved panels and nonlinear source strength functions
- Desingularized, triangular, constant source density panels (Jensen [15])

- Collocation patches with zero cross-flux in body surface, source points in interior (Söding [18])

b) <u>Free surface panels:</u>
These are singularity panels arranged in or above the free fluid surface in order to meet the linear or nonlinear free surface boundary condition:
- Planar, constant source strength panels (Rankine sources) in the free surface, linear case (Dawson [19])
- Panels raised above free surface, linear case (Söding [20])
- Panels above free surface, nonlinear case (Ni [14], Jensen [15])

These panels arranged above the free surface must be of sufficient number and resolution to capture the free surface deformation realistically, hence with several panels per wave length. Above 1000 panels in one half-space are not unusual for medium ship speeds.

Figure 4 shows a comparison between measured and calculated, nonlinear wave resistance curves $C_w = R_w / (0.5\rho V^2 S)$ by Jensen's method, Figure 5 demonstrates the noticeable differences between linear and nonlinear analysis. The overall agreement with experiments is encouraging and a good basis for design conclusions.

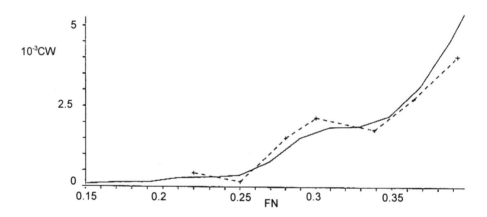

Figure 4: Wave resistance coefficient C_W for Series 60 hull, measured (——) and calculated (– – –) [15], plotted against Froude number F_n

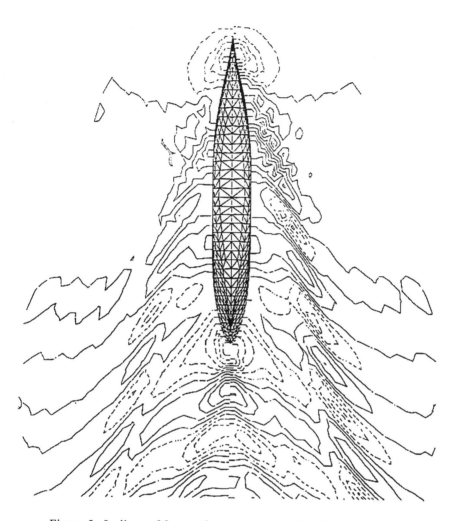

Figure 5: Isolines of free surface wave pattern for Series 60 hull, linear case (left), nonlinear case (right) [15]

4.2.2 Viscous Flow

Viscous flows around ships are modelled in similar ways as for other fluid dynamic shapes. Reynolds Averaged Navier Stokes Equation solvers are governing the state of the art. Ships are unique in their elongated, sometimes full and bluff shapes and in their very high Reynolds number ($Re_N = V \cdot L/\nu \approx 10^9$, turbulent flow). Little direct evidence can be carried over from aerodynamics of aircraft and automobiles. Yet ship viscous flow solvers have established themselves in their own right. The meetings held in Iowa City [21] and Tokyo [22] in recent years testify to the progress made in viscous flow calculations for ships.

The majority of the existing flow solvers is based on k-ε models, often combined with wall functions. Recently Reynolds stress transport models have also been tested and shown encouraging success (Sotiropoulos, Patel [23]). Figures 6 and 7 show some results from these calculations. The axial flow components in the propeller plane are compared in Figure 6 for a full tanker shape at model scale. The Reynolds stress transport model achieves a close resemblance with the measured wake, including the characteristic "hook" in the upper half of the propeller disk for isolines near 0.3. and 0.4. A look at Figure 8 explains why this is possible. The pattern of longitudinal vortex bands which detach themselves from the afterbody appear to be more realistically modelled here than in the k-ε-model. The ability to capture such complex secondary flows is remarkable.

For validations of ship viscous flow solvers comprehensive comparisons were made for many current CFD codes with results from experiments for a few well documented hull shapes at model scale. For the viscous flow about a ship without free-surface deformation (zero Froude number) the agreement with test results in frictional resistance was very good for most flow solvers wheras the viscous pressure drag varied more. Nonlinear potential flow solvers with free surface also gave trustworthy results in most cases. The new challenge of combining viscous flow solvers with free surface effects to investigate wave-viscous interference effects still requires further research.

There remain open questions for CFD solvers for ships, both for potential and viscous flows. They are mainly related to:

- High Froude numbers (wavebreaking, spray)
- Low Froude numbers (resolution of short wave length)
- Ship Reynolds numbers (resolution)
- Turbulence models
- Separation, secondary flows
- Integration accuracy

Despite these current specific limitations CFD codes are valuable prediction tools for ship flows in many applications. They support design decision in a rational way because they give an appreciation of cause and effect in shape variation and flow. They are beginning to be accurate and fast enough for exploratory work in early stage ship design.

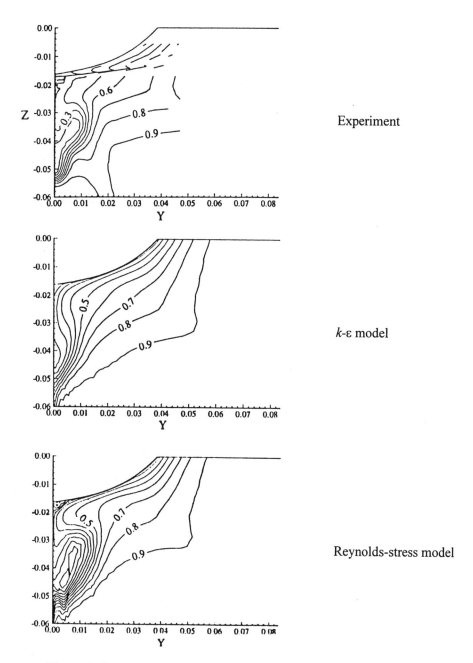

Figure 6: Isovelocity lines for axial wake in propeller plane ($x/L = 0.976$), measured and calculated from k-ε and *RSM* models, Wieghardt and Kux (1980), Sotiropoulos and Patel (1995)

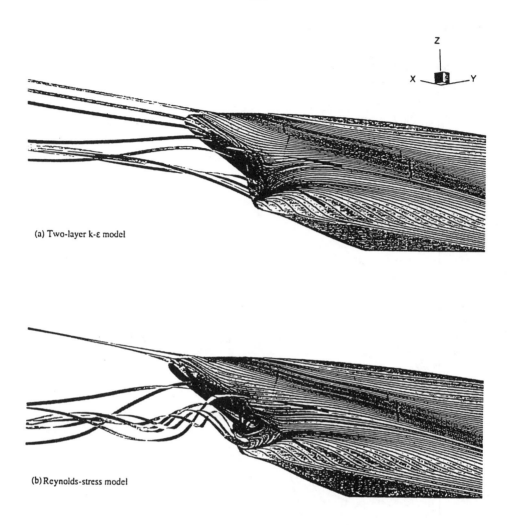

Figure 7: Limiting streamlines and stream ribbons, HSVA tanker calculated by k-ε model (top) and Reynolds stress model (bottom), Sotiropoulos and Patel (1995)

5 Examples

5.1 Systematic Variation of Center of Buoyancy

The volume centroid of the underwater hull form in naval architecture is called center of buoyancy. The longitudinal coordinate of this point is abbreviated by LCB, the vertical coordinate by VCB (i.e., longitudinal and vertical center of buoyancy). The location of this centroid is a governing influence on the general character of underwater hull shape and hence flow conditions, both for wavemaking and viscous resistance. This is why it was systematically varied in a study by S.-Y. Kim [23]. The systematic variation is used as an exploratory step before full optimization in order to demonstrate the sensitivity of the viscous resistance to hull form changes.

In this study a parent hull form, the model SSPA-720 from the Swedish ship model basin in Gothenburg, was chosen as the starting point. It is shown in Figures 8 and 9 by solid lines in the ship body plan (underwater cross sections forward and aft). This hull form corresponds to a conventional cargo ship of medium fullness ($C_B = 0.675$).

This parent form was systematically varied by a longitudinal centroid shift (by ±2% of ship length) combined with a vertical shift (by ± 1% of draft), which resulted in 9 combinations of form variation in total. Figures 8 and 9 document the parent and three of its variations. Figure 10 displays the results of boundary layer calculations for all nine variations, denoting the shifts forward and aft, up and down by symbols explained in the figure.

The momentum thickness at a station far aft in the afterbody (83.2% of ship length from the bow) is displayed as a function of the girth of this cross section between keel and design waterline. For low girth values, i.e. under the ship bottom, the momentum thickness is small, indicating mild energy losses and stable, attached flow. For high girth positions, i.e. along the ship sides beginning at the bilge, the momentum thickness is much greater, a warning signal for major energy loss upstream and a tendency for separation. The momentum thickness at the stern of the ship can be integrated to estimate the momentum loss in the ship wake and hence the approximate viscous drag (Squire-Young method).

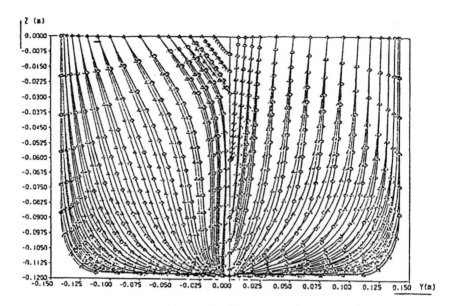

Figure 8: Hull form variation in parent SSPA-720 by longitudinal shift of centroid, Lackenby method (LCB shifted by ±2% of length)

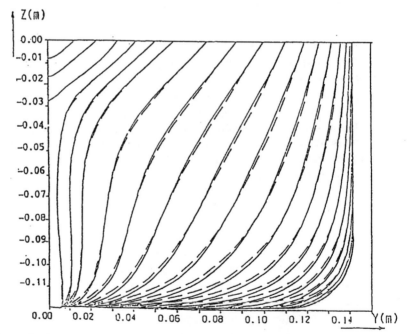

Figure 9: Hull form variation in parent SSPA-720 by vertical shift of centroid (VCB shifted by +1% of draught), dashed line after shift

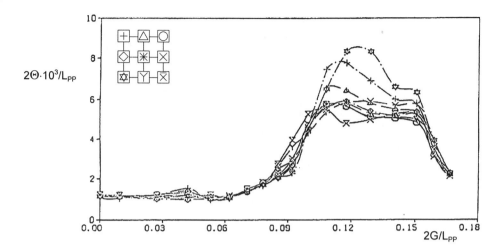

Figure 10: Momentum thickness of ship boundary layer as a function of station girth at $x = 0.832\ L$ from the bow

In the current case some of the shape variations clearly show a smaller momentum thickness than others. The shape with a centroid shift aft and down appears to be the most favourable with regard to viscous resistance.

5.2 Total Resistance Minimization for a Slender Hull

In a study by Larsson et al. [16] the minimization of the total resistance was applied, inter alia, to a slender ship form, the mathematically defined Wigley hull (Figure 11). In this investigation the hull form was represented by a B-spline surface. Cerrtain nodes in the B-spline mesh, denoted by arrows in Figure 11, were treated as free variables (about two times four at four ship stations in Figure 11). The other nodes at those stations were dependent variables following the change of the master nodes in a certain mode. Thus the overall hull shape was controlled by only eight free variables in this instance. The optimization was performed by an algorithm called the Method of Moving Asymptotes.

This slender hull shapes at a medium speed ($F_n = 0.25$) has a significant share of wave-making resistance. The flow was analyzed by the CFD code SHIPFLOW (developed by Larsson et al.), which combines a potential flow solver with free surface with a boundary layer regional solver and a Navier-Stokes solver for the stern domain. In this instance only the boundary layer solution in conjunction with the Squire-Young formula was used. Constraints were imposed on the displacement so that volume could be shifted, but not removed.

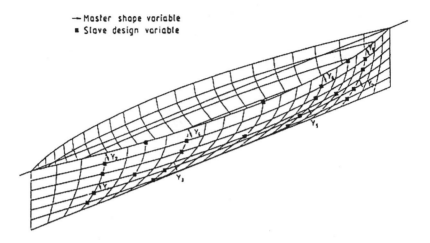

Figure 11: The Wigley hull with master and slave nodes

Figure 12 shows the history of the optimization runs. The results show fast convergence in only eight iterations in a monotonic trend. The wave resistance is reduced by 28%. The total resistance of the Wigley hull decreases by 7% despite a certain increase in wetted area. The potential for reducing wave making resistance in practice is often more significant than for viscous drag. Similar energy saving potentials exist for conventional cargo ships at the early design stage.

A word of caution is in place with regard to the reproducibility of optimization improvements in experiments or full scale ships. Recent results by Janson and Larsson [25] indicate that the advantages predicted by CFD are not always confirmed by measurements. This can be particularly true for those flow phenomena which are over-predicted by current flow solvers many of which neglect the effect of viscosity upon wave flows. The wave-viscous interaction effects should be taken into account in optimization studies and need to be better understood.

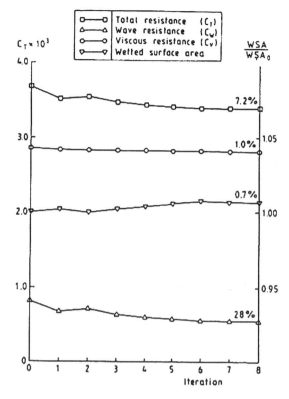

Figure 12: Optimization history [16]

6 Summary

Hull form resistance optimization can be performed using CFD analysis methods. The resulting optimization problem is of Nonlinear Programming type. Shape variation can be based on a systematic form parameter representation of hull forms, supported by modern polynomial spline surface representations. Shape variation, flow analysis by CFD, and design evaluation need to be closely linked. Systematic improvements in shape design are thereby feasible. They often yield worthwhile practical improvements. The accuracy and reliability of CFD solvers should be further improved. But design conclusions can be drawn from their results today well before all complex flow phenomena are predicted with absolute accuracy and precise detail.

References

[1] Froude, W.: On Experiments upon the Effect Produced on the Wave-Making Resistance of Ships by Length of Parallel Middle Body, Trans. Inst. Naval Architects, vol. 18, 1877.

[2] Froude, R. E.: On the Leading Phenomena of the Wave-Making Resistance of Ships, Trans. Inst. Naval Architects, vol. 22, 1881.

[3] Lord Kelvin: Ship Waves, Trans. IME, London, 1887.

[4] Michell, J. H.: The Wave Resistance of a Ship, Phil. Magazine, London, 1898.

[5] Taylor, D. W.: Calculations for Ships' Forms, and the Light Thrown by Model Experiments upon Resistance, Propulsion and Rolling of Ships, Trans. Intl. Eng. Congress, San Francisco, 1915.

[6] Weinblum, G.: Systematische Entwicklung von Schiffsformen, Jahrb. STG, Berlin, 1953.

[7] Wigley, C.: Ship Wave Resistance, Trans. NECI, vol. 47, 1931.

[8] Havelock, T. H.: Wave Resistance Theory and its Applications to Ship Problems,
Trans. SNAME, vol. 59, 1951.

[9] Inui, T./Takahei, T./Kumano, M.: Tank Experiments on the Wave-Making Resistance Characteristics of the Bulbous Bow, Part 1, Trans JSNA, vol. 108, 1960.

[10] Lin, W.-C./ Webster, W. C./Wehausen, J. V.: Ships of Minimal Total Resistance, Proc. Intl. Seminar in Wave Resistance, Univ. of Michigan, Ann Arbor, 1963.

[11] Hess, J. H./Smith, A. M. O.: Calculation of Non-Lifting Potential Flow about Arbitrary Three-Dimensional Bodies, Douglas Aircraft Rept. No. E.S.40622, March 1962.

[12] Nagamatsu, T./Sakamoto, T./Baba, E.: Study on the Minimization of Ship Viscous Resistance, Soc. Naval Arch. Japan, 1983.

[13] Larsson, L./Patel, V. C./Dyne, G. (eds.): Ship Viscous Flow, Proc. of 1990 SSPA-CTH-IIHR Workshop, Gothenburg, FLOWTECH Intl. AB, 1991.

[14] Ni, S. Y.: Higher Order Panel Methods for Potential Flow with Linear or Nonlinear Free Surface Boundary Condition, Dissertation, Chalmers Univ., Gothenburg, 1987.

[15] Jensen, G.: Calculation of Wave Resistance for Practical Ship Forms, in German, Jahrb. STG, 1988.

[16] Larsson, L./Kim, K.-J./Esping, B./Holm, D.: Hydrodynamic Optimization Using SHIPFLOW, Proc. PRADS '92 Conf., Newcastle, 1992.

[17] Webster, W.C.: The Flow about Arbitrary, Three-Dimensional Smooth Bodies, J. of Ship Research, Dec. 1975.

[18] Söding, H.: Fortschritte bei der Berechnung von Potentialströmungen, STG-Workshop on Ship Hydrodynamics, Potsdam, 1995.

[19] Dawson, C.W.: A Practical Computer Method for Solving Ship-Wave Problems, 2nd Intl. Conf. on Num. Ship Hydrodynamics, Berkeley, 1977.

[20] Jensen, G./Söding, H./Mi, Z.X.: Rankine Source Methods for Numerical Solutions of the Steady Wave Resistance Problem, 16th Symp. Naval Hydrodynamics, Berkeley, 1986.

[21] Anon.: Proc. Sixth Intl. Conf. on Numerical Ship Hydrodynamics, Iowa City, 1993.

[22] Anon.: Proc. CFD Workshop Tokyo, Ship Research Institute, Tokyo, 1994.

[23] Sotiropoulos, F./Patel, V.J.: Application of Reynolds-Stress Transport Models to Stern and Wake Flows, J. of Ship Research, vol. 39, no.4, 1995.

[24] Kim, S.-Y.: A Contribution to Drag Reduction of Ships by Systematic Form Variation, in German, Dissertation, TU Berlin, 1987.

[25] Janson, C.-E./Larsson, L.: A Method for the Optimization of Ship Hulls from a Resistance Point of View, 21st Symp. on Naval Hydrodynamics, Office of Naval Research, Trondheim, 1996.

Superplastic Protective Structures

Leonid I. Slepyan†‡ and Mark V. Ayzenberg‡

†Department of Solid Mechanics, Materials and Structures,
Tel Aviv University, 69978 Tel Aviv, leonid@eng.tau.ac.il
‡Institute for Industrial Mathematics, Beer Sheva, 84213, Israel,
ayzenbe@math.bgu.ac.il

Abstract

Superplastic protective structures (SPPS) designed from regular materials are dedicated for protection against impact, penetration or explosion. An increase in resistance of the structure to dynamic loading is provided by a high level of the total strain energy of SPPS under extension. Some principles of creating SPPS and some aspects of the effectiveness of their application are considered. The comparative role of two main factors – the strength and the limiting strain energy of the structure – is examined. In this context, the following processes are analyzed: collision of a tank filled by a fluid with a rigid obstacle, impact of a projectile onto a plate, a spherical shell under an internal explosion, and perforation of a composite armor.

1 Introduction

Consider collision of a tanker, filled with oil, with an obstacle. The cause of a breach in the hull is not so much the direct action of the collision force as internal pressure resulting from deformation of the hull. This pressure can reach a high level in the dynamics of oil, which being an almost incompressible fluid strongly resists a decrease of its volume under deformation of the hull during the collision. Thus, in this process extension of the material of the hull is the most important type of strain. Under these conditions, two factors define the resistance of the hull to the impact: the strength and the limiting strain energy of the material under extension, and the latter can be the main factor.

The role of these two characteristics depends on the type and parameters of the impact which, in turn, depend on the conditions of the impact and features of the bodies. To show this consider collision of two identical bodies. In the case of 'purely elastic' impact there is no strain energy after the impact. The duration of the impact is short, and the amplitude of the force is high. In contrast to this, the impact duration can be long and the force can be low in the case of 'purely inelastic' impact. However, in the latter case all the initial kinetic energy goes into the strain energy. In such a situation, a structure is better which can accumulate as much energy as possible before it is broken. SPPS are just the ones possessing the mentioned quality. This consideration in full measure concerns a tanker hull, as well as a tank with a fluid under collision.

The next example for the use of SPPS is a protective shell for an atomic power station, chemical plant, etc. Having a high level of strain energy under extension, it can resist possible explosions and protect the environment from pollution and/or contamination. The problem is to create a structure with an optimal combination of strength and limiting strain energy. This is a well-known fact, and many works have been devoted to the investigation of dynamic resistivity (see, for instance, [1], and [2]. As can be seen in these books, the main attention has been paid to dynamic compression. In contrast to this, we concentrate on dynamic extension, where SPPS are preferable (see [3], [4], [5], [6] and [7]). Some principles of SPPS design are presented by Cherkaev and Slepyan [7]. It is shown that the strain energy of a material or of a construction under extension can be increased by using special structures of ordinary elements. The possibilities are discussed for increasing the strain energy density in a sample and the total strain energy in a construction.

The theoretical local characteristics of the material provide some natural bounds to the ability of the construction to resist dynamic loading. However, in practice a construction is destroyed long before the limit of material resistance is achieved. The real limiting mechanism is *the instability* of the strains which manifests as the localization of strains. Therefore the obvious way to significantly increase the quality of the construction is to stabilize somehow the process of damaging. This involves the need to increase the energy density consumed in the material and to distribute the damage (the large strains) throughout a large part of the construction.

In the present paper, some principles of creating SPPS and some aspects of the effectiveness of their application are considered: the comparative role of the two main factors, the strength and the limiting strain energy of the structure, and other parameters which define the SPPS quality. Last, the role of an exstensible component in a composite armor is examined. In this context, the following processes are analyzed: collision of a tank filled by a fluid with a rigid obstacle, impact of a projectile onto a thin plate, a spherical shell under an internal explosion, and perforation of a composite armor with extensible plies.

2 Strength and Limiting Strain Energy

2.1 Fluid-filled tank under impact

Consider a flat-ended cylindrical tank under a longitudinal impact on a rigid barrier. Below R, L, h, ρ_0, K and v are the radius and length of the tank, the thickness of the shell, density and bulk modulus of the fluid, and the speed of the impact, respectively. The influence of the mass density of the shell on the process of the impact is neglected because of a small ratio of the shell weight to the fluid weight as assumed. Let us discuss the two following limiting cases:

(i) *The rigid tank.* In this limiting case, the total strain energy of the tank's material is expected to be negligible in comparison to that of the fluid. It is the case of a high level of stresses in the shell under tension. The fluid's strain energy density is $W = p^2/(2K)$, where p is pressure. Then the kinetic energy density is $T = \rho_0 v^2/2$, and the amplitude of the pressure in the step wave in the fluid is $p = \rho v c_0$, where $c_0 = \sqrt{K/\rho_0}$ is the sound velocity in the fluid. This pressure produces the tension stresses in the cylindrical shell: $\sigma = (R/h)\rho_0 v c_0$.

Under these conditions, the 'strength-limiting' thickness, h_e, which corresponds to the limiting strength in the material, $\sigma = \sigma^*$, is

$$h_e = R\frac{\rho_0 v c_0}{\sigma^*} = R\frac{\rho_0 v^2}{\sigma^*}\frac{c_0}{v}. \tag{1}$$

(ii) *The plastic tank.* Now consider the opposite limiting case, when the main absorber of the kinetic energy of the fluid is plastic strain of the tank's shell. This case corresponds to a low level of the stresses and a high level of the limiting strain energy of the shell. The energy of the fluid's compression is now considered to be negligibly small. Equating the strain energy of the tank to the fluid's kinetic energy one obtains the following 'energy-limiting' thickness of

$$h_e = R\frac{\rho_0 v^2}{4A^*}, \tag{2}$$

where A^* is the limiting strain energy or the specific work of plastic deformation.

It is obvious that h_e essentially differs from h_s: their ratio depends on the impact velocity, as well as on the ratio of the limiting strength of the first material to the specific strain energy of the second:

$$k = \frac{h_p}{h_e} = \frac{1}{4}\frac{v}{c_0}\frac{\sigma^*}{A^*} \tag{3}$$

This result shows that for a moderate impact velocity when $v \ll c_0$ (for oil $c_0 \approx 1300 m/s$), the plastic material should be much better: the use of this material allows the thickness of the shell to be decreased if the limiting strain energy, A^*, is not too small in comparison with the limiting stress, σ^*, of the rigid material (see Fig.1, where the straight line corresponds to the limiting thickness of the rigid shell, while the curves represent the thickness of the plastic shell).

Taking into account both the strain energy of the fluid and the cylindrical shell (assuming the latter to be the product: $A^* = \sigma^* \epsilon^*$) one has the energy balance equation

$$2\pi R h L \sigma \epsilon + \frac{\pi R^2 L p^2}{2KR^2} = \frac{\rho v^2}{2}\pi R^2 L, \quad p = \frac{h}{R}\sigma. \tag{4}$$

It follows from this that

$$\frac{h}{R} = \frac{2K\epsilon}{\sigma}\left(\sqrt{1 + \frac{\rho v^2}{4K\epsilon^2}} - 1\right) \tag{5}$$

with the asymptotes

$$\frac{h}{R} \sim \frac{\rho vc}{\sigma}\left(\frac{K\epsilon^2}{\rho v^2} \to 0\right), \quad \frac{h}{R} \sim \frac{\rho v^2}{4A^*}\left(\frac{\rho v^2}{K\epsilon^2} \to 0\right), \tag{6}$$

which correspond to the limiting cases (1) and (2) obtained above.

Formulas (5) and (6) are the basis for the criterion of effectiveness of the material. For this goal, a criterial coefficient can be introduced: specific weight, cost of unit volume of the material, etc., depending on what is to be minimized.

2.2 A concentrated impact

Consider an 'infinite' plate impacted by a cylindrical projectile with radius r, length l and mass M moving with velocity v along a normal to the plate. In estimating the plate's strength under these conditions the two limiting formulations are used, the same as above.

First consider a rigid plate, whose limiting strain energy (which can be produced by the impact) is negligibly small in comparison with the kinetic energy of the projectile. In this case, the projectile is assumed to be like a jet, and the plate to be intact. The impulse, I, and force, F, acting on the plate can be estimated as $I = Mv$ and $F = Mv^2/l$, where l/v is the duration of the impact. Assume that the strength of the plate is limited by shear stresses on the cylindrical boundary of the possible plug produced by the projectile. One has $F = 2\pi rh\tau^*$, where τ^* is the limiting shear stresses. The limiting thickness of the plate, h, is tentatively presented as

$$h = h_e = \frac{Mv^2}{2\pi rl\tau^*} = r\frac{\rho v^2}{2\tau^*}, \tag{7}$$

where ρ is the density of the projectile material. In the other extreme case the kinetic energy of the impactor is expected to be transformed into the plastic strain energy of the plate whose density is limited by A^*. The latter is distributed nonuniformly, it is diminishing to the periphery (from the impact spot). In this connection, let us introduce the effective radius of the area of considerable plastic deformations, R, which can be large in comparison to the radius of the contact area r. In this case, the projectile is expected to be rigid, and the limit thickness of a plastic plate, h_p, and the sought ratio of thicknesses are

$$h = h_p = \frac{Mv^2}{2\pi R^2 A^*}, \quad \frac{h_p}{h_e} = \frac{rl\tau^*}{R^2 A^*} \tag{8}$$

Thus the limiting strain energy can be the criterial value when not only its density, A^*, is large enough but also when the effective radius of the area involved in considerable plastic deformations is large. It can also be seen from (8) that the plastic plate is more effective at small l, that is, under sufficiently brief loading. The latter can also be characteristic for a shock wave, whose action will be estimated by considering a uniformly loaded spherical shell.

2.3 Spherical shell under explosion wave action

Consider a spherical shell of radius R, thickness h and density of the material ρ. The shell is under an internal spherical shock wave (the symmetrical problem is under consideration). The wave is assumed to be short enough to warrant the impulse-type formulation of the problem. Let I be the impulse per unit area.

First consider an elastic-brittle shell made of a material with the Young modulus E and limiting stress σ^*. The kinetic energy of the shell (per unit area) produced by the shock wave is $I^2/(2\rho h)$. The maximal strain energy is $(\sigma^*)^2 h/(2E)$. Thus the limiting thickness is

$$h = h_e = \frac{Ic}{\sigma^*} \quad (c = \sqrt{\frac{E}{\rho}}) \tag{9}$$

In the second limiting case, the specific kinetic energy goes into the plastic strain work which is limited by A^*, and one has

$$h = h_p = \frac{I}{\sqrt{2\rho A^*}}, \quad k = \frac{h_p}{h_e} = \frac{\sigma^*}{\sqrt{2EA^*}} = \sqrt{\frac{\sigma^*}{2E}}\sqrt{\frac{\sigma^*}{A^*}}. \tag{10}$$

This not unexpected result shows that it is sensible to increase not the strength of the material, but the specific limiting energy, because the first multiplier in the right hand part of (10) is usually rather small, and the plastic shell can be thinner. Of course, this conclusion does not concern a repeatedly loaded shell.

3 Some Ways of Creating SPPS

The main factor which limits the strain energy is instability of plastic strain caused by the localization of strain when the sample is under extension. Under the condition of the localization, only a small part of the structure is involved in a large strain, and the total strain energy turns out to be very small. An increase of the specific limiting strain energy can be achieved by a special 'microstructure' which prevents strain localization. In addition to this, the optimally designed structure must provide the most possible uniformity in the distribution of large strains – to increase the total limiting strain energy of the structure as a whole.

Possible types of SPPS, which possess these properties, are systems composed of friction elements, elements of moving plastic zones, elements with a thick net of cracking, etc. A good example of such composition is the so-called waiting element structure, which is multi-stable under extension [7]. In such a structure, a thick net of breaking is realized under a large extension. This leads to an essential increase in the total strain energy.

The next way is a transformation of the unstable extension on the 'macro-level' into a more stable type of strain on the 'micro-level'. A plastic helix is shown in [5] (also see [7]) to be the structure which provides this transformation. In this case, a significant increase of the limiting strain energy is achieved by replacing a straight, plastic rod (wire) by the proper helix of the same amount of the material. In opposite to the straight rod, the lengthening of the helix is realized by the wire twist, then by the bending, and at last, by the extension. Just at the last stage the tension force reaches maximum, and stability of large strain is obtained. Theoretical estimations confirmed by experiments show that the limiting strain energy of the proper plastic helix is up to 20 times greater than that of the straight rod.

The above-mentioned principles of SPPP creation can be used in many cases; one nontrivial application is shown below.

4 Composite, Metal-Fabric Armor

The tentative estimations obtained in Section 2 can be a basis for the detailed design of a structure. In this way, precise calculations allow to determine the true distribution of strains under the given conditions of impact. At the Institute for Industrial Mathematics (Beer-Sheva, Israel), software has been developed for simulating penetration into a composite armor. This software allows to optimize the armor as SPPS-related structure.

Composite armor is considered, subjected to high-speed impact and penetration by a small-arms bullet. The structure consists of a primary armor (hard metal plate) and of a secondary multilayered armor manufactured of strong but flexible, extensible fabric plies. The role of the primary armor is not only to decrease the impact energy, but also to maximally deform the bullet head shape. In contrast to this, the secondary armor must stop the bullet with decreased speed and deformed shape. This can be achieved thanks to the ability of the fabric to consume the remaining kinetic energy of the bullet. Such combination armors are used in practice for protection of light combat or cash-carrying vechicles, security doors, superstructures, cabins and control rooms in boats and small ships, etc.

Some results shown in Fig. 2, 3 relate to computer simulations treated with the aim to actually design a target of high stopping power. It is associated with either lesser penetration depth or lesser bullet residual velocity in the case of perforation. These armors consist of available materials at our disposal: aluminum alloy plate of $8mm$ thickness and $250BHN$ hardness as the primary armor, and two materials intended for the secondary armors, manufactured of special fabrics: (a) "rigid fabrics", $Kevlar$-49, and (b) "pliable fabrics", $Kevlar$-29, with the same ply areal density: $0.5\ kg/m^2$, thickness: $0.5\ mm$, and the tensile strength limit:

2.76 GPa. These fabrics are linear-elastic-brittle and have different Young moduli: (a) $E = 131 GPa$, and (b) $E = 62 GPa$ (see [8]).

Comparative estimations of the stopping power of these targets can be seeeen in Fig.2 for the case of an AK-47 bullet impacting onto the armors with the initial velocity 740 m/s. After perforation of the primary armor, the bullet loses about half of its initial kinetic energy: its residual velocity drops to 560.5 m/s. The secondary armor stops the bullet or is perforated by it depending on the ply number and deformability. In the discussed case, the stopping power of the pliable armor (b) proves better than that of the rigid one (a) because the former possesses the capacity to absorb more impact energy for its deformation before breaking. In Fig.3, the configuration can be seen of armors (of the same thickness, the ply number $N_p = 40$, and weight) realized under penetration, (b), and perforation, (a). The ply positions correspond to boundaries between light and dark stripes.

Acknowledgements: This research was supported by grants No. 9673-1-96 from the Ministry of Science, Israel, and No. 94-00349 from the United States – Israel Binational Science Foundation (BSF), Jerusalem, Israel.

References

[1] Chou Tsu-Wei *Microstructural Design of Fiber Composites*. Cambridge, Cambridge University Press (1992).

[2] Jones, N. and Wierzbicki, T. (Ed.) *Structural crashworthiness and failure*. Elsevier Applied Science. University Press (Belfast) Ltd (1993).

[3] Slepyan, L., Slepyan, V. and Scherbinin, V. The hanger system for a protective covering. The USSR Invention No. 4803025/33-30876, MKI5 E04V7/14, 06.03.90 (1990).

[4] Slepyan, L., Slepyan, V. and Scherbinin, V. The netted shell. The USSR Invention No. 4897633/33/528, MKI4 E04V7/10, 02.01.91 (1990).

[5] Slepyan, L. Some principles of prevention of plastic strain instability. Abstracts of VII All-Union Congr. Theor. Appl. Mech., Moscow (1991), p. 320.

[6] Slepyan, L. Impact and penetration. Rigid and extensible structures. In: *Structures under Shock and Impact - III*. Proc. 3rd Intern. Conf. Comp. Mech. Publ. (1994), 281-288.

[7] Cherkaev, A. and Slepyan, L. Waiting element structures and stability under extension. *Intern. Journal of Damage Mechanics* **4**, 1 (1995), 58-82.

[8] Schipholt, B.: Nomex and Kevlar aramid fibres for defence purposes. Inf. Bull. Du Pont de Normous International S. A. (1980), Genova.

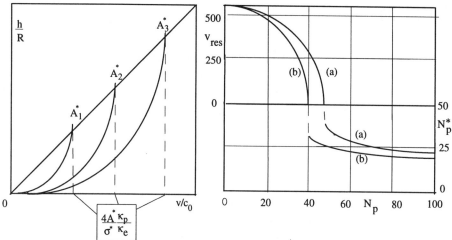

Figure 1: Limiting thickness of tank's shell vs. relative impact velocity v/c_0

Figure 2: Target stopping power vs. number of plies, N_{pl}, in the secondary armor:
(a) – Kevlar-49, (b) – Kevlar-29
v_{res} – residual velocity after perforation (m/s)
N_p^* – number of perforated plies

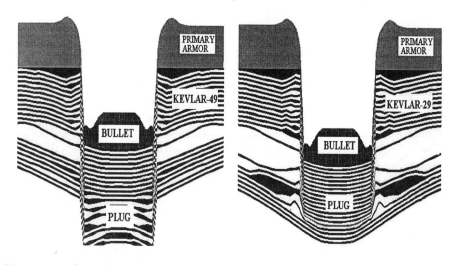

(a) target perforated at t = 36.5 mcs, v_{res} = 266.7 m/s

(b) bullet stopped at t = 82.5 mcs, 33 plies are broken through

Figure 3: Metal-fabrics composite targets (N_p = 40) vs. an AK-47 bullet

Trial Methods for Nonlinear Bernoulli Problem

Kari Kärkkäinen   Timo Tiihonen
University of Jyväskylä, Department of Mathematics,
P.O. Box 35, FIN-40351 Jyväskylä, Finland,
ktkar@math.jyu.fi, tiihonen@math.jyu.fi

Abstract

In this article we consider a free boundary problem which is related to formation of waves on a fluid surface (for example the ship waves). We study the possibility to construct 'trial' methods where one solves a sequence of standard flow problems formulated in different geometries that converge to the final free boundary. Furthermore, we use the shape optimization techniques to analyse the convergence of the fixed point iteration near a fixed point. For stream function case we conclude that the fast convergence can be obtained by using non-standard boundary conditions and we present numerical results to confirm the analysis.

1 Introduction

The main purpose of this paper is to introduce some numerical methods inspired by shape optimization techniques for free boundary problems. One possible application of this analysis will be the calculation of the free water surface around a moving ship, which is in particular interest of the authors.

The two most popular approaches for numerical solution of free fluid surfaces seem to be the use of linearised conditions or the convergence to a steady state in transient problem with moving boundary. In our approach we first formulate the free boundary problem as a shape optimization problem. The advantage of this procedure is that we can use the tools of 'shape calculus' to analyse the behaviour of the problem and to generate fast methods to compute the solution. The methods to be developed are based on solving a sequence of 'standard' flow problems and are in principle independent of the discretization method.

To understand the questions related to solving the nonlinear problem we formulate very simple problem that describes the inviscid irrotational stationary flow with free boundary. Although we neglect the viscous behaviour of the fluid we can still gain some insight of the behaviour of flow with free boundary.

The potential function formulation can be written in dimensionless form as

$$-\Delta u = 0 \quad \text{in } \Omega \tag{1a}$$

$$\frac{\partial u}{\partial n} = 0 \quad \text{on } \Sigma, \tag{1b}$$

$$\frac{\partial u}{\partial n} = 0 \quad \text{on } \Gamma_b, \tag{1c}$$

$$|\nabla u|^2 = 2 - \frac{1}{\text{Fr}^2}(z-1) =: \theta^2 \quad \text{on } \Sigma. \tag{1d}$$

Here Fr is the dimensionless Froude number $\text{Fr} = \frac{U^2}{gL}$ where g is gravity, U and L are velocity and length respectively. In $2D$ we can use the stream function formulation, which is

$$-\Delta u = 0 \quad \text{in } \Omega \tag{2a}$$

$$u = 1 \quad \text{on } \Sigma, \tag{2b}$$

$$u = 0 \quad \text{on } \Gamma_b, \tag{2c}$$

$$\left(\frac{\partial u}{\partial n}\right)^2 = \theta^2 \quad \text{on } \Sigma. \tag{2d}$$

These problems are very popular both in naval hydrodynamics [2] and in mathematics where quite number of results are known. Especially for linearised versions [8] or in absence of gravity [6], [1].

Here we are going to proceed as follows. In the section 2 we discuss the energy principle for this problem. In particular we analyse the second derivatives with respect to geometry of the energy from the point of view deriving solution algorithms. In the section 3 we present a shape optimization formulation, where the kinematic condition is fulfilled in least square sense. In the section 4 we introduce some solution algorithms inspired by the analysis of the shape optimization problem and present some numerical results.

2 Shape variational principle

We shall analyse the energy principle for (1) presented in [10] or [11]. For simplicity we confine ourselves in a bounded region Ω. We split the boundary of Ω to four different boundaries, $\partial\Omega = \bar{\Sigma} \cup \bar{\Gamma}_i \cup \bar{\Gamma}_b \cup \bar{\Gamma}_o$. We choose the functional space where we are going to work as $\mathcal{V} = \{v \in H^1(\Omega) \,|\, v = 0 \text{ at } \Gamma_o\}$. This is equivalent to imposing zero vertical velocity on artificial outflow boundary Γ_o. Let us now introduce an energy functional E,

$$E(\Omega, u) = \frac{1}{2}\int_\Omega |\nabla u|^2\, dx - \int_{\Gamma_i} \mathbf{n}_x u\, ds + \frac{1}{2}\int_\Omega \theta^2\, dx. \tag{3}$$

The variation of E with respect to variable u reads now as

$$dE(\Omega, u; v) = \int_\Omega \nabla u \cdot \nabla v\, dx - \int_{\Gamma_i} \mathbf{n}_x v\, ds, \qquad (4)$$

where v is any test function from test function space $\mathcal{V}$. So at any critical point we have $dE(\Omega, u; v) = 0$ for all $v \in \mathcal{V}$,

$$\int_\Omega \nabla u \cdot \nabla v\, dx = \int_{\Gamma_{in}} \mathbf{n}_x v\, ds \quad \text{for all } v \in \mathcal{V}. \qquad (5)$$

This implies (1a)-(1c) and specifies the inflow velocity on Γ_i.

Now we differentiate energy E with respect to Ω. This is done by introducing a transformation $T_t x = x + tV(x)$ with smooth velocity field V that maps Ω to $\Omega_t := T_t(\Omega)$ and by computing the derivative $dE(\Omega, \varphi; V) = \lim_{t \to 0} \frac{E(\Omega_t, \varphi) - E(\Omega, \varphi)}{t}$, see [9], [3], [5], [4] for further details.

We get an expression for the shape derivative of E, [7]

$$dE(\Omega, \varphi; V) = -\frac{1}{2} \int_\Sigma |\nabla u|^2 \langle V, \mathbf{n} \rangle\, ds + \frac{1}{2} \int_\Sigma \theta^2 \langle V, \mathbf{n} \rangle\, ds. \qquad (6)$$

We conclude by following proposition which has been already suggested in [11] and [10]:

Proposition 1. *Let (Ω, u) be a critical point of energy E satisfying $(\Omega, u) \in \mathcal{D} \times \mathcal{V}$, $\mathcal{D}$ is an appropriate set of domains. Then (Ω, u) is a solution of the free boundary problem (1).*

Thus, to find one solution to FBP, it is sufficient to find a critical point of E. In order to be able to do that we have to analyse the second order optimality conditions so that we know how E behaves near the critical points.

We now denote by P the Neumann to Dirichlet map which is defined by

$$P\mu = v\Big|_\Sigma$$

for v being solution of

$$-\Delta v = 0 \quad \text{in } \Omega,$$
$$\frac{\partial v}{\partial n} = \mu \quad \text{on } \Sigma,$$
$$v = 0 \quad \text{on } \Gamma_b.$$
$$\frac{\partial v}{\partial n} = 0 \quad \text{on } \partial\Omega \setminus (\Sigma \cup \Gamma_b).$$

By S we denote the inverse of P, i.e. the Dirichlet to Neumann map $S = P^{-1}$.

Following proposition is suggested in [7]

Proposition 2. *At any critical point of the energy E the shape Hessian of E has the expression*

$$d^2 E(\Omega, u; V, W) = \int_\Sigma \nabla_\Sigma \cdot (\langle V, \mathbf{n} \rangle \nabla_\Sigma u) \, P \left(\nabla_\Sigma \cdot (\langle W, \mathbf{n} \rangle \nabla_\Sigma u) \right) \, ds \\ + \int_\Sigma \left(\theta^2 H - \frac{\mathbf{n}_z}{\mathsf{Fr}^2} \right) \langle W, \mathbf{n} \rangle \langle V, \mathbf{n} \rangle \, ds, \quad (7)$$

where H is the mean curvature of Σ and $\mathbf{n}_z$ is the upward component of the normal vector. By ∇_Σ we denote the tangential derivative.

We observe that the Hessian is a continuous mapping from $H^{\frac{1}{2}}(\Sigma) \times H^{\frac{1}{2}}(\Sigma) \to \mathbb{R}$. This implies, among the other things, that straightforward discretisation of the problem leads to discrete optimization problem where the condition number is inversely proportional to grid size.

Closer look to the shape Hessian tells us how the energy E behaves near the solution. For big Froude numbers, say $\mathsf{Fr} \gg 1$, we can see that the last term $\frac{\mathbf{n}_z}{\mathsf{Fr}^2}$ is small and the first term majorises the Hessian if the curvature of the surface Σ is near zero. Thus the operator is positive definite. Then we have a local minimum in every critical point. So we can use optimization to achieve the critical point. But for small Froude numbers the last term gets bigger and so we lose the positive definiteness and the critical point turns to a saddle point. This means that the free boundary is not a minimiser of the 'energy' of the system.

Similar results can be obtained for stream function formulation.

3 Shape optimization formulations

As we mentioned in the previous section the variational principle works only for big Froude numbers for current problem. The case of small Froude numbers is the one which is more interesting from the viewpoint of applications. So here we consider problems where the functional is constructed so that any solution of our free boundary problem is a minimum point.

Now we restrict ourselves to $2D$ model and stream function formulation. We want to satisfy the kinematic condition in the least squares sense. That is, we look for Σ that minimises

$$J(\Sigma) = \frac{1}{2} \int_\Sigma (u - 1)^2 \, ds \quad (8)$$

under constraint that u is the stream function satisfying modified dynamic condition

$$\int_\Omega \nabla u \cdot \nabla v \, dx = \int_\Sigma [\alpha(u - 1) - \theta] v \, ds \quad \text{for all } v \in \mathcal{V}, \quad (9)$$

where $\mathcal{V} = \{v \in H^1(\Omega) \,|\, v = 0 \text{ on } \Gamma_b\}$. J is well defined for all u in $\mathcal{V}$ and for regular enough Σ. Parameter α is here an additional parameter which can be chosen freely as $u = 1$ at the solution. With α we can control the properties of the optimization problem. For example we have the following propositions [7]:

Proposition 3. *If we choose $\alpha = 0$ in (9) and we have $J(\Sigma) = 0$ the shape Hessian of the cost J reads*

$$d^2 J(\Sigma; V, W) = \int_\Sigma \left[P \left(\theta H \langle V, \mathbf{n} \rangle + \frac{\partial \theta}{\partial n} \langle V, \mathbf{n} \rangle \right) + \theta \langle V, \mathbf{n} \rangle \right] \\ \cdot \left[P \left(\theta H \langle W, \mathbf{n} \rangle + \frac{\partial \theta}{\partial n} \langle W, \mathbf{n} \rangle \right) + \theta \langle W, \mathbf{n} \rangle \right] ds \quad (10)$$

This Hessian is bounded in $L^2(\Sigma) \times L^2(\Sigma)$.

Proposition 4. *If we choose $\alpha = H - \frac{n_z}{Fr^2 \theta^2}$ in (9) and we have $J(\Sigma) = 0$ the shape Hessian of the cost J reads*

$$d^2 J(\Sigma; V, W) = \int_\Sigma \theta^2 \langle V, \mathbf{n} \rangle \langle W, \mathbf{n} \rangle \, ds \quad (11)$$

So here we have gained the best possible Hessian for the problem – θ^2 times identity operator in $L^2(\Sigma)$.

Another way to formulate the stream function model is to set up a cost functional for the dynamic boundary condition. This formulation leads again to an unbounded Hessian as in the potential flow case, so the better way is to formulate the problem by this kinematic condition. We should remark here that proposition 4 can not be extended for potential function formulation in $3D$ as we can not impose the nonlinear dynamic condition.

4 Algorithms and numerical results

As shown above, the free boundary problem can be formulated as a shape optimization problem (in several ways). There are two main practical difficulties in optimisation formulation. First the optimisation routines can not be easily combined with black box flow solvers. Secondly, efficient optimization algorithms require gradient information that is not always available. Hence we are led to study fixed point algorithms that are simplified versions of algorithms based on shape optimization.

In the case Fr > 1 the shape variational principle provides us a first candidate for a fixed point algorithm as it leads to minimisation problem with gradient depending explicitly on the solution of the state problem alone. Thus applying steepest descent method with constant step size will give an explicit fixed point type algorithm. The free boundary Σ is to be moved to the direction of the negative gradient $-\nabla E$ which can be easily evaluated once the state problem is solved. However, the iteration will not converge in the continuous case. This is because the Hessian of the problem is unbounded and the steepest descent method will diverge with any constant step length. In the discrete case the step length (and hence the convergence rate) has to be decreased as a function of grid refinement.

A possible remedy is to apply preconditioning, that is, to change the norm in the space of design variables. From proposition 2 we know that the shape Hessian maps H^l to H^{l-1}. Hence a good preconditioner should map H^{l-1} into H^l. One possible choice is $(\alpha I + S)^{-1}$ where α is a free parameter that can be chosen to speed up the convergence. This can be also viewed as an approximate Newton method.

Summarising the above we notice that the variational principles lead to fixed point schemes only if appropriate preconditioning is applied. Moreover, the preconditioning is essentially equivalent to solving an additional state problem with different boundary conditions.

Let us now analyse the shape optimization formulation of section 3. Due to proposition 3 and 4 the Hessian is bounded and hence, steepest descent methods should work without preconditioning. In contrast to variational principles we now have a freedom in selecting the boundary conditions in the state problem, even after the cost function has been selected. As noted in proposition 4, if we choose the boundary condition $\alpha u + \frac{\partial u}{\partial n} = \alpha + \theta$ on Σ for the stream function with $\alpha = H - \frac{\mathbf{n}_z}{Fr^2 \theta^2}$, the gradient of the cost function has the form $dJ_2(\Sigma; V) = \int_\Sigma (u-1)\theta \langle V, \mathbf{n} \rangle$ at the solution and the Hessian reads as $d^2 J(\Sigma; V, W) = \int_\Sigma \theta^2 \langle V, \mathbf{n} \rangle \langle W, \mathbf{n} \rangle$. Hence, the following algorithm is suggested:

Algorithm (Approximate shape Newton method)

1. Set $n = 0$. Choose Σ^0, $\alpha = H - \frac{\mathbf{n}_z}{Fr^2 \theta^2}$.

2. Solve equation $-\Delta u = 0$ in Ω with boundary conditions

$$u = 0 \quad \text{on } \Gamma_b,$$
$$\frac{\partial u}{\partial n} = 0 \quad \text{on } \Gamma_i \cup \Gamma_o,$$
$$\alpha u + \frac{\partial u}{\partial n} = \alpha + \theta \quad \text{on } \Sigma. \qquad (12)$$

3. Set $\Sigma^{n+1} = \Sigma^n - \mathcal{N}(u-1)/\theta$, where $\mathcal{N}$ is a regularization of the unit normal $\mathbf{n}$.

4. If y is small enough, then stop. Otherwise, set $n = n + 1$ and continue from 2.

Of course, H and $\mathbf{n}_z$ in definition of α can be computed only approximatively using the corresponding values at the current iterate. Moreover, to avoid the loss of regularity, H has to be made more regular using some smoothing process. This has the effect that the final observed converge rate can not be quadratic, see [6].

Although this algorithm was motivated using the value $H - \frac{n_z}{\mathsf{Fr}^2 \theta^2}$ for α, other values can be tried also, at least in the case $\mathsf{Fr} > 1$. The convergence of the algorithm relies on u'_V being 'small'. If we consider the case $\alpha = 0$ for example, we see that $u'_V = P((H\theta - \frac{n_z}{\mathsf{Fr}^2 \theta})\langle V, \mathbf{n}\rangle)$ which is small if the surface is flat ($H \approx 0$) and $\mathsf{Fr} \gg 1$. When $\mathsf{Fr} < 1$, the data is too large for mapping $\langle V, \mathbf{n}\rangle \mapsto u'_V$ to be contractive. Hence the algorithm diverges with $\alpha = 0$. On the other hand, with $\mathsf{Fr} < 1$ and α as in proposition 4, the state problem is indefinite and hence qualitatively different from 'standard' flow problems.

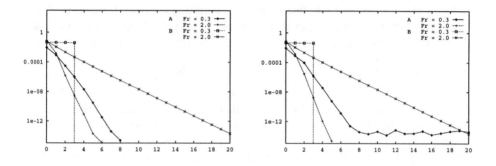

Figure 1: Convergence of the algorithm, 64 × 8 and 128 × 16 elements.

For fixed point method we first used the flow over an obstacle as a test case. We tested two values of α, namely $\alpha = 0$ (method B) and $\alpha = H - \frac{n_z}{\mathsf{Fr}^2 \theta^2}$ (method A) that should formally be the optimal value. The results are summarised in figure 1 for different Froude numbers. In this case the observed convergence rates seem to be independent of the grid size.

Finally algorithm was tested also in the case $\mathsf{Fr} < 1$. The results appeared to be strongly dependent on Froude number. For most values of $\mathsf{Fr} < 0.5$ the convergence was quite rapid. However, in some critical regions the convergence slowed down significantly, or was lost completely. This appears to have some relation to resonance effects related with (strongly reflecting) boundary conditions that were used in the tests.

5 Conclusions

The shape optimization techniques that are presented in this paper seem to be an efficient way in analysing free boundary problems. The energy formulation for potential flow in section 2 turned out to be indefinite for small Froude numbers, so optimisation will fail for that kind of formulation. However, in two dimensional case the free boundary problem was succesfully re-formulated in section 3 and with help of shape analysis the convergence of the numerical method was shown to be formally quadratic. In future we will study different formulations for inflow and outflow boundary conditions and try to extend the shape analysis for more complicated cases.

References

[1] H. Alt / L. Caffarelli: Existence and Regularity for a Minimum Problem with Free Boudary. J. Reine Angew. Math. 325, (1981), pp. 105–144.

[2] P. Bassani / U. Bulgarelli / E. Campana / F. Lalli. The Wave Resistance Problem in a Boundary Integral Formulation. Surveys on Mathematics for Industry 4, 3 (1994), pp. 151–194.

[3] M. Delfour. Shape Derivative and Differentiability of Min Max. In Delfour and Sabidussi [4], pp. 35–111.

[4] M. Delfour / G. Sabidussi, editors. Shape Optimization and Free Boundaries, Kluwer Academic Publishers, 1992.

[5] M. Delfour / J. Zolésio. Anatomy of the Shape Hessian. Ann. Mat. Pura Appl. 158, 4 (1991), pp. 315–339.

[6] M. Flucher / M. Rumpf. Bernoulli's Free-Boundary Problem, Qualitative Theory and Numerical Approximation. submitted, 1995.

[7] K. Kärkkäinen / T. Tiihonen. Trial Methods for a Nonlinear Bernoulli Problem. Report 6, University of Jyväskylä, Department of Mathematics, Laboratory of Scientific Computing, 1996.

[8] N. Kuznetsov / V. Maz'ya. On the Unique Solvability of the Plane Neumann-Kelvin Problem. Mat. Sb. 2, 1989, pp. 425–446.

[9] J. Sokolowski / J. Zolésio. Introduction to Shape Optimization. Springer-Verlag, 1992.

[10] J. Stoker. Water Waves. Interscience Publishers, INC. / New York, 1957.

[11] G. Whitham. Linear and Nonlinear Waves. John Wiley & sons, 1974.

A DIRECT (POTENTIAL BASED) BOUNDARY ELEMENT METHOD FOR THE LIFTING BODIES HYDRODYNAMIC CALCULATION

Bogdan Ganea
ICEPRONAV-SA , Str. Portului, Nr. 19A , 6200 Galati , ROMANIA

Abstract

A direct (potential based) boundary element method (BEM) developed in our company (Romanian Research and Design Institute for Shipbuilding) is briefly presented. It may be applied for the calculation of the potential flow paste the ship hull, also around the lifting bodies like: hydrofoil, nozzle, rudder, propeller. The lifting bodies hydrodynamic calculation has some modelling particularities, e.g. : the wake shape determination , and numerical difficulties, e.g.: the BEM integrals calculation. Therefore the BEM application to such kind of bodies substantially differs from standard BEM application.

1 Theoretical background

The mathematical model of the potential incompressible flow is:

$$\begin{aligned} \Delta \varphi &= 0 |_D \\ (\vec{V}_\infty + \vec{V}_i) \vec{n} &= 0 |_S \\ \vec{V}_i &= \nabla \varphi \end{aligned} \quad (1)$$

φ : perturbation potential ;
V_∞ : undisturbed inflow velocity ;
V_i : perturbation velocity ;
D : fluid domain ;
S : body surface ;
n : outward normal.

A way to solve it is to solve the equivalent integral equation:

$$2\pi\varphi(\vec{y}) - \int_S \varphi(\vec{x}) \frac{(\vec{x}-\vec{y})\vec{n}(\vec{x})}{|\vec{x}-\vec{y}|^3} dS - \int_{S_W} \delta\varphi(\vec{x}) \frac{(\vec{x}-\vec{y})\vec{n}(\vec{x})}{|\vec{x}-\vec{y}|^3} dS_W = -\int_S \frac{\vec{V}_\infty(\vec{x})\vec{n}(\vec{x})}{|\vec{x}-\vec{y}|} dS, \vec{y} \in S \quad (2)$$

Particular for a lifting body is the wake surface S_W, across it there is a potential jump:

$$\delta\varphi = \varphi_-^{TE} - \varphi_+^{TE} \quad . \quad (3)$$

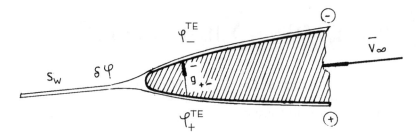

Figure 1: The wake surface (vortex sheet) at the trailing edge of a lifting body

Because dφ is not priori known, it must be determined from the Kutta-Jukovski condition. Noting the trailing edge back-face pressure difference:

$$\delta p^{TE} = (p_- - p_+)\big|_{TE} , \qquad (4)$$

due to the Bernoulli law, the Kutta-Jukovski condition is a non-linear equation:
$$\delta p^{TE} = 0 \qquad (5)$$
which may be solved by means of a specific method, e.g. the Newton iterative method. Because such kind of methods are very complex, for the shake of simplicity may be used the linear form:

$$\delta\varphi = \varphi_-^{TE} - \varphi_+^{TE} + \vec{V}_\infty \vec{g}_{+-}^{TE} , \qquad (5)$$

which is obtained from (5) by mathematical means.
The Kutta-Jukovski condition is specific to the lifting body hydrodynamic calculation and its correct treatment is very important in order to obtain good results.
Using (6), (2) becomes:

$$2\pi\varphi(\vec{y}) - \int_S \varphi(\vec{x}) \frac{(\vec{x}-\vec{y})\vec{n}(\vec{x})}{|\vec{x}-\vec{y}|^3} dS - \int_{S_w} (\varphi_-^{TE} - \varphi_+^{TE}) \frac{(\vec{x}-\vec{y})\vec{n}(\vec{x})}{|\vec{x}-\vec{y}|^3} dS_w =$$
$$= \int_{S_w} \vec{V}_\infty(\vec{x}) \vec{g}_{+-}^{TE} \frac{(\vec{x}-\vec{y})\vec{n}(\vec{x})}{|\vec{x}-\vec{y}|^3} dS_w - \int_S \frac{\vec{V}_\infty(\vec{x})\vec{n}(\vec{x})}{|\vec{x}-\vec{y}|} dS , \quad \vec{y} \in S \qquad (6)$$

To solve it, the body and wake surface will be discretizated in small flat regions (boundary elements - panels) on which surface the unknown potential φ is assumed to be constant. Now the integral equation (7) becomes a linear equations system, easy to solve numerically:

$$\sum_{j=1}^{N_{BEB}} [2\pi\delta_{ij} - I_{ij}^1 + f I_{ij}^2]\phi_j = \sum_{j=1}^{N_{BEB}} [I_{ij}^4 - I_{ij}^3] \ , \ i = \overline{1, N_{BEB}} \quad (7)$$

N_{BEB} : total number of body surface boundary elements;
I : integrals expressed on BE surface;
f : coupling function.

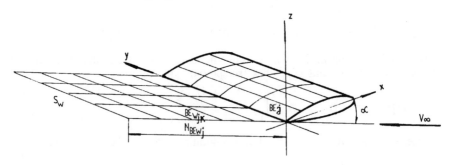

Figure 2: Body and wake surfaces discretization

A major problem is the I integrals calculation. There are two main types:

$$i^1 = \int_{BE} \frac{(\vec{x}-\vec{y})\vec{n}(\vec{x})}{|\vec{x}-\vec{y}|^3} dBE = -\int_{BE} \vec{n} \ \nabla(\frac{1}{r}) dBE \ , \quad (8)$$

$$i^3 = \int_{BE} \frac{1}{|\vec{x}-\vec{y}|} dBE = \int_{BE} \frac{1}{r} dBE \ . \quad (9)$$

Usually, BEM/FEM uses the Gauss numerical integration method but for the lifting bodies, due to its very particular geometry, it is not an appropriate way. Therefore I integrals are analytically calculated.

Knowing the potential φ, differentiating it, results the fluid velocity anywhere in fluid domain, particularly on the body surface. By means of Bernoulli law, may be calculated the pressure on the body surface and, integrating it, the hydrodynamic force.

Because for a lifting body the boundary layer is often thin, the viscosity effect is modelled by means of an empirical correction.

2 Numerical results

Using this general model and taking into account the particularities of each lifting body, it was calculated: finite span wings, nozzles, marine propellers. The results are presented in the figures bellow.

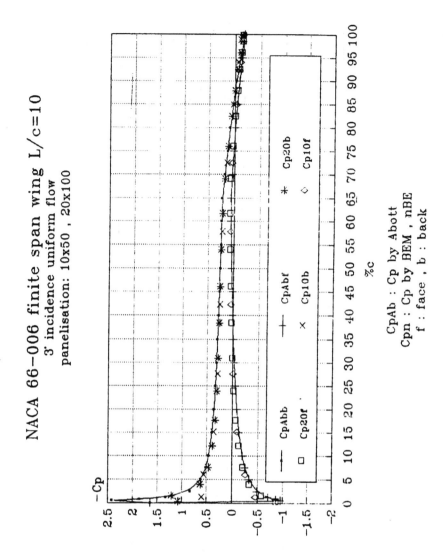

Figure 3: A finite span wing hydrodynamic calculation by present BEM against the method exposed by Abbot

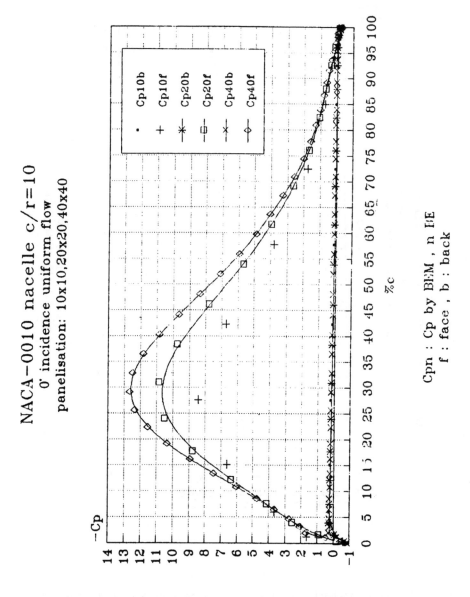

Figure 4 : A very particular nozzle (length/diameter=5), appreciated in [3] to be 'a very demanding test of a panel code'

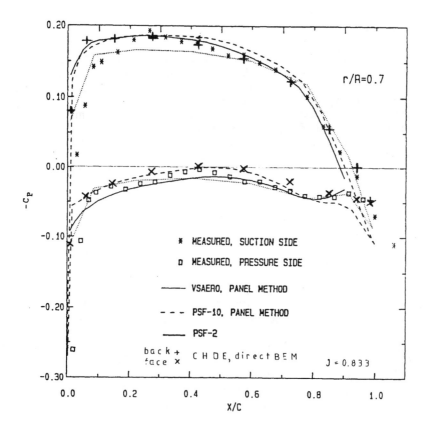

Figure 5: the DTRC4119 ITTC test propeller, J=0.833, r/R=0.7

The calculation follows the *20th ITTC Propulsor Committee Comparative Calculation by Surface Panel Method, 1992* recommendations.
Our results (CHDE program) are presented against the experimental ones and the computed ones by three well known codes.
The main assumptions are:
-non-linear Kutta-Jukovski condition;
-linear wake;
-without hub;
-without friction correction;
-discretization (spanwise×chordwise) 10×10 panels per propeller blade.

3 Conclusions

This paper dials with a direct BEM. Instead to compute as unknown singularities like: sources, dipoles, vortices, this method has as unknown the potential itself. Therefore it is simpler than the previous panel methods. Due to its low order singular integrals, it is proper to thin bodies hydrodynamic calculation.
Such kind of bodies are the lifting bodies like: hydrofoils, nozzles, marine propellers. Applying the proposed BEM, good results was obtained.
Briefly, can say that BEM is an intermediate stage between the obsolete methods based on lifting line or surface assumption (also potential methods but not three-dimensional methods) and the ideal RANS, using a more accurate fluid model but very complicate and CPU time consuming.

4 References

[1] Brebbia,C.,A./Telles,J.,C.,E./Wrobel,L.,C.: Boundary Element Techniques, Springer-Verlang, 1984
[2] Carafoli,E./Constantinescu,V.N: Dinamica fluidelor incompresibile (Incompressible Fluid Dynamics), Editura Academiei R.S.R., Bucuresti, 1981
[3] Kervin,J.,E./Kinnas,S.,A./Lee,J.-T./Shih,W.-Z.: A Surface Panel Method for the Hydrodynamic Analysis of Ducted Propellers, SNAME Transactions, Vol 85 (1987), pp. 93-122
[4] Koyama,K.: Comparative Calculation of Propellers by Surface Panel Method, 20th ITTC Propulsor Committee - Workshop on Surface Panel Method for Marine Propellers, August 23, 1992, Seoul
[5] Petrila,T./Gheorghiu,C.I.: Metode element finit si aplicatii (Finite Element Method and its Applications), Editura Academiei R.S.R., Bucuresti, 1987

Micutei Sirene R_+ .

Procedure for Free Surface Potential Flow Numerical Simulation Around Ship Model Hulls Using Finite Element Method (Galerkin Formulation)

Horatiu TANASESCU
ICEPRONAV S.A., 19A, Portului street, 6200 GALATI - ROMANIA

Abstract

Inviscid flow models remain, in the author conception, the most important for naval architecture despite the recently increased application of viscous flow tools. In this paper a physical, mathematical and numerical model for free surface potential flow around ship model hulls using finite element method is presented. Non-linear effects on the free surface are taken into account by an iterative procedure. From all application for inviscid flows, the work is focusing on the wave resistance problem only.

1 Introduction

According to William Froude's hypothesis (which stays at the foundation for conducting ship model tests), we have,

$$C_D(Re, Fn) = C_F(Re) + C_R(Fn) \tag{1}$$

where:

C_D - total resistance coefficient;

C_F - frictional resistance coefficient (viscous forces from the boundary layer, dominated by Reynolds number);

C_R - residual resistance coefficient (wave resistance, dominated by Froude number).

The present work is included within the efforts for evaluating of residual resistance $(C_R(Fn) = R_w/0.5\rho U^2 S)$.

2 Physical and Mathematical Model

Let us consider a ship hull model (piercing the free surface) moving steady horizontally in still water of infinite depth with a constant velocity U_∞ (upstream). We formulate the wave resistance problem in a Cartesian coordinate system fixed to the ship model (time-independent flow). The XY plane is at the design draft, the X axis is positive toward the stern and the vertical axis Z is positive upward. Assumptions:
- the fluid is incompressible and irrotational;
- we neglect viscous effects (boundary layers, turbulence, separation), surface tension, breaking waves, the effects of appendages and propellers.

By virtue of assumptions mentioned above, it is convenient to introduce a function $\phi(X, Y, Z)$, called the potential function or velocity potential such that,

$$\vec{U} = grad\phi = \nabla\phi \tag{2}$$

ϕ's partial derivative in any direction gives the velocity's component in that direction,

$$\phi_X = u, \phi_Y = v, \phi_Z = w \tag{3}$$

The equation of continuity (conservation of mass) for steady flow of an incompressible fluid

$$div\vec{U} = \nabla.\vec{U} = 0 \tag{4}$$

in the case of potential flow becomes $div\vec{U} = \nabla.\nabla\phi = \nabla^2\phi =$

$$\Delta\phi(X, Y, Z) = 0 \tag{5}$$

known as Laplace's equation. Thus potential ϕ fulfills Laplace equation (harmonic function) and is governed by Laplace's equation. This is an elliptic partial differential equation which requires boundary conditions at all boundaries of the computational domain (the wetted hull surface, the water free surface and at infinity). Laplace's equation together with the boundary conditions will determine the motion.

2.1 Boundary Conditions for Velocity Potential ϕ

- *on wetted hull surface*: water does not penetrate the hull (Neumann condition), $\vec{n}.\vec{U} = \vec{n}.\nabla\phi = 0$ or

$$\phi_n = 0 \tag{6}$$

- *on water free surface*; physical nature of the water free surface requires two boundary conditions:

Kinematical condition: water does not penetrate the water surface; thus at $Z = \zeta$,

$$\nabla\phi.\nabla\zeta = \phi_Z \tag{7}$$

or

$$\phi_X\zeta_X + \phi_Y\zeta_Y - \phi_Z = 0 \tag{8}$$

Dynamical condition: pressure on the water free surface must be atmospheric and independent of the position on it,

$$\frac{1}{2}(\nabla\phi)^2 + g\zeta = -\frac{1}{2}U_\infty^2 \tag{9}$$

or

$$g\zeta + \frac{1}{2}(\nabla\phi.\nabla\phi - U_\infty^2) = 0 \tag{10}$$

- *at infinity*: far away from the ship model, the flow is undisturbed parallel, having a speed equal with ship model speed U_∞ (regularity or decay condition),

$$\lim_{(X^2+Y^2+Z^2)\to\infty} \nabla\phi = (U_\infty, 0, 0) \tag{11}$$

- *radiation condition*: waves appear only in a sector behind the ship model (no upstream waves); this is fulfilled using infinite elements (polynomial-exponential decay shape functions);

- *equilibrium condition*: the ship model hull must be in equilibrium.

Combining of kinematical and dynamical boundary conditions (the equations (7) and (9)), eliminates the unknown ζ and leads to the following non-linear free surface boundary condition,

$$\frac{1}{2}\nabla\phi\nabla(\nabla\phi)^2 + g\phi_Z = 0 \tag{12}$$

with a first order linear approximation of the wave height ζ':

$$\zeta'(X,Y) = -\frac{U}{g}.\phi_X(X,Y,0) \tag{13}$$

The exact problem formulated above is non-linear because of the quadratic terms in the free-surface boundary conditions (8) and (10) and also because these conditions are to be applied on the initially unknown wavy surface. A solution method choosed here for the non-linear problem described is to linearize the free-surface boundary conditions around a known solution and solve the problem in an iterative manner. The free surface boundary condition in each iteration is linearized about the immediate former iteration. When the difference between two consecutive iterations vanishes the process converges and the exact solution is approached. First time only a start solution is obtained from a single-body, Neumann condition applied on design draft free surface.

Unknown pressures p on the hull and wavy surface $Z = \zeta(X,Y)$ will generate a potential ϕ and a wave elevation ζ, which fulfill the boundary conditions (8) and (10). Thus we can introduce the following functions:

$$K(p,\zeta) = \phi_X . \zeta_X + \phi_Y . \zeta_Y - \phi_Z = 0 \tag{14}$$

$$D(p,\zeta) = \zeta + \frac{1}{2g}[(\phi_X^2 + \phi_Y^2 + \phi_Z^2) - U_\infty^2] = 0 \tag{15}$$

By introducing small perturbations δp, in conformity with classical Michell's theory, small $\delta\phi$ and $\delta\zeta$ are induced,

$$p = p' + \delta p \rightarrow \phi = \phi' + \delta\phi . \zeta = \zeta' + \delta\zeta \tag{16}$$

where:
  ϕ - the exact solution;
  ϕ' - an approximate solution;
  $\delta\phi$ - the potential difference;
  $'$ - denotes a quantity from the immediate former iteration.

A great simplification of the boundary condition functions (14) and (15) is obtained by:
- expanding them in the first order Taylor series,

$$K(p,\zeta) \approx K(p',\zeta') + \Delta K(p,\zeta') + \Delta K(p',\zeta) \approx$$
$$\approx K(p',\zeta') + \frac{\partial}{\partial p}K(p,\zeta').\delta p + \frac{\partial}{\partial \zeta}K(p',\zeta).\delta\zeta \approx 0 \tag{17}$$

where:

$$K(p',\zeta') = \phi_X . \zeta'_X + \phi_Y . \zeta'_Y - \phi_Z \tag{18}$$

$$\Delta K(p,\zeta') = \underline{\delta\phi_X} . \zeta'_X + \underline{\delta\phi_Y} . \zeta'_Y - \underline{\delta\phi_Z} \tag{19}$$

$$\Delta K(p',\zeta) = \phi'_X . \delta\zeta_X + \phi'_Y . \delta\zeta_Y + $$
$$+ (\underline{\phi'_{XZ}} . \zeta'_X + \underline{\phi'_{YZ}} . \zeta'_Y - \underline{\phi'_{ZZ}}) . \delta\zeta \tag{20}$$

and

$$D(p,\zeta) \approx D(p',\zeta') + \Delta D(p,\zeta') + \Delta D(p',\zeta) \approx$$
$$\approx D(p',\zeta') + \frac{\partial}{\partial p}D(p,\zeta').\delta p + \frac{\partial}{\partial \zeta}D(p',\zeta).\delta\zeta \approx 0 \qquad (21)$$

where:

$$D(p',\zeta') = \zeta' - \frac{1}{2g}[U_\infty^2 - \\ -({\phi'_X}^2 + {\phi'_Y}^2 + \underline{{\phi'_Z}^2})] \qquad (22)$$

$$\Delta D(p,\zeta') = \frac{1}{g}(\phi'_X.\phi_X + \phi'_Y.\phi_Y + \underline{\underline{\phi'_Z.\phi_Z}}) \qquad (23)$$

$$\Delta D(p',\zeta) = \delta\zeta + \frac{1}{g}(\phi'_X.\underline{\underline{\phi'_{XZ}}} + \\ + \phi'_Y.\underline{\underline{\phi'_{YZ}}} + \phi'_Z.\underline{\underline{\phi'_{ZZ}}}).\delta\zeta \qquad (24)$$

The partial derivatives of $\delta\phi$ (single underlined terms) are non-linear; from this reason are neglected to linearization. Also in our particular case ϕ'_Z and the second order derivatives in the Z direction are expected to go to zero (double underlined terms).

Thus the free surface boundary conditions become:

$$\phi_X\zeta'_X + \phi_Y\zeta'_Y - \phi_Z + \phi'_X\delta\zeta_X + \phi'_Y\delta\zeta_Y = 0 \qquad (25)$$

and,

$$\delta\zeta = \frac{1}{2g}[U_\infty^2 - ({\phi'_X}^2 + {\phi'_Y}^2) + \\ + 2(\phi'_X.\phi_X + \phi'_Y.\phi_Y)] - \zeta' \qquad (26)$$

Inserting (26) in (25), the following final linearized free surface boundary condition is obtained:

$$\phi_X.\zeta'_X + \phi_Y.\zeta'_Y - \phi_Z + \\ \phi'_X\{[-\frac{\partial {\phi'_X}^2}{\partial X} - 2\frac{\partial}{\partial X}(\phi'_X.\phi_X)]\frac{1}{2g} - \frac{\partial \zeta'}{\partial X}\} + \\ \phi'_Y\{[-\frac{\partial {\phi'_Y}^2}{\partial Y} - 2\frac{\partial}{\partial Y}(\phi'_Y.\phi_Y)]\frac{1}{2g} - \frac{\partial \zeta'}{\partial Y}\} = 0 \qquad (27)$$

3 Numerical Model

The basic concepts associated with the application of the finite element method to solve the problem presented above are summarized in the following block diagram. Once the governing equation and boundary conditions has been defined, the procedure is conceptually straightforward. It is highly dependent on the way in which both the geometry of the flow domain and spatial variation of the variables is defined.

First time the code is executed in single-body, Neumann condition applied on design draft free surface mode. Then the computations are continued in linearized free surface mode, in an iterative way (using immediate former ϕ' and ζ'), always linearizing the free surface boundary condition about the immediate former solution. Simultaneously the finite elements and trim-sinkage are moved (grid point locations are made additional dependent variables in Galerkin formulation) and adjusted correspondingly. The modular type FORTRAN solver (3D) already conceived allows each section or subroutine can be examined in isolation so that different linearized versions free surface boundary conditions are easily possible to be tested.

4 Wave Resistance Computation

Once the velocity potential determined, the wave resistance can be expressed as follows,

$$R_w = \int\int_{S_w} p n_x dx \tag{28}$$

where S_w is the wetted ship hull surface and n_x, is the outward unit normal on that surface.

The fluid pressure p is given by Bernoulli equation,

$$p = -\frac{\rho}{2}(\nabla\phi.\nabla\phi - U^2) - \rho g Z \tag{29}$$

5 Conclusions

The boundary element method, unlike the finite element method, is based on the discretisation of the exterior boundary only; so the dimensionality of the problem is reduced by one. The main idea of this paper is to prove that the finite element method, if efficiently used, being more natural, is consequently, a more accurate alternative than any boundary element method.

BLOCK DIAGRAM SHOWING SPECIFIC APPLICATION OF THE FINITE ELEMENT METHOD - GALERKIN FORMULATION

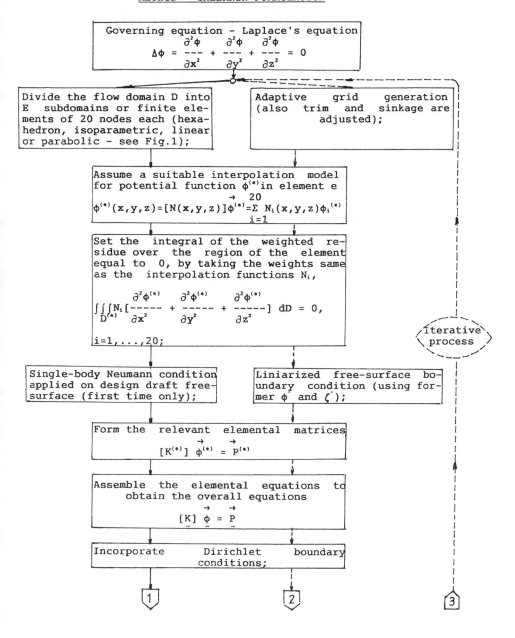

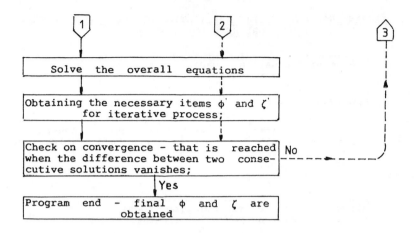

Fig.1 Hexahedron finite element with curved boundaries (location of nodes)

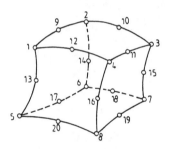

6 References

[1] Batchelor, G.K.,"An Introduction to FLUID DYNAMICS". Cambridge University Press, 1991;

[2] Fletcher, C.A.J.,"Computational Techniques for Fluid Dynamics", Springer-Verlag, 1987;

[3] Larsson, L., Broberg, L., Kim, K.J., Zhang, D.H.. "New Viscous and Inviscid CFD Techniques for Ship Flows", 5th International Conference on Numerical Ship Hydrodynamics, Hiroshima, 1989;

[4] Tanasescu, H.,"A New Procedure for Free Surface Potential Flow Numerical Simulation by Finite Element Method", National Colloquy on Fluid Mechanics, Ploiesti, 1995.

Oil industry

Mathematics for the Oil Industry
Non-mathematical review and views

Tore Gimse
SINTEF Applied Mathematics
P.O.Box 124, Blindern
N-0314 Oslo, Norway
Tore.Gimse@math.sintef.no

Abstract

In this paper we describe how mathematics is fundamental to all major engineering and scientific parts of the oil industry today. Although the applications may be very different, basic mathematical and numerical problems are shared between the disciplines. We point out the historical evolution in some of the fields, and indicate future directions for the development and use of mathematics in the oil industry.

Background - Acquiring Data

Petroleum fields are scattered all over the world. The size and location may vary significantly, from tens of meters just beneath the earth's surface in populated areas to huge reservoirs, stretching out several square kilometres, two or more kilometres underneath remote and stormy seas. The dimension and location of the resources cause major problems in the exploitation. Finding, producing and transporting the oil and gas is extremely costly, and involves highly advanced engineering and lots and lots of calculations and mathematics.

A typical North Sea reservoir is located some two kilometres under the sea floor and several hundred kilometres offshore. The sea depth is usually between one hundred and three hundred meters. Even though the reservoirs may be fairly large horizontally, the vertical dimensions may be just a few meters. The first problem is to determine exactly where resources are likely to be found. This is basically done by extensive seismic surveys which are combined with general and local geological experience. Fortunately, the geological structures in the North Sea area stretches out all the way to Greenland, where geological activity has lifted the structures above sea level. Field studies of giant outcrops in Greenland are compulsory for a North Sea geologist! The knowledge gained by such studies is then combined with seismic data. The interpretation of seismic measurements relies heavily on advanced mathematics. A seismic survey is conducted by ships travelling back and forth, shooting sound signals down, and detecting the reflected signals. By analysing these, media properties can be determined. The rock structures are usually layered, with different rock types in different layers. The rock types transmit and reflect the sound waves differently, yielding complex patterns for the received signals. To analyse these patterns, *inverse problems' mathematics* is fundamental. In general, the problem is severely underdetermined, but including additional information from outcrops and from test drillings, can tune the geological models remarkably. However, the seismic resolution is still on a scale that is often coarser than the desired scale needed later in the exploration and production process. In order to overcome this, an additional level of modelling has emerged among geologists and simulation people: introducing *geostatistics*. The aim of this discipline is to create models of the reservoirs based on the (coarse) seismic information, and (fine scale) information from outcrops and core samples. From the latter, one can build statistical models on possible facies distribution in the reservoirs, and matching this to the global knowledge from seismics fine scale geostatistical realisations of the reservoir can be generated. Usually, and in order to incorporate the uncertainty, several realisations are created.

The next step in the modelling process is to provide the simulation people with models that are practical for the purpose of simulation of fluid flow in the reservoir. Unfortunately, the geostatistical models are usually far to complex to be handled by present computing facilities. A geostatistical model can have a hundred million cells in it, while present computers can handle between one hundred thousand and one million grid blocks with a reasonable amount of CPU-time spent. Thus, there is a need to *homogenise* or *upscale* the realisations. In the process of upscaling, the aim is to preserve the small-scale effects in large-scale computations. Properties that are modelled in the geostatistical representations are usually porosity and permeability. Systematic small scale variations of these can have significant effect on a larger scale, and this should be captured in the upscaled model. Upscaling methods range from very simple averaging to relatively complex local flow analysis, where the presumed direction of flow plays an important role in the results. The latter involves solving numerous flow problems locally, and requires substantial computational efforts. Although upscaling is still a growing field of activity, present achievements indicate that future efforts should be focused on avoiding upscaling as often as possible. The ultimate goal should be to avoid both geostatistical modelling and upscaling as a whole, and rather acquire detailed seismic data, and be able to handle these data in the simulation tools. At present, however, this situation is purely a wish.

Production Devices

Once the reservoir is located (with a certain probability), it is necessary to drill wells to investigate the potential in some detail, and eventually produce the oil and gas that is available. Drilling technology itself is, of course, complicated at depths of several kilometres, and offshore conditions present even more challenging problems. Most of the offshore operations have been based on floating or stationary platforms. These are huge constructions in steel and concrete, built to withstand severe weather and climatic conditions over a long period of time. Structural analysis and load calculations are essential parts of the platform building industry. Today most of the methods and software for these purposes are well established, mostly relying on *finite element analysis*. However, the recent Sleipner accident, when a platform under construction (at the cost of some 150 mill. USD) suddenly disappeared into a deep Norwegian fjord, clearly indicates that methods and routines for such calculations are not sufficiently tested. The single cause of the accident has not been officially revealed, but obviously erroneous use and interpretation of calculations can impose dramatic consequences under such circumstances.

Once built and put into production, a platform is exposed to tremendous wave forces. Swells of amplitude 15-20 metres occur frequently, and the ability to withstand such forces is crucial. Usually the models for wave impact involve the linearized wave equations, but recently it has turned out that the full *non-linear problem* must be treated, in order to catch strange non-linear effects: The platforms are observed to be shivering at certain conditions, and such high frequency vibrations (albeit half a metre in amplitude) are properly predicted in non-linear models only.

The final problem concerning the platforms is where to position them and how to equip them with processing devices. Both these issues are related to what the platforms are expected to produce throughout their lifetime. The only way to make reasonable forecasts regarding this is to perform simulation studies of the processes that occur within the reservoir when production starts.

Reservoir Mathematics

The reservoir, of course, is the most vital part in the oil production process. However, it is also the most complex entity. In the North Sea most reservoirs are situated more than one kilometre beneath the sea floor, some several kilometres. The zones containing hydrocarbons may be just a few metres thick, stretching some few kilometres horizontally. The process of *reservoir simulation* includes the study of all processes related to the flow in a reservoir under production.

The first problem that arises is related to the dimensions and the shape of a reservoir. Vertically, the interesting parts may be less than ten meters, while it horizontally may be several kilometres. Most numerical methods divide the computational area in grid blocks, which are each assigned the relevant properties. For numerical methods to behave properly, one would prefer the grid blocks to be as regular as possible. However, the reservoir dimensions impose very different lengths in the vertical and horizontal directions. Furthermore, the grid blocks are rarely orthogonal. The reason for this is that the geological structures in the reservoir are often highly irregular. To represent the geology and geometry properly, skewed or irregular blocks are frequently applied. In addition, there are other structures in a reservoir which introduce additional problems: Geological activity has created faults which may impose non-neighbour connections between grid blocks. Fluid that exits one grid block may (physically) appear in a non-neighbour block. Fractures at faults or elsewhere may be sealing or open, and even though the structures are detectable, seismics do not reveal of which type they are.

To sum up these considerations, the computational area is usually highly irregular, the underlying geological data is scarcely known, and the constructed numerical grid blocks are unsuited for stable calculations. And yet we have not looked into the real reservoir problems, the liquids and gases floating there.

First of all, these fluids are subject to strange conditions; the average pressure in a reservoir may be some 200 bars, the temperature may usually exceed 70 degrees Celsius, and the hydrocarbons may consist of a vast number of components, each one with different chemical and physical properties. The relative amount of each component determines the main properties of a reservoir; light hydrocarbons will form gases. If both gas and oil is present, the gas will form a cap at the top of the reservoir. Below the oil layer, there is usually an aquifer.

The processes to be simulated span over a wide range, both in physical and mathematical complexity. In the start of a field's life, primary production is performed by pressure depletion. Some 20-30 percent of the total amount of oil is recoverable this way. In secondary production water is injected. This affects the reservoir conditions in two ways. First, some of the pressure is rebuilt, and secondly (most important), the water pushes out more oil. The efficiency of this process depends on the mobility ratio between the oil and the water. For unfavourable mobility ratios fingering will occur, and most of the oil may be bypassed by the injected water. Of course, since water is usually heavier than the oil, water should be injected at the bottom of the reservoir and the oil should be produced at the top. The effect of this then depends on the dip of the reservoir. For a completely flat reservoir water injection may be significantly less efficient. Instead of water, gas may be injected. This is usually gas that has been produced from the reservoir, but which is not shipped or used (gas re-injection). In general, the more gas that is present, the more complicated the modelling, since compressibility effects are then dominant, and gas can dissolve in oil depending on local pressures.

The governing principles for all reservoir simulation are *conservation of mass and momentum*. Continuity assumptions introduces Darcy's law for momentum balance, relating the average phase velocity and the pressure gradient. Several attempts have been made to derive Darcy's law from momentum balance mathematically, but the simplifying assumptions reduce the problem to an academic exercise. Thus, Darcy's law should be considered an experimental relation. Energy balance is usually not taken into consideration. The conservation equations may be formulated for the phases present, or for each component (or more frequently, each pseudo-component). If component balances are not considered, we usually have a black oil model. In these models, there are three components (gas, oil and water) and three phases (gas, oil and water), and the components may exist in different phases, depending on pressure. Two-phase models still are widespread, and may reveal important mechanisms and flow processes. The simpler processes, the faster and more detailed the simulations can be performed. This illustrates the continuous struggle between efficiency and accuracy. Introducing more "physics" always increases computational time, and thus reduces efficiency.

The basic flow equations and principles are well established. Yet a few fundamental mathematical questions remain unsolved. We will briefly mention two examples. The mathematical models are usually split into one equation for the reservoir pressure and one (or more) equations for the saturation(s). Numerically these equations are very often treated sequentially, first solving for the pressure implicitly and then updating saturations explicitly. Such an operator splitting approach is called an IMPES method. Although all experience indicates that the IMPES methods converge to a "correct" solution as the splitting time step goes to zero, no mathematical proofs of the convergence of general IMPES methods are known. The major underlying difficulty in working with this problem theoretically is the different structure of the equations. The pressure equation is generally elliptic, or slightly parabolic (due to compressibility), while the saturation equation is hyperbolic or slightly parabolic (due to capillary effects).

An other unsolved problem (although discussed and investigated for almost a decade) is the appearance of elliptic regions in the three-phase models. A three-phase model consists of a 2x2 system of conservation laws. Ignoring capillary effects, the equations possess a region in the physical phase space where the eigenvalues of the Jacobian are complex. This contradicts the physical interpretation of eigenvalues as wave speeds, and the mathematical and numerical behaviour within and in the vicinity of the elliptic region is weird and poorly understood. The relation between these effects and the physical meaning is also not well established. Although several structures of the solution involving elliptic regions are established, the underlying relevance and significance are not settled.

Transportation

Once the hydrocarbons in the reservoir approach the well, another regime of modelling is entered. Well inflow, well flow and pipeline flow is usually separated from the reservoir flow area. For the mathematical modelling in this regime, Darcy's law is no longer valid. Thus, Navier-Stokes equations are usually used for momentum balance. In addition, temperature often plays an important role. Even if the pipelines are axially symmetrical, three dimensional effects due to gravity are present. Furthermore, networks of pipelines and different pumping and controlling devices should be included in a full-scale model. Remembering that the pipeline systems can extend over several hundred kilometres, the computational difficulties in this part of the oil producing process are evident.

Among the important effects that should be modelled are formation of vax, emulsions and hydrates (ice and light hydrocarbons in crystals). These processes are highly dependent on the flow and temperature. Particularly in deep-water transportation (below 600 metres) at low temperatures this is known to be important.

The aim of performing correct modelling of pipeline transportation is to predict, and thereby take proper actions against, the above effects. In uncontrolled flow, plugs of e.g. hydrates can block the pipeline completely. Removing such plugs in deep waters is an almost impossible task, and the only possible solution may be to abandon that part of the pipeline completely. Such consequences, of course, are extremely costly.

To avoid or remove forming plugs, chemical substances may be injected in the flow to dissolve the material. The mathematical modelling of the related chemical processes are complex, but is necessary to include in the simulations.

Finally, the transportation modelling also includes engineering problems related to the installation of pipelines on the sea floor, load calculations on raisers from the sea floor to the surface and the pumping devices.

Processing

The final stage of the oil industry where mathematical modelling is a cornerstone is the downstream processing and refinement. A wide range of processes are used to produce petroleum products which are applicable in other industries and for consumers. Chemical models, of course, are central in all these processes. From a mathematical modelling point of view, such processes may be modelled on very different levels, from a molecular to a very large scale.

An important, and mathematically challenging, class of processes occurs in chemical reactors. The purpose of a chemical reactor is to enhance chemical reactions by catalysis. Inserted hydrocarbons are exposed to a catalyst in a reactor. The reactor may be of different types, fluid beds, fixed beds or trickle beds. The latter two impose modelling very similar to the reservoir simulation, since the reactor is in fact a porous medium, in which the hydrocarbons flow. Of course, the flow conditions and the environment are different, and a much more detailed level of modelling is necessary. Often very complex compositional models, including thermodynamics are used. For practical purposes one is usually restricted to one- or two-dimensional simulations.

The aim of mathematical modelling in this field is twofold. First, to optimise the output of the process economically. This includes taking into account the construction of the reactors, the input rates, catalyst distribution, quench etc. Secondly, safety and regular operation are important. In case of shut-down due to uncontrolled reactions, a clean-up may be very time consuming and expensive.

Technology Attitudes

As is evident from the above review, mathematical modelling and numerical simulations are important for several aspect of the oil industry. The oil industry, however, is fairly young, and is traditionally very engineering oriented. Thus, the need and advantage of using mathematical modelling as a production tool is not always evident to the oil companies. Discussing this, one may meet one or more of the following "ignorances":

Mathematical ignorance: It may turn out that fundamental mathematical problems are not of interest to the oil industry as long as the engineering solutions work properly. Furthermore, interesting, challenging and crucial mathematical issues may be disguised in physical effects that are added onto the models without any further investigation. The consequence may be that important effects are never observed in the computer simulations.

Model ignorance: The classical (and widespread) physical models are very seldom questioned or revised. As long as computer codes yield reasonable results, the underlying physical model is rarely discussed. A classical example of this, also related to mathematical ignorance, is the occurrence of elliptical regions in the three-phase models. All commonly used three-phase models possess these when investigated in detail. However, when it comes to practical applications, the strange effects are buried in other physical and numerical effects. Thus, one has never really re-investigated the basic three-phase models themselves.

Numerical ignorance: Problems and issues like numerical convergence, stability, grid orientation effects and numerical diffusion are rarely discussed among reservoir engineers. Again, the attitude is often that as long as the computer codes yield reasonable results, everyone is happy. Even among people that are aware of the issues, improvements or basic research in the direction of improved methods are rarely supported. As long as a majority of the industry uses the "wrong" methods, these seem to be accepted as the industrial standard.

Result ignorance: As we have seen in the above review, a lot of information enters the mathematical models of oil industry, and a lot of information is computed during a simulation process. However, when an engineer presents the results to the management, only a very limited amount of the available information in taken into account. In reservoir simulation, even if the complete saturation distribution and flow in the reservoir are computed, often only the production profiles (produced amount as a function of time) for the wells are presented. These "answers" are rarely related or discussed in the light of other data. Thus, important aspects of a production scenario are easily lost. Furthermore, uncertainty and statistical aspects of the results are very often ignored.

New technology ignorance: In many aspects, the petroleum industry is very conservative. The vast amount of today's management was educated a few decades ago, when mathematical modelling was significantly less sophisticated than today, and when computer facilities and numerical methods were extremely primitive compared with today's technologies. In their career, investments in "old" technology have committed the industry to continued use, instead of continuously developing modern tools. Of course, one also has a certain amount of faith in technology that has matured and is widespread.

Summarising the ignorances, they may all be due to lack of communication between industry and academia. Obviously, the academia has not managed to convince the industry of the necessity of more sophisticated methods, more thorough analysis of fundamental problems, and more careful use of existing and new technologies. On the other hand, the industry itself has not been very active in bringing up these issues and their fundamental problems.

Future

Based on the above section, one might think that the future of mathematics in oil industry does not look too bright. However, there are certainly signs of a slow change. First of all, there is an increased awareness in the industry that some fundamental insight may lack. The use of relatively simple streamline tools to study fundamental problems related to the physical flow in the reservoir has increased, and has proven successful. On the other side, more advanced production processes require more advanced modelling methods, and there is a growing interest in "new" approaches to classical problems. Finally, several major oil companies are reducing the in-house research activity, and are rather outsourcing their projects.

To some extent, historically, academia has also misused the oil industry. When oil prices rose dramatically in the seventies, the oil industry became "rich", and a vast number of research projects were launched and conducted virtually beyond the control or influence of the oil industry. Researchers proposed projects with little real relevance to the oil industry and got funding. After some years, when the oil industry accountants asked for results they were often disappointed. Academia thus acquired a rumour of being of little use, over-selling, and hardly controllable. Basic research shall probably not be controlled, but applied research certainly should have some relevance to the problems in the industry. Defining the relevance should not be left to the researchers alone.

Another aspect of the future of mathematics in the oil industry is how new software tools are designed and implemented. At present, most software is based on relatively old FORTRAN codes, which have been modified and extended over several years. This development has resulted in large packages, capable of handling complex physical problems, but with inherent old fashioned numerical methods and stiff data structures. Future developments should to a large extent be based on object oriented design, building modules which can be integrated in dedicated simulators on request. Such developments have been initiated in some other fields (e.g., fluid mechanics), but have not yet received any broad support in the traditional oil industry.

Developments in this new framework will be a great challenge, both to the industry, and to the co-operating partners. It requires state-of-the-art knowledge in both reservoir engineering, mathematical modelling and software design. It is not obvious how the work should start. Since the industry relies heavily on existing tools, these should probably be integrated as separate modules. By doing this, the next generation tools can develop virtually (but not conceptually) from the existing software, and practical use and supervision will be ensured throughout the development process.

We should also keep in mind that the field of mathematical modelling is a very young one, and that the future may reveal fundamentally new methods and approaches. Even if the present production devices and methods are relatively "advanced", that may perhaps not be the case for the corresponding mathematics; historically, in ancient Egypt, the pyramid constructors could hardly compute the area of an arbitrary triangle.

Discretization on general grids for general media

I. Aavatsmark, T. Barkve, Ø. Bøe, T. Mannseth *

Norsk Hydro Research Center, N-5020 Bergen, Norway
E-mail: ivar.aavatsmark@nho.hydro.com

Abstract

Discretization methods are presented for control-volume formulations on quadrilateral, triangular and polygonal grid cells in two space dimensions. The methods are applicable for any system of conservation laws where the flow density is defined by a gradient law, like Darcy's law for porous-media flow. A strong feature of the methods is the ability to handle media inhomogeneities in combination with full-tensor anisotropy.

1 Introduction

In many flow phenomena, the flow density is given by a gradient law $q = -K\nabla u$, like Fourier's law of heat conduction, Darcy's law for porous-media flow, or Ohm's law of electric conduction. For a general anisotropic, inhomogeneous medium, the conductivity field K is represented by a space-dependent symmetric tensor. Off-diagonal elements in the tensor exist if the coordinate directions are not aligned with the principal directions of K. Discretization methods for conservation laws are generally not designed to handle both inhomogeneity and general anisotropy. The purpose of this study is development of methods to improve this situation for general grids in two space dimensions.

Large discontinuities in medium properties require construction of numerical schemes with a proper definition of the effective conductivity across cell interfaces. The methods presented in this contribution will produce a generalization of the harmonic average commonly applied for a diagonal tensor K and orthogonal grids.

To satisfy local continuity in flux between grid cells with strong discontinuities in conductivity, control-volume methods are especially well suited. For time-dependent problems with large solution gradients, it is important that the methods can be combined with a fully implicit time stepping.

Although having wider applicability, the methods will be introduced based on model equations for multi-phase flow in porous media. Important effects of rock anisotropy and inhomogeneities, along with complex model geometry, motivate the

---

* Now with *RF-Rogaland Research, Thormøhlens gate 55, N-5008 Bergen, Norway*

use of the presented methods in reservoir simulation. Inactive grid cells and internal no-flow cell faces, commonly featured in reservoir simulation, will be handled by the methods with almost no increase in complexity.

The presented methods were developed by the authors in [1], [2], [3], [4], [5]. Parallel and independent development is found in [8], [9], [18], [19], [20]. Related methods are found in [6], [7], [10]. The subclass of $\boldsymbol{K}$-orthogonal methods is discussed in [13], [14], [15], [16], [17], see also [11], [12].

2 Flux calculations

The flux across an interface S is defined by

$$f = -\int_S \boldsymbol{w} \cdot \nabla u \, dS, \tag{1}$$

where $\boldsymbol{w} = \boldsymbol{K}\boldsymbol{n}$ and $\boldsymbol{n}$ is a unit normal at the interface. The vectors $\boldsymbol{w}$ and ∇u will in general have different magnitude and direction on each side (termed left (L) and right (R)) of the interface, but their inner product is continuous. In general, it is not possible to connect the cell centers of the two neighboring cells of an interface with a broken line, which on each side of the interface is parallel to $\boldsymbol{w}$, confer Figs. 1–3. Grids for which such a connection is possible, are called $\boldsymbol{K}$-orthogonal. Hence, in general, the discrete flux formula should involve more than two grid points:

$$f = \sum_{i=1}^m t_i u_i, \quad m \geq 2. \tag{2}$$

The coefficients t_i will be termed transmissibilities. For $\boldsymbol{K}$-orthogonal grids only, $m = 2$.

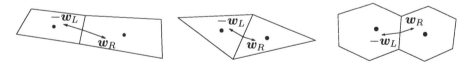

Fig. 1. Quadrilateral grid cells

Fig. 2. Triangular grid cells

Fig. 3. Polygonal grid cells

To derive an expression for the transmissibilities, *interaction regions*, shown by the dashed lines in Figs. 4–6, are introduced. As the grid cells, the interaction regions are non-overlapping and cover the domain of the differential equation. Based on the interaction regions, a unified approach for the calculation of the transmissibilities is presented.

The interaction regions divide the cell edges in two segments, and the flux will be computed on the edge segments lying inside an interaction region. For each interaction region, a linear expression will be assumed for the unknown variable u

inside each cell. At the cell center, u will be assumed to be equal to the numerical cell value. This gives two additional degrees of freedom for the variation of u in each cell, represented by two coefficients in the linear expression. These coefficients will be determined by imposing continuity conditions at cell interfaces, both for the unknown u and for the flux f.

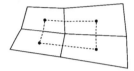 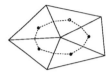

Fig. 4. Interaction region for quadrilateral grid cells

Fig. 5. Interaction region for triangular grid cells

Fig. 6. Interaction region for polygonal grid cells

In two or three space dimensions, requirement of full continuity in both u and f at all cell interfaces will lead to an overdetermined system of equations for the unknown coefficients in the linear representation. Therefore, a reduced set of conditions must be imposed. Several alternatives of reduced conditions exist, giving different discretizations.

In [1][2], two discretizations, termed *O-* and *U-method*, are introduced for quadrilateral grids. In [3][4], this technique is extended to triangular and polygonal grids. For both methods, full continuity will always be ensured for the flux.

The edges of the interaction regions intersect the edges of the cells in points termed *dividing points*. In the O-method, the continuity conditions are imposed at the dividing points. At each dividing point of an interaction region, the potential u and the flux f are required to be continuous. With n cells in an interaction region, this gives $2n$ conditions for the $2n$ unknown coefficients in the linear expressions for u. Having determined these coefficients, the flux for each edge segment can be calculated, giving an expression of the form (2).

In the U-method, the continuity conditions depend on the edge segment for which the flux is to be computed. This edge is termed the central edge. At the central edge the continuity conditions are as for the O-method. At the the two neighboring edges of the central edge, full continuity in u and f is required. The remaining edges in the interaction region are left without continuity conditions, and hence, only the four grid cells adjacent to the three edges with continuity conditions are involved. For these four grid cells, having 8 degrees of freedom, there are $2 + 2 \cdot 3$ continuity conditions.

Among all methods in which the flux across an edge is constructed from continuity conditions in the interaction region, the U-method uses the minimum number of cell nodes, whereas the O-method uses the maximum number of cell nodes.

For polygonal cells with triangular interaction regions, the interaction region involves only three cells. Hence, for this grid, the U-method reduces to an O-method. For triangular cells, the O-method involves more cells than the U-method, which only involves four cells. This is indicated in Fig. 5, where only four of the cell centers are marked with filled circles. The remaining cell centers in the interaction region are marked with open circles. For quadrilateral cells, the O- and the U-method involve the same number of cell nodes, but the flux molecule is different.

For the whole interface, the flux discretization molecule will involve the grid cells from the two interaction regions which divide the interface in two edge segments. Hence, in Eq. (2), for quadrilateral grids $m = 6$, whereas for polygonal grids with triangular interaction regions $m = 4$. For triangular grids using the U-method, $m = 6$.

The number of grid cells in the cell molecule, to which the control-volume formulation leads, is nine for quadrilateral grids. For polygonal grids with triangular interaction regions, the number of cells in the cell molecule equals the number of corners in the polygon (usually six) plus one. Using the U-method for triangular grids, the cell molecule involves ten points.

3 Properties of the methods

Grids without orthogonality conditions are flexible and can easily be adjusted to medium discontinuities. However, the polygonal grids are less flexible than quadrilateral and triangular grids. By comparing sequentially refined grids, convergence of the method is confirmed also for inhomogeneous, anisotropic test cases. A reduced grid orientation effect is observed, compared to K-orthogonal grids (i.e., grids with two-point flux molecules) [1][2][4].

For triangular and polygonal grids, large anisotropy in the conductivity may cause a transmissibility to change sign, and the method may break down. The break-down typically occurs for a ratio k_{max}/k_{min} larger than 10 for equilateral triangular cells and larger than 100 for equilateral hexagonal cells, where k_{max} and k_{min} are the eigenvalues of the matrix K. In general, this will impose restrictions on the shape of grid cells, discussed further in [4].

For quadrilateral grids, the transmissibilities will have no change in sign in the homogeneous case. However, for inhomogeneous media such a change in sign may occur. In general, this imposes minor restrictions on the shape of the grid cells along discontinuities.

4 References

[1] I. AAVATSMARK, T. BARKVE, Ø. BØE, T. MANNSETH: Discretization on non-orthogonal, curvilinear grids for multi-phase flow. *Proc. 4th European Conference on the Mathematics of Oil Recovery*, vol. D, Røros 1994, 17 pp.

[2] I. AAVATSMARK, T. BARKVE, Ø. BØE, T. MANNSETH: Discretization on non-orthogonal, quadrilateral grids for inhomogeneous, anisotropic media. *J. Comput. Phys.* **127** (1996), 2–14.

[3] I. AAVATSMARK, T. BARKVE, Ø. BØE, T. MANNSETH: Discretization on unstructured grids for inhomogeneous, anisotropic media. Part I: Derivation of the methods. To appear in *SIAM J. Sci. Comput.*

[4] I. AAVATSMARK, T. BARKVE, Ø. BØE, T. MANNSETH: Discretization on unstructured grids for inhomogeneous, anisotropic media. Part II: Discussion and numerical results. To appear in *SIAM J. Sci. Comput.*

[5] I. AAVATSMARK, T. BARKVE, Ø. BØE, T. MANNSETH: A class of discretization methods for structured and unstructured grids in anisotropic, inhomogeneous media. *Proc. 5th European Conference on the Mathematics of Oil Recovery*, Leoben 1996, edited by Z.E. Heinemann and M. Kriebernegg, pp. 157–166.

[6] P.I. CRUMPTON, G.J. SHAW, A.F. WARE: Discretisation and multigrid solution of elliptic equations with mixed derivative terms and strongly discontinuous coefficients. *J. Comput. Phys.* **116** (1995), 343–358.

[7] L.J. DURLOFSKY: A triangle based mixed finite element—finite volume technique for modeling two phase flow through porous media. *J. Comput. Phys.* **105** (1993), 252–266.

[8] M.G. EDWARDS, C.F. ROGERS: Multigrid and renormalization for reservoir simulation, *Proc. 4th European Multigrid Conference*, Amsterdam 1993, edited by P.W. Hemker and P. Wesseling, Birkhäuser, Basel 1994, pp. 189–200.

[9] M.G. EDWARDS, C.F. ROGERS: A flux continuous scheme for the full tensor pressure equation, *Proc. 4th European Conference on the Mathematics of Oil Recovery*, vol. D, Røros 1994, 15 pp.

[10] I. FAILLE: Control volume method to model fluid flow on 2D irregular meshing. *Proc. 2nd European Conference on the Mathematics of Oil Recovery*, Arles 1990, edited by D. Guérillot and O. Guillon, Éditions Technip, Paris 1990, pp. 149–156.

[11] P. FORSYTH: A control-volume, finite-element method for local mesh refinement in thermal reservoir simulation. *SPE Reservoir Engineering* **5** (1990), 561–566.

[12] L.S.-K. FUNG, A.D. HIEBERT, L.X. NGHIEM: Reservoir simulation with a control-volume finite-element method. *SPE Reservoir Engineering* **7** (1992), 349–357.

[13] Z.E. HEINEMANN, C.W. BRAND: Gridding techniques in reservoir simulation. *Proc. First Intl. Forum on Reservoir Simulation*, Alpbach 1988, pp. 339–425.

[14] Z.E. HEINEMANN, C.W. BRAND, M. MUNKA, Y.M. CHEN: Modeling reservoir geometry with irregular grids. *SPE Reservoir Engineering* **6** (1991), 225–232.

[15] B. HEINRICH: *Finite difference methods on irregular networks*. Birkhäuser, Basel 1987.

[16] R. HERBIN: An error estimate for a finite volume scheme for a diffusion-convection problem on a triangular mesh. *Numer. Methods Partial Differential Equations* **11** (1995), 165–173.

[17] P.S. VASSILEVSKI, S.I. PETROVA, R.D. LAZAROV: Finite difference schemes on triangular cell-centered grids with local refinement. *SIAM J. Sci. Stat. Comput.* **13** (1992), 1287–1313.

[18] S. VERMA, K. AZIZ, J. FAYERS: A flexible gridding scheme for reservoir simulation. *Proceedings of the SPE Annual Technical Conference and Exhibition*, vol. II, Dallas 1995, pp. 657–672.

[19] S. VERMA, K. AZIZ: Two- and three-dimensional flexible grids for reservoir simulation. *Proc. 5th European Conference on the Mathematics of Oil Recovery*, Leoben 1996, edited by Z.E. Heinemann and M. Kriebernegg, pp. 143–156.

[20] A.F. WARE, A.K. PARROTT, C.. ROGERS: A finite volume discretisation for porous media flows governed by non-diagonal permeability tensors. *Proceedings of CFD95*, Banff, Canada, 1995, 8 pp.

Identification of Mobilities for the Buckley-Leverett Equation by Front Tracking

Vidar Haugse,
Department of Petroleum Engineering and Applied Geophysics,
Norwegian University of Science and Technology,
N-7034 Trondheim, Norway, vidarh@ipt.unit.no

Abstract

Displacement experiments of oil by water are widely used to investigate the effects of two-phase flow. These one dimensional experiments are used to determine the fractional flow and the total mobility functions which are used as input for reservoir simulations. A new method has been developed to find fractional flow and total mobility functions that will reproduce the discrete measurements. Both constant rate and constant pressure drop experiments are investigated. The calculated fractional flow function is piecewise linear in the range of observed fractional flow values. It is assumed that that the core is homogeneous. Gravity, compressibility, and capillary effects are neglected.

1 Introduction

Multi–phase flow in porous media is modelled by Darcy's law. This empirical relation relates the macroscopic velocity of each phase to the pressure gradient. The proportionality factor is determined by three factors: absolute permeability is the ability of a porous medium to conduct flow (rock property), viscosity is a fluid property, and relative permeability is a measure of the reduction in the ability of a phase to flow due to the presence of other fluids (rock/fluid property).

An overview of relative permeability measurements are given in [4]. Flooding experiments were first analysed by Johnson et al. [5]. Jones and Rozelle [6] developed a graphical technique to determine relative permeability from measurements at discrete times. For a constant pressure drop experiment, interpolation is used to find total velocity at times of measurements. These velocities are used to determine relative permeabilities by a graphical technique. Others have approximated relative permeabilities by multi–parameter families of functions, and determined the parameters by minimising the error (see [7] and [8]).

A new method to calculate relative permeabilities from measured data will be presented, where the fractional flow and total mobility functions are considered as the unknown functions. Relative permeabilities are easily calculated from fractional flow and total mobility data. We are able to construct a fractional flow function and a total mobility function that will reproduce the measured data exactly.

The mathematical approach to one dimensional scalar conservation laws is based on front tracking. It is assumed that the flux function is piecewise linear and that the initial data is piecewise constant. This Cauchy problem may be solved exactly by tracking the positions of the shocks defined by the initial data, and solving the Riemann-problems defined by collisions between shocks. The method was first presented by Dafermos [2], and later analysed by Holden et al. [3]. Front tracking is also used to solve the saturation equation in a commercial reservoir simulator [1].

2 Equations

The core is assumed to be homogeneous. Gravity, compressibility, and diffusive forces will be neglected, and viscosities of the two phases are assumed to be constant. The velocity u_i of a phase i is described by Darcy's law

$$u_i = -\frac{kk_{r_i}(s)}{\mu_i}p_x(x), \quad i = w, o,$$

where k is permeability, μ_i is viscosity of phase i, k_{r_i} is relative permeability of phase i, s is the saturation of water, and p is pressure. Conservation of mass for incompressible water and oil gives

$$\phi s_t + (u_w)_x = 0, \quad \phi(1-s)_t + (u_o)_x = 0,$$

where ϕ is porosity. Total velocity is found to be constant in space by adding the conservation equations

$$(u_o + u_w)_x = u_x = 0. \tag{1}$$

Total mobility is defined by $\lambda_T = \lambda_w + \lambda_o$, where $\lambda_i(s) = k_{r_i}(s)/\mu_i$, $i = w, o$. Using equation (1), pressure is given by the formula

$$p(x,t) = p(0,t) - \frac{u(t)}{k}\int_0^x \frac{dx}{\lambda_T(s(x,t))}. \tag{2}$$

By using the expression for the fractional flow curve for water, $f(s) = \frac{u_w}{u} = \frac{\lambda_w}{\lambda_T}$, the conservation equation for water may be written

$$\phi s_t + u f(s)_x = 0. \tag{3}$$

Equation (2) and (3) are decoupled by introducing new dimensionless variables

$$x_D = \frac{x}{L}, \quad t_D = \frac{\int_0^t u(\tau)d\tau}{\phi L}. \tag{4}$$

Note that dimensionless time t_D is equal to the number of pore volumes injected. Equation (3) is now simplified to a scalar hyperbolic conservation law

$$s_{t_D} + f(s)_{x_D} = 0,$$

and the pressure drop in the core is given by

$$p(x_D, t_D) = p(0, t_D) - \frac{u(t_D)L}{k}\int_0^{x_D} \frac{dx_D}{\lambda_T(s(x_D, t_D))}.$$

3 Front Tracking Solution with Constant Rate

We will analyse a unsteady-state displacement of oil with water. Boundary conditions are given by

$$s(x,0) = s_R, \quad s(0,t) = s_L > s_R, \quad p(0,t) = p_L, \quad u(t) = u_T > 0. \quad (5)$$

Assume that measurements are made at times $t_0, t_1, \cdots, t_M$, where $0 = t_0 < t_1 < \cdots < t_M$ and t_1 is the time of breakthrough. Let Δp_i be the pressure drop over the core at time t_i, and f_i the average fractional flow value at the outlet during time interval $[t_i, t_{i+1}]$. We will assume that $f_i < f_{i+1}$, for $i = 0, \cdots, M-2$. The required measurements are indicated in Figure 1.

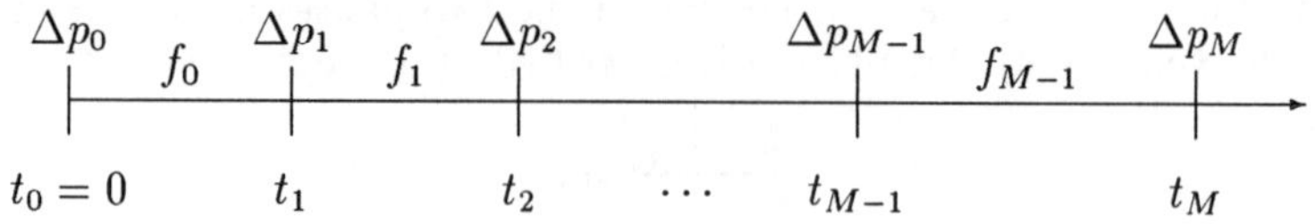

Figure 1: Measurements of pressure drop Δp_i and average fractional flow f_i are made at times t_i.

Define dimensionless time $t_{D_i} = u_T t_i/(\phi L)$. We will make the following assumptions on the concave envelope of the fractional flow function f: The concave envelope of f is continuous, increasing, and piecewise linear in saturation, for fractional flow values in the range $f_0 < f < f_{M-1}$. Breakpoints[1] are given by (s_i, f_i). No assumptions are made on f for fractional flow values larger than f_{M-1}.

The structure of the solution is simple in this case: The saturation profile is piecewise constant for saturations corresponding to fractional flow values in the interval $[f_0, f_{M-1}]$. Let s_i be the saturation corresponding to f_i. The dimensionless velocity of the front between f_{i-1} and f_i is given by the Rankine–Hugoniot condition

$$v_{i-1,i} = \frac{f_i - f_{i-1}}{s_i - s_{i-1}}.$$

Note that the average fractional flow values are reproduced if the producing fractional flow in the interval $[t_{i-1}, t_i]$ is given by the constant f_i. We will therefore assume that the front between s_{i-1} and s_i reach the outlet $x_D = 1$ at dimensionless time t_{D_i}, that is

$$\frac{f_i - f_{i-1}}{s_i - s_{i-1}} t_{D_i} = 1.$$

Accordingly, $v_{i-1,i} = 1/t_{D_i}$. Note that saturations may be calculated sequentially

$$s_0 = s_R, \quad s_i = s_{i-1} + (f_i - f_{i-1})t_{D_i}, \quad i = 1, \cdots, M-1. \quad (6)$$

Let λ_{T_i} be the total mobility corresponding to s_i. We will calculate λ_{T_i}, for $i = 1, \cdots, M-1$, by using the measured pressure drops. The saturation profile is piecewise constant for saturations between s_0 and s_{M-1}. Thus, total mobility will

---
[1] We will use the following definition: The point (s_b, f_b) is a *breakpoint* of the function f, if f' is discontinuous at s_b and $f(s_b) = f_b$.

also be piecewise constant in the corresponding part of space at a given time. The pressure drop may be rewritten as

$$\Delta p = \Delta p^{(M)} + \Delta p^{(M-1)} + \cdots + \Delta p^{(1)} + \Delta p^{(0)},$$

where $\Delta p^{(M)}$ is the pressure drop due to saturations not observed at the outlet, and $\Delta p^{(i)}$ is the pressure drop due to total mobility λ_{T_i}. Since the saturation profile is a similarity solution, $s(x_D, t_D) = s(x_D/t_D) = s(\xi)$, we have that

$$\Delta p^{(M)}(t_D) = \frac{u_T L}{k} \int_0^{t_D/t_{D_M}} \frac{dx_D}{\lambda_T(s(x_D/t_D))} = \frac{u_T L}{k} t_D \int_0^{1/t_{D_M}} \frac{d\xi}{\lambda_T(s(\xi))} \equiv \frac{u_T L}{k} t_D I_M.$$

For $\Delta p^{(i)}$, $i = 1, \cdots, M-1$ we find that

$$\Delta p^{(i)}(t_D) = \frac{u_T L}{k} \int_{t_D/t_{D_{i+1}}}^{t_D/t_{D_i}} \frac{dx_D}{\lambda_{T_i}} = \frac{u_T L}{k} t_D \frac{1/t_{D_i} - 1/t_{D_{i+1}}}{\lambda_{T_i}} \equiv \frac{u_T L}{k} t_D I_i.$$

Initial total mobility λ_{T_0} is calculated from $\Delta p_0 = u_T L/(k\lambda_{T_0})$. The integral I_M and the total mobilities λ_{T_i}, for $i = M-1, \cdots, 1$ are calculated sequentially by matching the observed pressure drops Δp_i. Since $\Delta p_i = \Delta p^{(M)}(t_{D_i}) + \cdots + \Delta p^{(i)}(t_{D_i})$, we have

$$\frac{\Delta p_i k}{u_T L t_{D_i}} = I_M + I_{M-1} + \cdots + I_i, \qquad i = 1, \cdots, M. \tag{7}$$

Theorem 1 *Let the fractional flow function f be calculated from equation (6), and the total mobility from equation (7). Then these functions will reproduce the measured data Δp_i and f_i.*

4 Solution with Constant Pressure Drop

We will now study the displacement of oil with the following boundary and initial conditions

$$s(x, 0) = s_R, \qquad s(0, t) = s_L > s_R, \qquad p(0, t) = p_L, \qquad p(L, t) = p_R < p_L. \tag{8}$$

Since total velocity is not constant in time, dimensionless time will now depend on the pressure solution.

Let u_0 be the total velocity at time 0, $\bar{u}_i$ be the measured average total velocity during time interval $[t_i, t_{i+1}]$, $\bar{u}_i = \frac{1}{t_{i+1} - t_i} \int_{t_i}^{t_{i+1}} u(t) dt$, and f_i the average effluent fractional flow value in the same time interval, $f_i = \frac{1}{\bar{u}_i(t_{i+1} - t_i)} \int_{t_i}^{t_{i+1}} u(t) f(L, t) dt$. The available measured quantities are illustrated in Figure 2.

Dimensionless time t_{D_i} is the number of pore volumes injected at time t_i

$$t_{D_i} = \frac{1}{\phi L} \sum_{j=1}^{i} (t_j - t_{j-1}) \bar{u}_{j-1}.$$

Since the saturation equation only depends on dimensionless time, s_i is given by the result in the previous section, see equation (6).

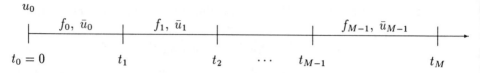

Figure 2: Measurements of average total velocity $\bar{u}_i$ and average fractional flow f_i are made at times t_i.

Initial mobility λ_{T_0} is calculated from the equation $\frac{\Delta p\,k}{L} = \frac{u_0}{\lambda_{T_0}}$. At time t_i we have the following expression for the pressure drop

$$\frac{\Delta p\,k}{L} = u_i t_{D_i}(I_M + I_{M-1} + \cdots + I_i), \qquad i = 1, \cdots, M, \quad (9)$$

where u_i is the unknown total velocity at time t_i. These velocities are not measured, and must be calculated from the observed initial velocity u_0 and average total velocities $\bar{u}_i$, for $i = 0, \cdots, M-1$. Pressure drop for $0 < t < t_1$ is described by

$$\frac{\Delta p\,k}{L} = u(t)\left(t_D(t)(I_M + I_{M-1} + \cdots + I_1) + \frac{1 - t_D(t)/t_{D_1}}{\lambda_{T_0}}\right).$$

By using the expression for initial mobility, the total velocity may be written

$$u(t) = \frac{u_0}{1 + b_0 t_D(t)}, \qquad 0 < t < t_1, \quad (10)$$

where $b_0 = \lambda_{T_0}(I_M + \cdots + I_1) - 1/t_{D_1}$. Total velocity for $t_i < t < t_{i+1}$ may be expressed by the integral equation

$$u(t) = \frac{u_i}{1 + b_i(t_D(t) - t_{D_i})}, \qquad t_i < t < t_{i+1}, \quad (11)$$

where

$$b_i = \frac{(I_M + \cdots + I_{i+1})/t_{D_i} - (I_M + \cdots + I_i)/t_{D_{i+1}}}{(I_M + \cdots + I_i)t_{D_i}(1/t_{D_i} - 1/t_{D_{i+1}})}, \qquad i = 1, \cdots, M-1.$$

Differentiation of equation (11) with respect to time yields

$$u'(t) = -\frac{b_i}{u_i \phi L}u(t)^3, \qquad u(t_i) = u_i,$$

with solution

$$u(t) = \frac{u_i}{\sqrt{1 + \frac{2b_i u_i}{\phi L}(t - t_i)}}, \qquad t_i < t < t_{i+1}, \quad (12)$$

where $b_i > -\phi L/(2u_i(t_{i+1} - t_i))$ to avoid singularities. Note that $u(t)$ is a strictly decreasing function of b_i.

The unknown variables b_i and u_{i+1}, for $i = 0, \cdots, M-1$, may know be calculated sequentially. Assume that u_i has been found. The variable b_i must be chosen so that the measured average velocity is matched

$$\frac{1}{t_{i+1} - t_i}\int_{t_i}^{t_{i+1}} u(\tau)d\tau = \frac{\phi L}{(t_{i+1} - t_i)b_i}\left(\sqrt{1 + \frac{2b_i u_i}{\phi L}(t_{i+1} - t_i)} - 1\right) = \bar{u}_i, \quad (13)$$

It is easily verified that there exist a unique solution b_i if $0 < \bar{u}_i < 2u_i$. The total velocity u_{i+1} is then given by equation (12). The results in this section are summarised in the following theorem.

W_i (ml)	N_p (ml)	Δp (psi)
0.00	0.00	138.6
7.00	7.00	97.5
11.20	7.84	91.9
16.28	8.43	87.9
24.27	8.93	83.7
39.2	9.30	78.5
62.3	9.65	74.2
108.9	9.96	70.0
155.6	10.11	68.1
311.3	10.30	65.4

f	s	λ_T	k_{r_o}	k_{r_w}
.000000	.350000	.074149	.774859	.000000
.800000	.529819	.096197	.201052	.074649
.883858	.559977	.102037	.123841	.087481
.937422	.587978	.106545	.069674	.096881
.975218	.617433	.111522	.028881	.105495
.984848	.629555	.119828	.018973	.114472
.993348	.646586	.128727	.008948	.124035
.996788	.658616	.138076	.004635	.133503
.998780	.668568	.145160	.001851	.140633

Table 1: Data from a constant rate waterflood experiment.

Table 2: Calculated results from data in Table 1.

Theorem 2 *Let the fractional flow function f be calculated from equation (6), let b_i be the solution of (13), and let total velocity u_{i+1} be calculated from equation (12). Equation (13) has a unique solution if $0 < \bar{u}_i < 2u_i$. Finally, let the total mobility be calculated from equation (9). The fractional flow and total mobility functions given by this construction will reproduce the measured data $\bar{u}_i$ and f_i.*

5 Examples

We will present two examples. The first is a constant rate waterflood example from Jones and Rozelle [6]. Data for the second example is generated by solving the flow equations with given relative permeabilities and calculating average total velocity and fractional flow in some given time intervals. Relative permeabilities from the "measured" data are compared with the true relative permeabilities.

5.1 Example 1

The input data from Jones and Rozelle [6] are as follows:
$\mu_o = 10.45$ cp $\quad \mu_w = 0.97$ cp $\quad s_0 = 0.35$ $\qquad L = 12.705$ cm
$k = 35.4$ mD $\quad \phi = 0.215$ $\quad u = 1.95 \cdot 10^{-3}$ cm/s

Production data are given in Table 1. W_i is total water volume injected, and N_p is cumulative production of oil. The fractional flow values given are calculated from W_i and N_p: $f = 1 - \Delta N_p/\Delta W_i$. Pressure drop is monotonically decreasing, since the viscosity of the oil is an order of magnitude larger than the viscosity of water. Also, there is no clear breakpoint in the pressure drop at the time of breakthrough.

Calculated relative permeability curves are close to the results given by Jones and Rozelle, and pressure drop and production are matched exactly at the points of measurement.

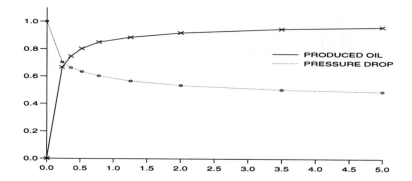

Figure 3: Normalised pressure drop $\Delta p/\Delta p_0$ and normalised cumulative oil production vs. dimensionless time. Markers indicate measured points.

5.2 Example 2

The algorithm in section 4 is tested on data from a given model. The total velocity and the producing fractional flow at the outlet are calculated and average data for some given time intervals are found. From these discrete "measurements", a total mobility and a fractional flow function are constructed by the proposed algorithm.

The input data are

$\mu_o = 1.0$ cp $\quad\quad \mu_w = 1.0$ cp $\quad\quad L = 10.0$ cm
$k = 100$ mD $\quad\quad \phi = 0.3 \quad\quad\quad \Delta p = 1.0$ atm.
$k_{r_w}(s) = (s - 0.1)^2 \quad k_{r_o}(s) = (0.9 - s)^2 \quad s_0 = 0.1.$

Production for this model is shown in Figure 4. Notice the breakpoint in total velocity at the time of breakthrough. Average total velocity and fractional flow values are calculated for the time intervals given in Table 3.

t_i (s)	$\bar{u}_i$ (cm/s)	f_i
373.16	0.0053280	0.00000
400.00	0.0045936	0.86075
500.00	0.0047227	0.88825
700.00	0.0049687	0.92742
1000.0	0.0052569	0.95855
1500.0	0.0055404	0.97849
2000.0	0.0057516	0.98847

Table 3: Averaged total velocities $\bar{u}_i$ and averaged fractional flow values f_i in some given time intervals.

From these "measured" data, fractional flow and total mobility values are calculated. The results are given in Figure 5. It is seen that the difference between the correct and calculated relative permeabilities is small.

Previous publications on determination of relative permeability have only presented relative permeability curves from data measured in the laboratory, where the "true" relative permeability curves are unknown. The author suggests that algorithms should be tested on data calculated from given relative permeability curves before they are used to analyse laboratory experiments.

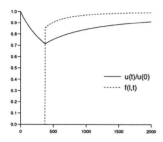

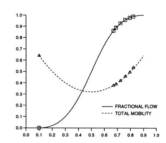

Figure 4: Normalised total velocity $u(t)/u_0$ and producing fractional flow vs. time [s].

Figure 5: Fractional flow and total mobility function vs. saturation. Markers indicate measured points.

Acknowledgements

The author would like to thank Prof. Helge Holden and Knut-Andreas Lie for many valuable comments.

References

[1] F. Bratvedt, K. Bratvedt, C.F. Buchholz, L. Holden, H. Holden, and N.H. Riesbro. A new front-tracking method for reservoir simulation. *SPE Res. Eng.*, pages 107–116, February 1992.

[2] C. Dafermos. Polygonal approximations of solutions of the initial value problem for a conservation law. *J. Math. Anal. Appl.*, 38:33–41, 1972.

[3] H. Holden, L. Holden, and R. Høegh-Krohn. A numerical method for first order nonlinear scalar conservation laws in one-dimension. *Comput. Math. Applic.*, 15:595–602, 1988.

[4] M. Honarpour and S.M. Mahmood. Relative permeability measurements: An overview. *J. Petr. Tech.*, pages 963–966, August 1988.

[5] E.F Johnson, D.P. Bossler, and V.O Naumann. Calculation of relative permeability from displacement experiments. *Trans. AIME*, 216:370–372, 1959.

[6] S.C. Jones and W.E. Rozelle. Graphical techniques for determining relative permeability from displacement experiments. *Trans. AIME*, 216:807–817, 1978.

[7] A. Mikelić and Z. Tutek. Identification of mobilities for the Buckley-Leverett equation. *Inverse Problems*, 6:767–787, 1990.

[8] O. Vignes. Application of optimization methods in oil recovery problems. Dr. ing. thesis, Norwegian Institute of Technology, University of Trondheim, 1993.

Flow of Waxy Crude Oils

Angiolo Farina* Luigi Preziosi+

*Dipartimento di Ingegneria Aeronautica e Spaziale
Politecnico di Torino
Corso Duca degli Abruzzi 24, 10129 - Torino, Italy

+Dipartimento di Matematica
Politecnico di Torino
Corso Duca degli Abruzzi 24, 10129 - Torino, Italy

Abstract

The crystallization of wax in crude oils causes severe difficulties in pipelining. At low temperature the waxy crudes are non-Newtonian fluids and their flow properties are time-dependent showing "thyxotropic" phenomena. Starting from the experimental data, we have developed a mathematical model to describe the low temperature behavior of a waxy crude in a laboratory experimental loop. The correspondent initial boundary value problem, which is of Stefan type, has been studied and solved numerically.

1 Introduction

The transport of waxy crude oils through pipelines can be troublesome, particularly during winter. This arises from the adverse flow characteristics influenced by the presence of relatively large amounts of wax [1].

The rheology of crude oils is believed to be influenced by waxes as well as by the thermal and mechanical history of the crude [2], [3].

Wax will begin to crystallize as soon as the equilibrium temperature and pressure is reached (Cloud Point). When the product is cooled to a temperature lower than the Cloud Point, the crystals agglomerate leading to the formation of a gel structure [4] which is responsible for the appearance of an yield-stress (a minimum stress that must be overcome before any fluid movement takes place).

The flow properties of waxy crudes are further complicated by their critical dependence upon mechanical and thermal "history". The apparent viscosity can be greatly reduced by mechanical shear, which can disintegrate large wax agglomerates [2].

This paper deals with the physical modeling of the low temperature behavior of waxy crudes and with the related mathematical and numerical problems. In particular, section 2 is devoted to the deduction of the model. Section 3 deals with the mathematical problem in planar geometry and Section 4 with the results of the simulations performed in planar geometry.

2 The physical model

Waxy crude oils show, at low temperature, the presence of a yield-stress [4]. According to this experimental evidence we describe them as Bingham fluids (see [5] p. 97).

A Bingham fluid is a non-Newtonian fluid which behaves like a rigid body when the shear stress τ is less than a threshold value τ_o, while for $\tau > \tau_o$ the relationship between the stress τ and the shear rate γ is linear

$$\tau = \tau_o + \eta\gamma \tag{1}$$

where η is the Bingham viscosity.

Assuming that the temperature is uniform, constant and below the so-called Pour Point the density of crystallized wax is constant in time and space. So the non-Newtonian behavior has to be attributed to the agglomeration of wax crystals only. We have, therefore, introduced a parameter α defined as the ratio between the density of agglomerated wax and the density of crystallized wax. So α is a non dimensional parameter ranging in the interval $[0, 1]$.

We will assume that the only quantity which is influenced by the agglomeration factor is the yield-stress τ_o:

$$\tau_o = \tau_o(\alpha) \tag{2}$$

where τ_o, as function of α, is C^1 and non-decreasing.

The evolution equation for α has to include both the spontaneous aggregation of wax crystals and the agglomerate fragmentation due to the mechanical shear. A first step is represented by the following evolution equation

$$\dot{\alpha} = k_1(1-\alpha) - k_2 \alpha |W| \tag{3}$$

where W is the power dissipated by the viscous forces per unit volume and k_1, k_2 are constant parameters whose values are known, unfortunately, with very poor precision.

Other evolution equations could be considered, but in order to open a proper discussion more precise experimental results are needed.

We remark that (3) implies $\alpha = \alpha(t)$. This limits the applicability of the model to implants characterized by pumps mixing the fluid over time intervals much smaller than the evolution time scale of α. This situation actually occurs in laboratory experimental loops.

3 The mathematical problem

Due to the presence of a yield-stress τ_o, a Bingham flow in a canal or a pipe may present a central rigid core and a fluid region near the wall.

Let us consider the dynamics in planar geometry. Consider a waxy crude oil flowing between two parallel plates at a distance $2L$. Let x be the coordinate along the direction of motion and y be the coordinate perpendicular to the plates. The velocity has the form $\vec{v} = v(y,t)\vec{e}_x$ since we are concerned with laminar flow.

The canal is divided into two regions: a fluid region $[-L; -s(t)] \cup [s(t); L]$ and a rigid core $[-s(t); s(t)]$, divided by the non-material interfaces $y = -s(t)$ and $y = s(t)$.

Due to symmetry we will only consider the upper half space $y \geq 0$. The velocity v satisfies, in the fluid regions, the equation

$$\rho v_t = \eta v_{yy} + f_o \qquad s(t) < y < L;\ t > 0 \qquad (4)$$

where ρ is the oil density (typically $\rho = 0.8\ \text{gr/cm}^3$) and

$$f_o = -\frac{\partial P}{\partial x} \qquad (5)$$

is the pressure gradient assumed to be a known constant. The boundary and initial conditions are

$$v(L,t) = 0 \qquad t > 0 \qquad (6)$$
$$v_y(s(t),t) = 0 \qquad t > 0 \qquad (7)$$
$$v(y,0) = v_o(y) \qquad s_o \leq y \leq L \qquad (8)$$
$$s(0) = s_o \qquad 0 < s_o < L. \qquad (9)$$

Equation (6) and (7) express respectively the no-slip condition at $y = L$ and the absence of strain rate at the boundary $s(t)$. Operating as in [6] pp. 90-92, that is, evaluating the momentum balance of a unit length portion of the rigid core, we get the additional free boundary condition

$$\rho v_t(s(t),t) = \left(f_o - \frac{\tau_o}{s(t)}\right) \qquad t > 0. \qquad (10)$$

The evolution equation for α is

$$\dot{\alpha} = k_1(1-\alpha) - k_2 \frac{\alpha}{L}\eta \left[\int_{s(t)}^{L} (v_y)^2\, dy + v(s(t),t)\tau_o(\alpha)\right] \qquad t > 0 \qquad (11)$$

whose initial condition is

$$\alpha(0) = \alpha_o \qquad 0 \leq \alpha_o \leq 1. \qquad (12)$$

Thus in planar geometry the time evolution of the system is governed by equations (4), (6) - (12), in the following referred to as problem (P_p).

It has been shown that (P_p) has a unique classical solution [7]. The proof of the existence is based on Schauder's theorem while uniqueness follows applying fixed point techniques. In particular, considering the time derivative of the flow v (i.e. the acceleration)

$$z = v_t(y, t) \tag{13}$$

(P_p) can be transformed in the following problem of Stefan type

$$\rho z_t - \eta z_{yy} = 0 \qquad t > 0; \ s(t) < y < L \tag{14}$$

$$z(L, t) = 0 \qquad t > 0 \tag{15}$$

$$z(s(t), t) = \frac{1}{\rho}\left(f_o - \frac{T_o(\alpha)}{s(t)}\right) \qquad t > 0 \tag{16}$$

$$z(y, 0) = \frac{\eta}{\rho} v_o''(y) + \frac{f_o}{\rho} \qquad s_o \leq y \leq L \tag{17}$$

$$z_y(s(t), t) = \frac{1}{\eta s(t)} T_o(\alpha) \dot{s}(t) \qquad t > 0 \tag{18}$$

$$s(0) = s_o \qquad 0 < s_o < L \tag{19}$$

$$\dot{\alpha} = k_1(1 - \alpha) - \frac{1}{L}k_2\alpha\left[\eta \int_s^L (v_y)^2 \, dy + T_o(\alpha) v(s, t)\right] \qquad t > 0 \tag{20}$$

$$\alpha(0) = \alpha_o \qquad 0 \leq \alpha_o \leq 1. \tag{21}$$

Then, after having defined a set Σ formed by Lipschitz functions defined in the interval $[0, 1]$, we have looked for the solution α in this set. In fact, once a certain $\tilde{\alpha} \in \Sigma$ was fixed, we have shown that the problem (14) - (19) (with α given by $\tilde{\alpha}$) has a unique classical solution and we have proved that the linear Cauchy problem

$$\begin{aligned}\dot{\alpha} &= k_1(1 - \alpha) - k_2\alpha |W(\tilde{\alpha})| \\ \alpha(0) &= \alpha_o\end{aligned} \tag{22}$$

($W(\tilde{\alpha})$ being the power density evaluated with $\tilde{\alpha}$) defines a map from Σ to Σ. Applying then Schauder's theorem we have deduced the existence of a classical solution for the complete problem (14) - (21).

It has been shown in [8] that, if the parameters k_1, k_2 satisfy certain hypotheses, the initial conditions on s and α satisfy the following relationship

$$s_o = \frac{T_o(\alpha_o)}{f_o} \tag{23}$$

and the initial flow is

$$v_o(y) = \frac{f_o}{2\eta}(L^2 - y^2) + \frac{f_o s_o}{\eta}(y - L) \qquad s_o < y < L \tag{24}$$

then the solution can be obtained, neglecting small perturbations, performing a quasi-steady approximation. To prove that we have used, essentially, perturbation techniques.

The quasi-steady approximation (studied in [8]) consists in considering the evolution of the parameter α and of the amplitude of the rigid core s much slower than that one characterizing the flow v. In this approximation (also called quasi-stationary approximation) the equation for v and s are

$$s = \frac{\tau_o(\alpha)}{f_o} \qquad (25)$$

$$v = \frac{f_o}{\eta}(L-y)\left[\frac{1}{2}(L+y) - \frac{\tau_o(\alpha)}{f_o}\right]. \qquad (26)$$

Therefore dynamics is completely governed by the equation for α, which rewrites

$$\dot{\alpha} = k_1(1-\alpha) - k_2 \frac{f_o^2}{\eta L}\alpha \left[\frac{1}{3}\left(L - \frac{\tau_o(\alpha)}{L}\right)^3 + \frac{\tau_o(\alpha)}{2L}\left(L - \frac{\tau_o(\alpha)}{L}\right)^2\right] \qquad (27)$$

still to be joined with the initial condition (12).

4 Numerical methods

Before performing a numerical study of the problem in planar geometry we have to estimate the parameters involved in the model, in fact, even if one assumes a simple linear form for $\tau_o(\alpha)$

$$\tau_o(\alpha) = \tau_M \alpha \qquad \alpha \in [0,1] \qquad (28)$$

still the parameters k_1, k_2 and τ_M are to be determined.

To get their estimates we have used the experimental data presented in [3]. The values found refer, then, to a sample of Jackson-Hutton crude oil cooled from $45\,°C$ to $10\,°C$ at an average cooling rate of $5\,°C/hour$ and subject to a constant shear rate of $7.3\ s^{-1}$. The values obtained are: $k_1 = 1.25 \times 10^{-4}\ s^{-1}$, $k_2 = 10^{-6}\ Pa^{-1}$ and $\tau_M = 5\ Pa$; which, as already mentioned, are nothing more than simple estimates of the order of magnitude. We can note, anyway, that these values of k_1 and k_2 are such that the hypotheses of the quasi-steady approximation are verified.

In order to perform the numerical simulation we have considered equations (13) - (21). This, in fact, allows to use an evolution equation for $s(t)$. The system (13) - (21) is then solved by a Chebychev collocation method in space and a fourth order Runge-Kutta method in time.

Since the integration interval and, therefore, the collocation nodes, are time dependent, time derivatives are handled in a pseudo-Lagrangian way.

The results of the numerical integration are given in Fig.1 - 3 where the time behavior of s (expressed in m) is reported. Time is expressed in seconds. To obtain these results we have considered the following values for the physical parameters involved in the model: pressure gradient $f_o = 25\,KPa/Km$, oil density $\rho = 0.8\,gr/cm^3$, Bingham viscosity $\eta = 0.6\,Pa\,s$ and width of the canal $L = 1m$.

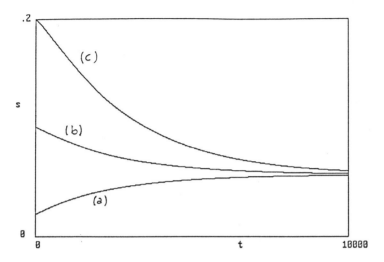

Figure 1. Initial conditions of s and α are linked by relation (23). The initial values of α are respectively: (a) $\alpha_o = 0.1$; (b) $\alpha_o = 0.5$; (c) $\alpha_o = 1$.

In Fig.1 the initial conditions on s and α are linked by the relation (23) and the initial flow is given by (24). In particular the curves (a), (b) and (c) refer respectively to the following initial values of α: 0.1; 0.5 and 1. So we are in the conditions which justify the quasi-stationary approximation and therefore the dynamics is governed by (27). In this case the system has a unique stable stationary point, given by $s_{eq} \simeq 0.06m$, and, as shown by the simulations, $s(t)$ tends to this value.

Figures 2 and 3 represent the behavior of s when the initial condition of s and α are not linked by (23). In particular in Fig.2 the initial condition on s is fixed ($s(0) = 0.5m$) while $\alpha(0)$ varies and is equal to 0.1, 0.5 and 0.8 respectively for the curves (a), (b) and (c). Finally in fig.3 the initial condition on α is equal for each one of the three curves ($\alpha(0) = 0.3$), while $s(0)$ varies.

We can note that, after a short transitory, the dynamics of the system is ruled by that one emerging by the quasi-steady approximation tending to the limit value s_{eq}. Of course, the transient period depends on the values of the parameters involved.

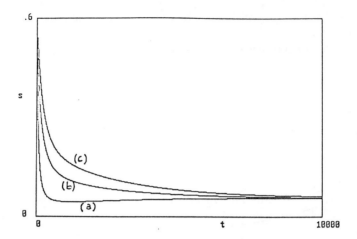

Figure 2. Initial conditions of s and α are not linked by (23). The initial value of s is fixed for each one of the three curves ($s_o = 0.5m$). Initial values of α: (a) $\alpha_o = 0.1$; (b) $\alpha_o = 0.5$; (c) $\alpha_o = 0.8$.

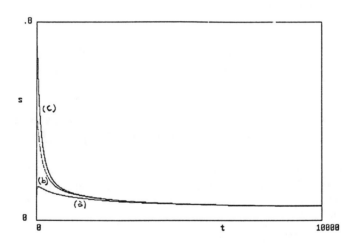

Figure 3. Initial conditions of s and α are not linked by (23). For each one of the three curves $\alpha_o = 0.5$. Initial values of s: (a) $s_o = 0.1$; (b) $s_o = 0.5$; (c) $s_o = 0.7$.

References

[1] K.M. Agarwal, R.C. Purohit, M. Surianarayanan, G.C. Joshi, R. Krishna, "Influence of waxes on the flow properties of Bombay high crude", Fuel **68**, (1989), pp. 937-938.

[2] L.T. Wardhaugh, D.V. Boger, "Measurement of the unique flow properties of waxy crude oils", Chem Eng Res Des **65**, (1987), pp. 74-83.

[3] L.T. Wardhaugh, D.V. Boger, "Flow characteristics of waxy crude oils: application to pipeline design", AIChE Journal **37**, (1991), pp. 871-884.

[4] J. Denis, J.P. Durand, "Modification of wax crystallization in petroleum products", Revue de l'Insitut Française du Pétrole **46**, (1991), pp. 637-649.

[5] R.B. Bird, W.E. Stewart, E.N. Lightfoot, *Transport Phenomena*, Wiley, New York (1960).

[6] L. Rubinstein, *The Stefan problem*, American Mathematical Society translation, vol.27, Amer. Math. Soc. Providence (1971).

[7] A. Farina, "Waxy crude oils: some aspects of their dynamics", to appear in Mathematical Models and Methods in Applied Sciences.

[8] A. Farina, A. Fasano, "Flow characteristics of waxy crude oils in laboratory experimental loops" to appear in Mathemaical and Computer Modelling.

Modelling of Bending Effects in Oil–Water Microemulsions

Yuriy Povstenko
Pidstryhach Institute for Applied Problems
of Mechanics and Mathematics,
Ukrainian National Academy of Sciences,
Naukova 3–b, 290601 Lviv–MCP, Ukraine
E–mail: Kalyniak@ippmm.lviv.ua
and Pedagogical University of Częstochowa,
Armii Krajowej 13/15, 42-201 Częstochowa, Poland

Abstract

Presence of surfactants and cosurfactants in microemulsions results in a very low surface tension. For this reason, the bending effects become of particular significance. In this paper a fluid/fluid interface is considered as a two-dimensional continuum described by two independent vectors – a velocity vector $\mathbf{v}_s$ and an intrinsic angular velocity vector $\mathbf{w}_s$; in such media a couple–stress tensor appears simultaneously with a stress tensor. The theory of non–equilibrium thermodynamics is applied to the system under consideration to obtain the linear phenomenological relations. It results in eight viscosity coefficients being attributed to the interface. Two equations of Navier–Stokes type for the velocity vector and the angular velocity vector are obtained.

1 Introduction

Many technological problems deal with the statics and dynamics of multiphase fluids and require a detailed knowledge of the essential features of fluid/fluid interfaces. Mixtures of oil and water can be stabilized by addition of surfactants and cosurfactants, thus, forming stable dispersions of oil–in–water or water–in–oil (microemulsions). These microemulsions are used in variuos industrial applications and are currently of great interest due to their importance in processes for increasing recovery of oil from undeground reservoirs [1–3].

Presence of surfactants and cosurfactants results in a very low (nearly vanishing) surface tension [4, 5]. This is of considerable significance in oil recovery because only a large decrease of the surface tension enables to the injected water to displace the oil droplets from the small pores [6]. If an interface has a very low surface tension and/or high curvature (conditions met by microemulsions), the bending effects become of particular significance. Molecules of surfactant adsorbed on the interface contain a polar group (soluble in water) and a hydrocarbon tail (soluble in oil) [3, 5] and, thus, have non–symmetric structure. On the one hand, the interface tends to bend making the electrical double layer, formed at the water side of an interface, diffuse. On the other hand, the surfactant lipophiles, residing at the oil side of the interface, also tend to repel one another, forcing the interface to bend [7]. Bending effects in microemulsions have attracted considerable theoretical attention of a number of researches (see [3–11], among others).

Gibbs [12] was the first to recognize that the change in free energy accompanying deformation of an interface between two phases involves terms describing the change in area and the change in curvature. However, subsequently he eliminated the term with the sum of the principle curvatures by selecting an appropriate position of the dividing surface and neglected the term with their difference in comparison with the term proportional to the surface tension. Buff [13] took into account the mean curvature of an interface and Murphy [14] complemented Buff's results by a term with the Gaussian curvature. Although extending the equilibrium theory to large surface curvature both Buff and Murphy did not include the surface rheology in their consideration.

By analogy with bulk fluids Boussinesq [15] introduced the concept of interfacial viscosity (the coefficients of shear and dilatational viscosity). The modern theory of viscous surface was presented by Scriven [16] (see also the previous works of Erickson [17] and Oldroyd [18] and the subsequent detailed presentation of Aris [19]) but this theory has not been generalized to surfaces of large curvature.

Gudrich [20] developed a theory with five viscosity coefficients (in the directions tangential to the interfacial region and in the direction normal to it) and interpreted the coefficients of surface dilatational and surface shear viscosity as excess transport properties. Lindsay and Straughan [21] proposed the constitutive equation for the surface stress tensor as a function not only of the rate–of–strain tensor and the first fundamental form but also of the second fundamental form of the surface. In the series of articles Edwards and Wasan [22] extended the static theory of Murphy to the dynamic case, to provide a surface rheological theory which is general to highly curved fluid surfaces. In parallel with the surface excess shear viscosity and the surface excess dilatational viscosity Edwards and Wasan defined the first moments and the second moments of these surface excess rheological parameters and developed the surface excess linear momentum and first moment of momentum equations for dynamic surfaces with large surface curvature.

In the present paper the curvature effect on the surface rheology is taken into account in another way extending Sriven's [16] approach to a case of the Cosserat surface, i. e. the surface to which two independent vectors, a velocity vector and an intrinsic angular velocity vector, are attributed (in a case of three-dimensional fluid such a theory was presented in [23, 24]). In addition to the terms with the surface shear viscosity and the surface dilatational viscosity, in the theory developed here the rheological equation for a stress tensor includes also the terms with the surface rotational viscosity and the surface cross–resultant viscosity. In the rheological equation for a surface couple-stress tensor four viscous coefficients describe the pure twisting, twisting, bending and cross–couple viscosities, respectively. Earlier Georgescu [25-27] investigated the two–dimensional reology with an asymmetric stress tensor but this study was incomplete because a couple-stress tensor necessary arising in such media was not considered.

2 Balance Equations

Consider a material volume V cut by an interface S. This interface divides the material volume into two volumes V_1 and V_2. The two bulk phases in contact and their interface are identified by subscripts 1, 2 and s, respectively. Any extensive quantity Ψ characterizing a volume V can be written as the sum

$$\Psi = \int_{V_1} \rho_1 \psi_1 \, dV + \int_{V_2} \rho_2 \psi_2 \, dV + \int_S \rho_s \psi_s \, dS \tag{1}$$

where ρ_1 and ρ_2 are the mass densities, ψ_1 and ψ_2 are the densities of Ψ per unit mass in V_1 and V_2; ρ_s is the surface mass density, ψ_s is treated either in accordance with Gibbs' approach as the surface excess density associated with the dividing surface [13] or presentation (1) can be substantiated by the measure theory [28, 29].

The local form of the general balance equation for points that lie in the surface S reads [30, 31]

$$\rho_s \dot{\psi}_s = \Theta_s - \boldsymbol{\nabla}_s \cdot \mathbf{J}_s + \mathbf{n}_1 \cdot \mathbf{J}_1 + \mathbf{n}_2 \cdot \mathbf{J}_2$$
$$+ (\psi_1 - \psi_s)\rho_1(\mathbf{v}_1 - \mathbf{v}_s) \cdot \mathbf{n}_1 + (\psi_2 - \psi_s)\rho_2(\mathbf{v}_2 - \mathbf{v}_s) \cdot \mathbf{n}_2. \tag{2}$$

Here Θ and $\mathbf{J}$ are the production and the flux of Ψ, $\mathbf{n}_k$ is the outer to V_k unit normal to S, $(k = 1,2)$; $\boldsymbol{\nabla}_s$ indicates the surface gradient operator, the superposed dot denotes the material time derivative.

Identifying the quantity ψ with the velocity vector $\mathbf{v}$, the production Θ with the body force vector $\mathbf{f}$ and the flux $\mathbf{J}$ with the stress tensor σ (with the opposite sign) we obtain the two-dimensional equation of the linear momentum balance

$$\rho_s \dot{\mathbf{v}}_s = \mathbf{f}_s + \boldsymbol{\nabla}_s \cdot \sigma_s - \mathbf{n}_1 \cdot \sigma_1 - \mathbf{n}_2 \cdot \sigma_2$$
$$+ (\mathbf{v}_1 - \mathbf{v}_s)\rho_1(\mathbf{v}_1 - \mathbf{v}_s) \cdot \mathbf{n}_1 + (\mathbf{v}_2 - \mathbf{v}_s)\rho_2(\mathbf{v}_2 - \mathbf{v}_s) \cdot \mathbf{n}_2. \tag{3}$$

The two-dimensional moment-of-momentum balance equation corresponds to the following choice: $\psi = \alpha \mathbf{w}$, $\Theta_s = \mathbf{m}_s - (\epsilon_s : \sigma_s)\mathbf{n}_1 + \epsilon_s \cdot \sigma_s \cdot \mathbf{n}_1$, $\mathbf{J} = -\mu$, where $\mathbf{w}$ is the intrinsic angular velocity vector, μ is the couple-stress tensor and $\mathbf{m}$ is the body couple vector, ϵ denotes the surface alternator dyadic. Coefficient α does not depend on time and is connected with the moment of inertia. Thus,

$$\alpha_s \rho_s \dot{\mathbf{w}}_s = \mathbf{m}_s + \boldsymbol{\nabla}_s \cdot \mu_s - \mathbf{n}_1 \cdot \mu_1 - \mathbf{n}_2 \cdot \mu_2 - (\epsilon_s : \sigma_s)\mathbf{n}_1 + \epsilon_s \cdot \sigma_s \cdot \mathbf{n}_1$$
$$+ (\alpha_1 \mathbf{w}_1 - \alpha_s \mathbf{w}_s)\rho_1(\mathbf{v}_1 - \mathbf{v}_s) \cdot \mathbf{n}_1 + (\alpha_2 \mathbf{w}_2 - \alpha_s \mathbf{w}_s)\rho_2(\mathbf{v}_2 - \mathbf{v}_s) \cdot \mathbf{n}_2. \tag{4}$$

In the local basis formed by the tangential vectors $\mathbf{a}_\alpha$, ($\alpha = 1, 2$) and the normal vector $\mathbf{n} \equiv \mathbf{n}_1$ the surface stress tensor $\boldsymbol{\sigma}_s$ and the surface couple-stress tensor $\boldsymbol{\mu}_s$ have the following dyadic representation

$$\boldsymbol{\sigma}_s = \sigma^{\alpha\beta}\mathbf{a}_\alpha\mathbf{a}_\beta + \sigma^{\alpha n}\mathbf{a}_\alpha\mathbf{n} = \boldsymbol{\sigma}_s^{\|} + \boldsymbol{\sigma}_s^{\perp},$$

$$\boldsymbol{\mu}_s = \mu^{\alpha\beta}\mathbf{a}_\alpha\mathbf{a}_\beta + \mu^{\alpha n}\mathbf{a}_\alpha\mathbf{n} = \boldsymbol{\mu}_s^{\|} + \boldsymbol{\mu}_s^{\perp} \quad (5)$$

with $\boldsymbol{\sigma}_s^{\|}$ denoting a projection of $\boldsymbol{\sigma}$ on a tangential plane, $\boldsymbol{\sigma}_s^{\perp} = \boldsymbol{\sigma}_s - \boldsymbol{\sigma}_s^{\|}$.

It is assumed that the surface couple-stress tensor $\boldsymbol{\mu}_s$ is of viscous origin, while the surface stress tensor $\boldsymbol{\sigma}_s$ includes both the eqilibrium surface tension σ_s and the viscous terms:

$$\boldsymbol{\sigma}_s = \sigma_s \mathbf{I}_s + \boldsymbol{\sigma}_s^* = \sigma_s \mathbf{I}_s + \sigma_s^* \mathbf{I}_s + \mathrm{Dev}_s \boldsymbol{\sigma}_s^{*\|S} + \boldsymbol{\sigma}_s^{*\|A} + \boldsymbol{\sigma}_s^{*\perp}, \quad (6)$$

$$\boldsymbol{\mu}_s = \boldsymbol{\mu}_s^* = \mu_s^* \mathbf{I}_s + \mathrm{Dev}_s \boldsymbol{\mu}_s^{*\|S} + \boldsymbol{\mu}_s^{*\|A} + \boldsymbol{\mu}_s^{*\perp} \quad (7)$$

where the asterisk denotes the viscous constituent part of a tensor, the superscripts S and A mean symmetrization and antisymmetrization, $\mathbf{I}_s$ is the unit surface tensor, $\mathrm{Dev}_s \boldsymbol{\sigma}_s$ indicates the surface deviator of the surface tensor $\boldsymbol{\sigma}_s$.

Multiplying the linear momentum equation (3) by the velocity vector $\mathbf{v}_s$ and the moment-of-momentum equation (4) by intrinsic angular velocity vector $\mathbf{w}_s$ and summing the results we get the balance equation of the surface kinetic energy. Substracting this equation from the balance of the total energy e_s

$$\rho_s \dot{e}_s = \mathbf{f}_s \cdot \mathbf{v}_s + \mathbf{m}_s \cdot \mathbf{w}_s - \boldsymbol{\nabla}_s \cdot (\mathbf{J}_s^q - \boldsymbol{\sigma}_s \cdot \mathbf{v}_s - \boldsymbol{\mu}_s \cdot \mathbf{w}_s)$$

$$+\mathbf{n}_1 \cdot (\mathbf{J}_1^q - \boldsymbol{\sigma}_1 \cdot \mathbf{v}_1 - \boldsymbol{\mu}_1 \cdot \mathbf{w}_1) + \mathbf{n}_2 \cdot (\mathbf{J}_2^q - \boldsymbol{\sigma}_2 \cdot \mathbf{v}_2 - \boldsymbol{\mu}_2 \cdot \mathbf{w}_2)$$

$$+ (e_1 - e_s)\rho_1(\mathbf{v}_1 - \mathbf{v}_s) \cdot \mathbf{n}_1 + (e_2 - e_s)\rho_2(\mathbf{v}_2 - \mathbf{v}_s) \cdot \mathbf{n}_2 \quad (8)$$

we arrive at the internal energy balance

$$\rho_s \dot{u}_s = -\boldsymbol{\nabla}_s \cdot \mathbf{J}_s^q + \boldsymbol{\sigma}_s^T : \boldsymbol{\gamma}_s + \boldsymbol{\mu}_s^T : \boldsymbol{\kappa}_s + \mathbf{n}_1 \cdot \mathbf{J}_1^q + \mathbf{n}_2 \cdot \mathbf{J}_2^q$$

$$-\mathbf{n}_1 \cdot \boldsymbol{\sigma}_1 \cdot (\mathbf{v}_1 - \mathbf{v}_s) - \mathbf{n}_2 \cdot \boldsymbol{\sigma}_2 \cdot (\mathbf{v}_2 - \mathbf{v}_s) - \mathbf{n}_1 \cdot \boldsymbol{\mu}_1 \cdot (\mathbf{w}_1 - \mathbf{w}_s) - \mathbf{n}_2 \cdot \boldsymbol{\mu}_2 \cdot (\mathbf{w}_2 - \mathbf{w}_s)$$

$$+ [e_1 - e_s + (\mathbf{v}_1 - \mathbf{v}_s) \cdot \mathbf{v}_s + (\alpha_1 \mathbf{w}_1 - \alpha_s \mathbf{w}_s) \cdot \mathbf{w}_s]\rho_1(\mathbf{v}_1 - \mathbf{v}_s) \cdot \mathbf{n}_1$$

$$+ [e_2 - e_s + (\mathbf{v}_2 - \mathbf{v}_s) \cdot \mathbf{v}_s + (\alpha_2 \mathbf{w}_2 - \alpha_s \mathbf{w}_s) \cdot \mathbf{w}_s]\rho_2(\mathbf{v}_2 - \mathbf{v}_s) \cdot \mathbf{n}_2 \quad (9)$$

with $\mathbf{J}^q$ being the heat flux,

$$\boldsymbol{\gamma}_s = \boldsymbol{\nabla}_s \mathbf{v}_s + \mathbf{I}_s \times \mathbf{w}_s, \qquad \boldsymbol{\kappa}_s = \boldsymbol{\nabla}_s \mathbf{w}_s. \quad (10)$$

It should be mentioned that the tensors $\boldsymbol{\gamma}_s$ and $\boldsymbol{\kappa}_s$ have the same structure as the surface tensors $\boldsymbol{\sigma}_s$ and $\boldsymbol{\mu}_s$ (5).

3 Thermodynamic Analysis

The entropy balance equiation reads

$$\rho_s \dot{s}_s = \Theta_s^s - \nabla_s \cdot \mathbf{J}_s^s + \mathbf{n}_1 \cdot \mathbf{J}_1^s + \mathbf{n}_2 \cdot \mathbf{J}_2^s$$
$$+ (s_1 - s_s)\rho_1(\mathbf{v}_1 - \mathbf{v}_s) \cdot \mathbf{n}_1 + (s_2 - s_s)\rho_2(\mathbf{v}_2 - \mathbf{v}_s) \cdot \mathbf{n}_2 \quad (11)$$

where s_s is the surface entropy density, Θ_s^s and $\mathbf{J}_s^s$ are the surface entropy production and flux, respectively.

The applicability bounds of local equilibrium in two-dimensional systems are narrower than those in a case of three dimensions because of sharp gradient of thermodynamic state in the normal direction and possible dependence of surface thermodynamic state on that of bulk phases in contact. However, the use of the local equilibrium principle in a case of interface is a quite feasible approximation [32]. We assume that local equilibrium exists in every surface mass particle and the surface may be considered as a locally autonomous system. Mathematically, this assumption means that the Gibbs equation, written for the surface, has the following form:

$$\frac{du_s}{dt} = T_s \frac{ds_s}{dt} + \sigma_s \frac{d}{dt}\left(\frac{1}{\rho_s}\right) \quad (12)$$

where σ_s is the equilibrium surface tension, T_s the temperature.

Using equations (9), (11) and (12) the surface entropy flux $\mathbf{J}_s^s$ and the surface entropy production Θ_s^s may be written as

$$\mathbf{J}_s^s = \frac{\mathbf{J}_s^q}{T_s}, \quad (13)$$

$$\Theta_s^s = \mathbf{J}_s^q \cdot \nabla_s \left(\frac{1}{T_s}\right) + \frac{1}{T_s}(\nabla_s^T : \gamma_s + \mu_s^T : \nabla_s - \sigma_s \nabla_s \cdot \mathbf{v}_s)$$
$$+ \mathbf{n}_1 \cdot \mathbf{J}_1^q \left(\frac{1}{T_s} - \frac{1}{T_1}\right) + \mathbf{n}_2 \cdot \mathbf{J}_2^q \left(\frac{1}{T_s} - \frac{1}{T_2}\right)$$
$$- \frac{1}{T_s}[\mathbf{n}_1 \cdot \sigma_1 \cdot (\mathbf{v}_1 - \mathbf{v}_s) + \mathbf{n}_2 \cdot \sigma_2 \cdot (\mathbf{v}_2 - \mathbf{v}_s)$$
$$+ \mathbf{n}_1 \cdot \mu_1 \cdot (\mathbf{w}_1 - \mathbf{w}_s) + \mathbf{n}_2 \cdot \mu_2 \cdot (\mathbf{w}_2 - \mathbf{w}_s)]$$
$$+ \frac{1}{T_s}[e_1 - e_s + (\mathbf{v}_1 - \mathbf{v}_s) \cdot \mathbf{v}_s + (\alpha_1 \mathbf{w}_1 - \alpha_s \mathbf{w}_s) \cdot \mathbf{w}_s + \frac{\sigma_s}{\rho_s} - T_s(s_1 - s_s)]\rho_1(\mathbf{v}_1 - \mathbf{v}_s) \cdot \mathbf{n}_1$$
$$+ \frac{1}{T_s}[e_2 - e_s + (\mathbf{v}_2 - \mathbf{v}_s) \cdot \mathbf{v}_s + (\alpha_2 \mathbf{w}_2 - \alpha_s \mathbf{w}_s) \cdot \mathbf{w}_s + \frac{\sigma_s}{\rho_s} - T_s(s_2 - s_s)]\rho_2(\mathbf{v}_2 - \mathbf{v}_s) \cdot \mathbf{n}_2. \quad (14)$$

It should be noted that the surface entropy production contains two types of terms. The first type terms involve inhomogeneities within the surface giving rise to heat flow, stresses and couple stresses; the second ones describe the entropy production due to discontinuites of temperature, velocity and angular velocity across the interface. We shall not present a detailed analysis of the general phenomenological equations and the Onsager reciprocal relations between the phenomenological coefficients but now confine attention to the non–interrelated linear laws between the thermodynamic fluxes and generalized thermodynamic forces which occur in the entropy production. In doing so we make use of the two–dimensional isotropy of the surface. From (14) it follows that

$$\sigma_s^* = \kappa_s(\nabla_s \cdot \mathbf{v}_s)\mathbf{I}_s + 2\mu_s \mathrm{Dev}_s(\nabla_s \mathbf{v}_s)^{\|S}$$

$$+ \nu_s(\nabla_s \times \mathbf{v}_s - 2\mathbf{w}_s) \cdot \mathbf{n}\epsilon_s + 2\mu_s^\perp(\nabla_s \mathbf{v}_s + \mathbf{I}_s \times \mathbf{w}_s)^\perp, \tag{15}$$

$$\mu_s^* = \zeta_s(\nabla_s \cdot \mathbf{w}_s)\mathbf{I}_s + 2\eta_s \mathrm{Dev}_s(\nabla_s \mathbf{w}_s)^{\|S} + \tau_s(\nabla_s \times \mathbf{w}_s) \cdot \mathbf{n}\epsilon_s + 2\eta_s^\perp(\nabla_s \mathbf{w}_s)^\perp, \tag{16}$$

$$\mathbf{J}_s^q = -\Lambda_s \nabla_s T_s, \tag{17}$$

$$\mathbf{n}_k \cdot \mathbf{J}_k^q = H_k(T_k - T_s), \tag{18}$$

$$\mathbf{n}_k \cdot \sigma_k^\| = \varphi_k^\|(\mathbf{v}_s - \mathbf{v}_k)^\|, \qquad X_k = \varphi_k^\perp(\mathbf{v}_s - \mathbf{v}_k) \cdot \mathbf{n}_k, \tag{19}$$

$$\mathbf{n}_k \cdot \mu_k^\| = \phi_k^\|(\mathbf{w}_s - \mathbf{w}_k)^\|, \qquad \mathbf{n}_k \cdot \mu_k \cdot \mathbf{n}_k = \phi_k^\perp(\mathbf{w}_s - \mathbf{w}_k) \cdot \mathbf{n}_k \tag{20}$$

where $k = 1, 2$,

$$X_k = \rho_k\left[e_k - e_s - (\mathbf{v}_k - \mathbf{v}_s) \cdot \mathbf{v}_s - (\alpha_k \mathbf{w}_k - \alpha_s \mathbf{w}_s) \cdot \mathbf{w}_s + \frac{\sigma_s}{\rho_s} - T_s(s_k - s_s) - \mathbf{n}_k \cdot \sigma_k \cdot \mathbf{n}_k\right],$$

κ_s, μ_s, ν_s, $\mu_s^\perp$, ζ_s, η_s, τ_s and $\eta_s^\perp$ are the dilatational, shear, rotational, cross-resultant, pure twisting, twisting, bending, and cross-couple viscosities, respectively; Λ_s is the surface heat conductivity, H_k describe convective heat transfer, the phenomenological coefficients φ_k and ϕ_k represent sliding effects, etc. If $H_k \to \infty$, $\varphi_k^\perp \to \infty$, $\varphi_k^\| \to \infty$, $\phi_k^\perp \to \infty$, $\phi_k^\| \to \infty$ the phenomenological equations provide the continuity conditions for the temperature T, the velocity vector $\mathbf{v}$ and the angular velocity $\mathbf{w}$ at the surface.

4 Navier–Stokes Type Equations

Introducing the surface stress tensor σ_s and surface couple-stress tensor μ_s from Eqs. (15) and (16) into equations of the linear momentum balance (3) and moment-of-momentum balance (4) we find two equations of Navier-Stokes type for the velocity vector $\mathbf{v}_s$ and the angular velocity vector $\mathbf{w}_s$

$$(\kappa_s + \mu_s)\nabla_s(\nabla_s \cdot \mathbf{v}_s) - (\mu_s + \nu_s)\nabla_s \times \nabla_s \times \mathbf{v}_s + 2\mu_s(\mathbf{b} : \nabla_s \mathbf{v}_s)\mathbf{n}$$

$$-(\mu_s + \nu_s - 2\mu_s^\perp)\left[\nabla_s^2(\mathbf{v}_s \cdot \mathbf{n}) + \nabla_s \cdot (\mathbf{b} \cdot \mathbf{v}_s)\right]\mathbf{n} - (\mu_s - \nu_s + 2\mu_s^\perp)\mathbf{b} \cdot (\nabla_s \mathbf{v}_s) \cdot \mathbf{n} + \nabla_s \sigma_s$$

$$+ 2H\left[\sigma_s\mathbf{n} + (\mu_s - \nu_s)(\nabla_s \mathbf{v}_s) \cdot \mathbf{n} + (\kappa_s - \mu_s)(\nabla_s \cdot \mathbf{v}_s)\mathbf{n}\right] + 2\nu_s \epsilon_s \cdot \nabla_s(\mathbf{w}_s \cdot \mathbf{n})$$

$$+ 2\mu_s^\perp\left[\mathbf{nn} \cdot (\nabla_s \times \mathbf{w}_s) - \mathbf{b} \cdot \epsilon_s \cdot \mathbf{w}_s\right] = \rho_s \dot{\mathbf{v}}_s - \mathbf{f}_s + \mathbf{n}_1 \cdot \sigma_1 + \mathbf{n}_2 \cdot \sigma_2$$

$$- (\mathbf{v}_1 - \mathbf{v}_s)\rho_1(\mathbf{v}_1 - \mathbf{v}_s) \cdot \mathbf{n}_1 - (\mathbf{v}_2 - \mathbf{v}_s)\rho_2(\mathbf{v}_2 - \mathbf{v}_s) \cdot \mathbf{n}_2, \tag{21}$$

$$(\zeta_s + \eta_s)\nabla_s(\nabla_s \cdot \mathbf{w}_s) - (\eta_s + \tau_s)\nabla_s \times \nabla_s \times \mathbf{w}_s + 2\eta_s(\mathbf{b} : \nabla_s \mathbf{w}_s)\mathbf{n}$$

$$-(\eta_s + \tau_s - 2\eta_s^\perp)[\nabla_s^2(\mathbf{w}_s \cdot \mathbf{n}) + \nabla_s \cdot (\mathbf{b} \cdot \mathbf{w}_s)]\mathbf{n} - (\eta_s - \tau_s + 2\eta_s^\perp)\mathbf{b} \cdot (\nabla_s \mathbf{w}_s) \cdot \mathbf{n}$$

$$+2H\left[(\eta_s - \tau_s)(\nabla_s \mathbf{w}_s) \cdot \mathbf{n} + (\zeta_s - \eta_s)(\nabla_s \cdot \mathbf{w}_s)\mathbf{n}\right] + 2\nu_s \mathbf{nn} \cdot (\nabla_s \times \mathbf{v}_s - 2\mathbf{w}_s)$$

$$+2\mu_s^\perp(\nabla_s \times \mathbf{v}_s - \mathbf{w}_s) \cdot \mathbf{I}_s = \alpha_s \rho_s \dot{\mathbf{w}}_s - \mathbf{m}_s + \mathbf{n}_1 \cdot \mu_1 + \mathbf{n}_2 \cdot \mu_2$$

$$- (\alpha_1 \mathbf{w}_1 - \alpha_s \mathbf{w}_s)\rho_1(\mathbf{v}_1 - \mathbf{v}_s) \cdot \mathbf{n}_1 - (\alpha_2 \mathbf{w}_2 - \alpha_s \mathbf{w}_s)\rho_2(\mathbf{v}_2 - \mathbf{v}_s) \cdot \mathbf{n}_2 \quad (22)$$

where b denotes the dyadic of the second fundamental form of the surface, H is the mean curvature.

In the absence of couple-stresses neglecting the inertia and interchange terms Eq. (21) simplifies and coincides with the linear momentum balance equation obtained by Scriven [16].

The two-dimensional rheology is also of particular significance to imposing approproate boundary conditions on the three-dimansional hydrodynamic equations. The Navier-Stokes type equations (21) and (22) together with the phenomenological equations (19), (20) provide such boundary conditions.

References

[1] Healy R N/Reed R L/Stenmark D G: Multiphase microemulsion systems, Soc. Petroleum Eng. J. 16, (1976), pp. 147–155.

[2] Hwan R/Miller C A/Fort T: Determination of microemulsion phase continuity and drop size by ultracentrifugation, J. Colloid Interface Sci. 68, 1 (1979), pp. 221-231.

[3] Miller C A/Neogi P: Thermodynamics of microemulsions: combined effect of dispersion entropy of drops and bending energy of surfactant films, AIChE J. 26, 2 (1980), pp. 212–220.

[4] Pouchelon A/Chatenay D/Meunier J/Langevin D: Origin of low interfacial tensions in systems involving microemulsion phases, J. Colloid Interface Sci. 82, 2 (1981), pp. 418–422 .

[5] De Gennes P G/Taupin C: Microemulsions and the flexibility of oil/water interfaces, J. Phys. Chem. 86, 13 (1982), pp. 2294–2304.

[6] Ruckenstein E: Stability, phase equilibrium, and interfacial free energy in microemulsions, Micellization, Solubilization, and Microemulsions. Vol. 2, 1977, pp. 755–778.

[7] Huh C: Equilibrium of a microemulsion that coexists with oil and brine, Soc. Petroleum Eng. J. 23, 5 (1983), pp. 829–847.

[8] Ruckenstein E: An explanation for the unusual phase behavior of microemulsions, Chem. Phys. Lett. 98, 6 (1983), pp. 573–576.

[9] Miller C A: Interfacial bending effects and interfacial tensions in microemulsions, J. Dispersion Sci. Techn. 6, 2 (1985), pp. 159–173.

[10] Neogi P/Kim M/Friberg S E: Micromechanics of surfactant microstructures, J. Phys. Chem. 91, 3 (1987), pp. 605–611.
[11] Neogi P/Friberg S E: Curved surfaces in surfactant aggregates, J. Colloid Interface Sci. 127, 2 (1989), pp. 492–496.
[12] Gibbs J W: Scientific Papers. Vol. 1, 1906/1961.
[13] Buff F P: Curved fluid interfaces. I. The generalized Gibbs–Kelvin equation, J. Chem. Phys. 25, 1 (1956), pp. 146–153.
[14] Murphy C L: Thermodynamics of Low Tension and Highly Curved Interfaces. Ph.D. thesis, University of Minnesota, Minneapolis, 1966.
[15] Boussinesq J: Sur l'existence d'une viscosité superficielle, dans la mince couche de transition séparant un liquide d'un autre fluide contigu, Ann. Chim. Phys. 29 (1913), pp. 349–357.
[16] Scriven L E: Dynamics of a fluid interface. Equation of motion for Newton surface fluid, Chem. Eng. Sci. 12, 2 (1960), pp. 98–108.
[17] Ericksen J L: Thin liquid jets, J. Ration. Mech. Anal. 1, 4 (1952), pp. 521–538.
[18] Oldroyd S G: The reology of some two-dimensional disperse systems, Proc. Cambridge Phil. Soc. 53, 2 (1957), pp. 514–524.
[19] Aris R: Vectors, Tensors and the Basic Equations of Fluid Mechanics, 1962, pp. 226–244.
[20] Goodrich F C: The theory of capillary excess viscosities, Proc. Roy. Soc. London A374, 1758 (1981), pp. 341–370.
[21] Lindsay K A/Straughan B: A thermodynamic viscous interface theory and associated stability problems, Arch. Ration. Mech. Anal. 71, 4 (1979), pp. 307–326.
[22] Edwards D A/Wasan D T: Surface rheology, J. Rheol. 32 (1988), pp. 429-484.
[23] Aero E A/Bulygin A N/Kuvshinskii E V: Asimmetric hydrodynamics, J. Appl. Math. Mech. 29, 2 (1965), pp. 333–344.
[24] Eringen A C: Theory of micropolar fluids, J. Math. Mech. 16 (1966), pp. 1–18.
[25] Georgescu L: A Navier-Stokes type equation for the phase interface of a liquid, Surface Sci. 15, 1 (1969), pp. 177–181.
[26] Georgescu L: Phenomenological equations for the multicomponent phase interface, Rev. Roum. Phys. 20, 8 (1975), pp. 781–790.
[27] Georgescu L: Contributii la studiul fazelor de interfata si al fenomenolor de transport neliniare pe baza termodinamicii proceselor ireversibile, Stud. Cerc. Fiz. 28, 1 (1976), pp. 25–56.
[28] Gurtin M E/Williams W O: An axiomatic foundation for continuum thermodynamics, Arch. Ration. Mech. Anal. 26, 2 (1967), pp. 83–117.
[29] Fisher G M C/Leitman M J: On continuum thermodynamics with surfaces, Arch. Ration. Mech. Anal. 30, 3 (1968), pp. 225–262.
[30] Ghez R: A generalized Gibbsian surface, Surface Sci. 4, 2 (1966), pp. 125–140.
[31] Moeckel G P: Thermodynamics of an interface, Arch. Ration. Mech. Anal. 57, 3 (1975), pp. 255–280.
[32] Defay R/Prigogine I/Sanfeld A: Surface thermodynamics, J. Colloid Interface Sci. 58, 3 (1977), pp. 498–510.

Optimization in industry

Optimal Power Dispatch via Multistage Stochastic Programming *

M.P. Nowak[1] and W. Römisch[1]

Abstract. The short-term cost-optimal dispatch of electric power in a generation system under uncertain electricity demand is considered. The system comprises thermal and pumped-storage hydro units. An operation model is developed which represents a multistage mixed-integer stochastic program and a conceptual solution method using Lagrangian relaxation is sketched. For fixed start-up and shut-down decisions an efficient algorithm for solving the multistage stochastic program is described and numerical results are reported.

1 Introduction

Mathematical models for cost-optimal power scheduling in hydro-thermal systems often combine several difficulties such as a large number of mixed-integer variables, nonlinearities, and uncertainty of problem data. Typical examples for the latter are uncertain prices in electricity trading, the future electric power demand, and future inflows into reservoirs of hydro plants. Incorporating the uncertainties directly into an optimization model leads to stochastic programming problems. In the context of power scheduling such models are developed e.g. in [4], [5], [9], [11].

In the present paper we consider a short-term optimization model for the dispatch of electric power in a hydro-thermal generation system over a certain time horizon in the presence of uncertain demand. The generation system comprises (coal-fired and gas-burning) thermal and pumped-storage hydro units (without inflows) which is typical for the eastern part of Germany. Short- and long-term energy contracts are regarded (and modelled) as (particular) thermal units. The operation of such a generation system is very complex, because it creates a link between a decision in a given time interval and the future consequences of this decision. Even for optimal on-line power scheduling future costs created by actual decisions have to be taken into account. Since a longer time horizon (e.g. one

---

[1]Humboldt University Berlin, Institute of Mathematics, 10099 Berlin, Germany

*This research is supported by the Schwerpunktprogramm "Echtzeit-Optimierung großer Systeme" of the Deutsche Forschungsgemeinschaft (DFG)

week) is often needed due to the pumping cycle of the hydro storage plants, the stochastic nature of the demand cannot be ignored. The optimization model thus represents a multistage stochastic program containing mixed-integer (stochastic) decisions which reflect the on/off schedules and production levels of the generating units for all time intervals of the horizon. The increase of scenarios in the stochastic programming model corresponds to a decrease of information on the power demand.

The stochastic model will be developed and discussed in some detail in section 2 (for more information we refer to [3]). In section 3 we sketch a conceptual decomposition method by applying Lagrangian relaxation to the loosely coupled multistage stochastic program, and in section 4 an efficient algorithm for solving the stochastic program for fixed on/off decisions is described and numerical results are reported.

2 Stochastic Model

The mathematical model represents a mixed-integer multistage stochastic program with linear constraints. Let T denote the number of (hourly or shorter) time intervals in the optimization horizon and $\{\mathbf{d}^t : t = 1,\ldots,T\}$ the stochastic demand process (on some probability space $(\Omega, \mathcal{A}, P)$) reflecting the stochasticity of the electric power demand. It is assumed that the information on the demand is complete for $t = 1$ and that it decreases with increasing t. This is modelled by a filtration of σ-fields

$$\mathcal{A}_1 = \{\emptyset, \Omega\} \subseteq \mathcal{A}_2 \subseteq \ldots \subseteq \mathcal{A}_t \subseteq \ldots \subseteq \mathcal{A}_T \subseteq \mathcal{A},$$

where $\mathcal{A}_t$ is the σ-field generated by the random vector $(\mathbf{d}^1,\ldots,\mathbf{d}^t)$. Let I and J denote the number of thermal and pumped-storage hydro units in the system, respectively. According to the stochasticity of the demand process the decisions for all thermal and hydro units

$$\{(\mathbf{u}_i^t, \mathbf{p}_i^t) : t = 1,\ldots,T\} \quad (i = 1,\ldots,I)$$
$$\{(\mathbf{s}_j^t, \mathbf{w}_j^t) : t = 1,\ldots,T\} \quad (j = 1,\ldots,J)$$

are also stochastic processes being adapted to the filtration of σ-fields. The latter condition means that the decisions at time t only depend on the demand vector $(\mathbf{d}^1,\ldots,\mathbf{d}^t)$ (nonanticipativity). Here, $\mathbf{u}_i^t \in \{0,1\}$ and $\mathbf{p}_i^t$ denote the on/off decision and the production level for the thermal unit i and time interval t, respectively, and $\mathbf{s}_j^t$, $\mathbf{w}_j^t$ are the generation and pumping levels, for the pumped-storage plant j during time interval t, respectively. Further, let $\mathbf{l}_j^t$ denote the water level (in terms of electrical energy) in the upper reservoir of plant j at the end of interval t.

The objective function is given by the expected value of the total fuel and start-up costs of the thermal units

$$\mathbb{E}\left[\sum_{i=1}^{I}\sum_{t=1}^{T} FC_i(\mathbf{p}_i^t, \mathbf{u}_i^t) + SC_i(\mathbf{u}_i(t))\right] \tag{2.1}$$

where $\mathbb{E}$ denotes the expectation, FC_i the fuel cost function and SC_i the start-up costs for the operation of the i-th thermal unit. It is assumed that the functions FC_i are monotonically increasing and piecewise linear convex with respect to $\mathbf{p}_i^t$ and that $SC_i(\mathbf{u}_i(t))$ is determined by $(\mathbf{u}_i^t, \ldots, \mathbf{u}_i^{t_{s_i}})$ where $t - t_{s_i}$ is the preceding down-time of the unit i (see e.g. [12] for typical start-up cost functions). All the (stochastic) variables mentioned above have finite lower and upper bounds reflecting unit capacity limits and reservoir capacities of the generation system:

$$p_{it}^{min} \mathbf{u}_i^t \leq \mathbf{p}_i^t \leq p_{it}^{max} \mathbf{u}_i^t, \; i=1,\ldots,I, \; t=1,\ldots,T$$
$$0 \leq \mathbf{s}_j^t \leq s_{jt}^{max}, 0 \leq \mathbf{w}_j^t \leq w_{jt}^{max}, 0 \leq \mathbf{l}_j^t \leq l_{jt}^{max}, j=1,\ldots,J, \; t=1,\ldots,T \quad (2.2)$$

The constants p_{it}^{min}, p_{it}^{max}, s_{jt}^{max}, w_{jt}^{max} and l_{jt}^{max} denote the minimal/maximal outputs and maximal water levels in the upper reservoir, respectively. During the whole time horizon reservoir constraints have to be maintained for all pumped storage plants. These are modelled by the equations:

$$\mathbf{l}_j^t = \mathbf{l}_j^{t-1} - \mathbf{s}_j^t + \eta_j \mathbf{w}_j^t, \quad t=1,\ldots,T, \; j=1,\ldots,J \quad (2.3)$$
$$\mathbf{l}_j^0 = l_j^{in}, \; \mathbf{l}_j^T = l_j^{end}, \quad j=1,\ldots,J.$$

Here, l_j^{in} and l_j^{end} denote the initial and terminal water level in the upper reservoir, respectively, and η_j is the efficiency of the j-th pumped-storage plant. Moreover, there are minimum down times τ_i and possible must-on/off constraints for each thermal unit i. Minimum down times are imposed to prevent the thermal stress and high maintenance costs due to excessive unit cycling. They are described by the inequalities:

$$\mathbf{u}_i^{t-1} - \mathbf{u}_i^t \leq 1 - \mathbf{u}_i^\tau, \; \tau = t+1,\ldots,\min\{t+\tau_i-1,T\}, \; i=1,\ldots,I, \; t=2,\ldots,T. \quad (2.4)$$

Load coverage for each time interval t of the horizon is described by the equations:

$$\sum_{i=1}^{I} \mathbf{p}_i^t + \sum_{j=1}^{J} \left(\mathbf{s}_j^t - \mathbf{w}_j^t\right) = \mathbf{d}^t, \; t=1,\ldots,T. \quad (2.5)$$

In order to compensate sudden load increases or unforeseen events on-line, some spinning reserve level $\mathbf{r}^t$ for the thermal units is required leading to the constraints:

$$\sum_{i=1}^{I} \left(p_{it}^{max} \mathbf{u}_i^t - \mathbf{p}_i^t\right) \geq \mathbf{r}^t, \; t=1,\ldots,T. \quad (2.6)$$

Altogether, (2.1)-(2.6) represents a multistage stochastic program with $2(I+J)T$ stochastic decision variables. For large power generation systems like that of VEAG Vereinigte Energiewerke AG in the eastern part of Germany (with $I = 25$, $J = 7$ and $T = 168$ which corresponds to an hourly discretization of one week) these models involve an enormous number of stochastic decisions.

Numerical approaches for solving (2.1)-(2.6) are mostly based on designing discretization schemes (scenario trees) for the probability distribution of the random demand vector $(\mathbf{d}^1, \ldots, \mathbf{d}^T)$ and lead to large-scale mixed-integer programs with a huge number of variables. In general, such problems are too large from the viewpoint of even the latest solution techniques for multistage stochastic programs with discrete distributions. However, it is possible to make use of the fact that the problem (2.1)-(2.6) is loosely coupled via the constraints (2.5) and (2.6) with respect to the operation of different units.

3 Lagrangian relaxation

For deterministic models of the form (2.1)-(2.6) (i.e. for $\mathcal{A}_T = \{\emptyset, \Omega\}$) the authors of [10] came to the conclusion that for solving realistic problems a clear consensus is presently tending toward the Lagrangian relaxation approach over other methodologies. The approach is based on introducing a partial Lagrangian for constraints linking the operation of different units and on solving the nondifferentiable concave dual maximization problem by modern nonsmooth optimization methods. This approach is also suggested for certain multistage stochastic models in [11] by relying on the same arguments as in the deterministic case, namely, that the dual problems decompose into single unit subproblems and the duality gap becomes small under certain circumstances (see [1], [3]).

In order to describe the Lagrangian relaxation approach for the model (2.1)-(2.6), let $\{\boldsymbol{\lambda}^t : t = 1, \ldots, T\}$ and $\{\boldsymbol{\mu}^t : t = 1, \ldots, T\}$ be stochastic processes in $L^1(\Omega, \mathcal{A}, P; I\!\!R^T)$ adapted to the filtration and consider the (partial) Lagrangian:

$$L(\mathbf{u}, \mathbf{p}, \mathbf{s}, \mathbf{w}; \boldsymbol{\lambda}, \boldsymbol{\mu}) = I\!\!E \sum_{t=1}^{T} \left\{ \sum_{i=1}^{I} \{FC_i(\mathbf{p}_i^t, \mathbf{u}_i^t) + SC_i(\mathbf{u}_i(t))\} \right.$$

$$\left. + \boldsymbol{\lambda}^t \left(\mathbf{d}^t - \sum_{i=1}^{I} \mathbf{p}_i^t - \sum_{j=1}^{J} (\mathbf{s}_j^t - \mathbf{w}_j^t) \right) + \boldsymbol{\mu}^t \left(\mathbf{r}^t - \sum_{i=1}^{I} (\mathbf{u}_i^t p_{it}^{max} - \mathbf{p}_i^t) \right) \right\}. \quad (3.1)$$

The dual problem then reads

$$\max \{ D(\boldsymbol{\lambda}, \boldsymbol{\mu}) : \boldsymbol{\mu} \geq 0, \boldsymbol{\lambda} \}, \quad (3.2)$$

where $D(\boldsymbol{\lambda}, \boldsymbol{\mu})$ denotes the infimum of $L(\mathbf{u}, \mathbf{p}, \mathbf{s}, \mathbf{w}; \boldsymbol{\lambda}, \boldsymbol{\mu})$ subject to $(\mathbf{u}, \mathbf{p}, \mathbf{s}, \mathbf{w})$ satisfying the constraints (2.2)-(2.4). Since (2.2)-(2.4) represent exclusively single unit constraints, the Lagrangian dual function can be written as:

$$D(\boldsymbol{\lambda}, \boldsymbol{\mu}) = \sum_{i=1}^{I} D_i(\boldsymbol{\lambda}, \boldsymbol{\mu}) + \sum_{j=1}^{J} \hat{D}_j(\boldsymbol{\lambda}) + I\!\!E \sum_{t=1}^{T} \left[\boldsymbol{\lambda}^t \mathbf{d}^t + \boldsymbol{\mu}^t \mathbf{r}^t \right].$$

where the functions D_i and $\hat{D}_j$ are defined by

$$D_i(\boldsymbol{\lambda}, \boldsymbol{\mu}) = \min_{(\mathbf{u}_i, \mathbf{p}_i)} I\!\!E \sum_{t=1}^{T} \left[FC_i(\mathbf{p}_i^t, \mathbf{u}_i^t) + SC_i(\mathbf{u}_i(t)) - (\boldsymbol{\lambda}^t - \boldsymbol{\mu}^t) \mathbf{p}_i^t - \boldsymbol{\mu}^t \mathbf{u}_i^t p_{it}^{max} \right] \quad (3.3)$$

$$\hat{D}_j(\lambda) = \min_{(s_j, w_j)} \mathbb{E} \sum_{t=1}^{T} \left[-\lambda^t \left(s_j^t - w_j^t \right) \right] \tag{3.4}$$

and the corresponding minimization is carried out subject to the constraints (2.2), (2.4) and (2.2), (2.3), respectively.

To solve the dual problem, an iterative bundle-type method ([6], [7]) is used for updating the Lagrange multipliers (λ, μ) and maximizing the concave function D, respectively. For given (λ, μ) the value of $D(\lambda, \mu)$ is computed by solving the single thermal unit subproblem (3.3) by dynamic programming (as described in [11]) and the single hydro subproblem (3.4) by a fast descent method developed in [8]. After solving the dual problem a heuristic approach is applied to obtain primal decisions $(\mathbf{u}, \mathbf{p}, \mathbf{s}, \mathbf{w})$ which are feasible for (2.5) and (2.6). This approach consists in a modification of the search for reserve-feasible solutions in [12] for the case of hydro-thermal systems.

4 Economic Dispatch

Having the binary variables $\mathbf{u}_i^t$ fixed, one has to solve the minimization problem with respect to $\mathbf{p}_i^t$, $\mathbf{s}_j^t$ and $\mathbf{w}_j^t$, i.e. the economic dispatch problem. It means that the objective function

$$\mathbb{E} \sum_{t=1}^{T} \sum_{i=1}^{I} FC_i(\mathbf{p}_i^t, \mathbf{u}_i^t) \tag{4.1}$$

has to be minimized subject to the constraints (2.2),(2.3),(2.5) and (2.6).

Taking the right-hand side of $\sum_{i=1}^{I} \mathbf{p}_i^t = \mathbf{d}^t - \sum_{j=1}^{J} (\mathbf{s}_j^t - \mathbf{w}_j^t)$ as a parameter $\mathbf{v}^t$ the problem for one time period and one realization of $\mathbf{v}^t$ reads:

$$\text{minimize} \sum_{i=1}^{I} FC_i(\mathbf{p}_i^t, \mathbf{u}_i^t) \text{ s.t. } \sum_{i=1}^{I} \mathbf{p}_i^t = \mathbf{v}^t, \ \mathbf{u}_i^t p_{it}^{min} \leq \mathbf{p}_i^t \leq \mathbf{u}_i^t p_{it}^{max}, \ i=1,..,I \tag{4.2}$$

Since $FC_i(\mathbf{p}_i^t, \mathbf{u}_i^t)$ are piecewise linear with respect to $\mathbf{p}_i^t$, the optimal value function $\phi^t(\mathbf{v}^t)$ of (4.2) is piecewise linear, too. Sorting the segments (see figure 1) of the cost functions of all thermal units, the computation of $\phi^t(\mathbf{v}^t)$ consists of a look up in a list. Then the problem (4.1) consists in minimizing

$$\mathbb{E} \sum_{t=1}^{T} \phi^t \left(\mathbf{d}^t - \sum_{j=1}^{J} (\mathbf{s}_j^t - \mathbf{w}_j^t) \right) \tag{4.3}$$

subject to (2.2), (2.3).

This problem can be solved by a modification of the algorithm described in [8]. The crucial point in this descent algorithm consists in selecting a direction from a prescribed subset of descent directions. Since the objective function is not linear as in [8], the algorithm has to regard the kinks of the piecewise linear function. At

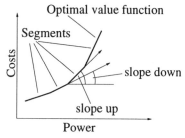
Figure 1: overall cost function

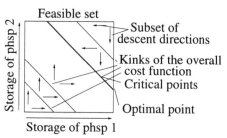
Figure 2: piecewise linear function

these kinks two different slopes (up and down in figure 1) have to be considered. Since the kinks of the objective function (see figure 2) do not coincide with the sufficiently large subset of directions of [8], one has to avoid critical points.

Under the additional assumption $l_i^{in} = l_i^{end}$, $i = 1, \ldots, I$ the point $\mathbf{s}_i^t \equiv \mathbf{w}_i^t \equiv 0$ is feasible. With this point as a starting point and choosing the direction of steepest descent the algorithm converges to an optimal solution.

5 Computational results

The algorithm described above is implemented in $C++$. The performance of the corresponding code ECDISP has been compared with CPLEX 4.0 [2] on several examples. Here we report computation times for both codes on an example including 25 thermal power units, 8 pumped hydro storage plants, 192 stages and 1 scenario. This corresponds to a linear program with 14200 columns, 17856 rows, 46256 nonzero elements of the matrix. The computation time of ECDISP is 50.95 seconds. For CPLEX we display the computation times (in seconds) for different methods and pricing strategies.

CPLEX function	Pricing strategy primal/dual					
	-1	0	1	2	3	4
Simplex/primal	1232.47	1188.4	1918.15	2664.14	2440.7	1696.9
Simplex/dual		1086.18	946.24	1103.48	1466.54	1083.8
baropt	94.78					
hybbaropt/primal	114.71	114.32	114.36	486.55	114.45	114.35
hybbaropt/dual		115.08	114.69	693.03	1424.86	114.84
hybnetopt/primal	957.66	910.39	1298.03	2252.83	1960.93	1162.68
hybnetopt/dual		1393.82	1253.76	1412.06	1833.96	1392.3

All these computations are done on a SPARCstation IPX (4/50) with 64 MB Main Memory and 40 MHz CPU-frequency.

Further comparisons with several numbers of scenarios are performed, too. At this time the code ECDISP is compared with the *baropt* function of CPLEX only.

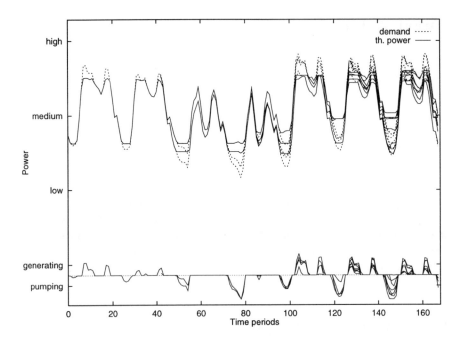

Figure 3: Stochastic solution

The number of nodes is the number of nodes in the scenario tree. Here, advantage denotes the advantage of using ECDISP versus CPLEX.

scenarios	nodes	columns	rows	nonzeros	ECDISP	CPLEX	advantage
2	252	18632	23436	60708	11.02	64.35	5.84
6	504	37248	46872	121408	37.17	228.76	6.15
10	723	53422	67239	174155	61.59	289.79	4.71
14	966	71372	89838	232686	116.81	534.78	4.58
18	1064	78592	98952	256272	103.33	504.41	4.88
22	1260	93064	117180	303476	128.18	794.93	6.20

The amount of memory needed by CPLEX exceeds the memory of the workstation if examples are computed with more than 22 scenarios, but ECDISP can even handle problems with more than 500 scenarios.

For testing the algorithm on a stochastic power dispatch model a scenario tree for approximating the stochastic demand process is generated as follows:

$$d^t_{sampled} = d^t_{given} + \alpha(d^{t-1}_{sampled} - d^{t-1}_{given}) + \beta\varepsilon$$

where d^t_{given} is the given load, $d^t_{sampled}$ a sample of the demand tree for the time interval t, ε is a standard normal (i.e. $N(0,1)$) random variable, and α, β are sampling parameters. A numerical example for the demand scenario and the corresponding stochastic schedule is given in figure 3.

Acknowledgement

We wish to thank G. Scheibner, G. Schwarzbach and J. Thomas (VEAG Vereinigte Energiewerke AG) for the fruitful and productive cooperation.

References

[1] D.P. Bertsekas, G.S. Lauer, N.R. Sandell, and T.A. Posbergh. Optimal short-term scheduling of large-scale power systems. *IEEE Transactions on Automatic Control*, 28(1983)1-11.

[2] CPLEX Optimization, Inc. *Using the CPLEX Callable Library*, 1989-1995.

[3] D. Dentcheva and W. Römisch. Optimal power generation under uncertainty via stochastic programming. Preprint Nr. 96-35, Humboldt-Universität Berlin, Institut für Mathematik, 1996.

[4] N. Gröwe, W. Römisch, and R. Schultz. A simple recourse model for power dispatch under uncertain demand. *Annals of Operations Research*, 59(1995)135-164.

[5] J. Jacobs, G. Freeman, J. Grygier, D. Morton, G. Schultz, K. Staschus, and J. Stedinger. SOCRATES: A system for scheduling hydroelectric generation under uncertainty. *Annals of Operations Research*, 59(1995)99-133.

[6] K.C. Kiwiel. Proximity control in bundle methods for convex nondifferentiable minimization. *Mathematical Programming*, 46(1990)105-122.

[7] C. Lemarechal. Lagrangian decomposition and nonsmooth optimization: bundle algorithm, prox iteration, augmented lagrangian. In: F. Gianessi (Ed.) *Nonsmooth Optimization Methods and Applications*, Gordon and Breach 1992,201-216.

[8] M.P. Nowak. A fast descent method for the hydro storage subproblem in power generation. Working Paper 96-109, International Institute for Applied Systems Analysis, September 1996.

[9] M.V.F Pereira and L.M.V.G. Pinto. Multi-stage stochastic optimization applied to energy planning. *Mathematical Programming*, 52(1991)359-375.

[10] G.B. Sheble and G.N. Fahd. Unit commitment literature synopsis. *IEEE Transactions on Power Systems*, 9(1994)128-135.

[11] S. Takriti, J.R. Birge, and E. Long. A stochastic model for the unit commitment problem. *IEEE Transactions on Power Systems*, 11(1996)1397-1506.

[12] F. Zhuang and F.D. Galiana. Towards a more rigorous and practical unit commitment by lagrangian relaxation. *IEEE Transactions on Power Systems*, 3(1988)763-773.

Solving the Unit Commitment Problem in Power Generation by Primal and Dual Methods*

D. Dentcheva[1], R. Gollmer[1], A. Möller[1],
W. Römisch[1] and R. Schultz[2]

Abstract. The unit commitment problem in power plant operation planning is addressed. For a real power system comprising coal- and gas-fired thermal and pumped-storage hydro plants a large-scale mixed integer optimization model for unit commitment is developed. Then primal and dual approaches to solving the optimization problem are presented and results of test runs are reported.

1 Introduction

The unit commitment problem in electricity production deals with the fuel cost optimal scheduling of on/off decisions and output levels for generating units in a power system over a certain time horizon. The problem typically involves technological and economic constraints. Depending on the shares of nuclear, conventional thermal, hydro and pumped-storage hydro power in the underlying generation system fairly different cost functions and side conditions arise in mathematical models for unit commitment. In the present paper we consider a power system comprising coal- and gas-fired thermal units and pumped-storage hydro plants. This reflects the energy situation in the eastern part of Germany. Our work grew out of a cooperation with the power company VEAG Vereinigte Energiewerke AG Berlin. In our unit commitment model, the objective function is given by start-up and operation costs of the thermal units. Pumped storage plants do not cause direct fuel costs. Their operation, nevertheless, has an impact on the total fuel costs in the system. Constraints of the unit commitment model comprise output bounds for the units of the generation system, load coverage over the whole time horizon as well as provision for a spinning reserve, minimum down times for thermal units and water balances in the pumped-storage plants. Typical optimization horizons vary from several days up to several months.

---

*This research is supported by a grant of the German Federal Ministry of Education, Science, Research and Technology (BMBF).
[1] Humboldt-Universität Berlin, Institut für Mathematik, 10099 Berlin
[2] Konrad-Zuse-Zentrum für Informationstechnik Berlin, Heilbronner Straße 10, 10711 Berlin

From the extensive literature in unit commitment we mention here [1], [3], [5], [11], [16], and [17], and refer to the comprehensive literature synopsis [15]. The papers [5], [9], [16] reflect recent developments of efficient algorithms based on modern mathematical techniques.

2 Modelling

Our mathematical model for the unit commitment problem is a mixed-integer optimization problem with linear constraints. In the following sections we will tackle this problem with both primal and dual solution methods, which will lead to different model specifications and variants. Here, we describe features of the model that are common to both situations.

The optimization horizon is partitioned into (hourly, half-hourly or shorter) time intervals, whose total number is denoted by T. By I and J we denote the number of thermal and pumped-storage hydro units in the system, respectively. Then the following variables occur

$u_i^t \in \{0, 1\}$ — on/off decisions for the thermal unit $i \in \{1, \ldots, I\}$ and time interval $t \in \{1, \ldots, T\}$,

p_i^t — output level for the thermal unit i during interval t,

s_j^t, w_j^t — level for generation and pumping, respectively, for the pumped-storage plant $j \in \{1, \ldots, J\}$ during interval t,

l_j^t — water level, in terms of energy, in the upper reservoir of plant $j \in \{1, \ldots, J\}$ at the end of interval t.

The objective function reads

$$F(\boldsymbol{u}, \boldsymbol{p}) := \sum_{t=1}^{T} \sum_{i=1}^{I} FC_i(p_i^t, u_i^t) + \sum_{t=1}^{T} \sum_{i=1}^{I} SC_i(\boldsymbol{u}_i(t)). \qquad (2.1)$$

Here, FC_i denotes the fuel cost function for the operation of the i-th thermal unit. With respect to p_i^t the function is monotonically increasing and often assumed to be convex (linear, piecewise linear, quadratic). Non-convex setups for fuel costs are not considered in the present paper. Start-up costs $SC_i(\boldsymbol{u}_i(t)) = SC_i(u_i^t, \ldots, u_i^{t_{s_i}})$ of the i-th thermal unit are determined by the preceding downtime $t - t_{s_i}$ and will be further specified later on.

The constraints of the unit commitment model are formulated as linear equations and inequalities. The same feasible set can be described by nonlinear expressions, too. Here, we prefer the linear variant in order to apply methodology from mixed-integer linear programming if also the objective is given by linear terms.

All variables mentioned above have finite lower and upper bounds which reflect the bounded output of all units in the generation system.

$$p_{it}^{min} u_i^t \leq p_i^t \leq p_{it}^{max} u_i^t, \ i = 1, \ldots, I, \ t = 1, \ldots, T \qquad (2.2)$$
$$0 \leq s_j^t \leq s_{jt}^{max}, \ j = 1, \ldots, J, \ t = 1, \ldots, T \qquad (2.3)$$
$$0 \leq w_j^t \leq w_{jt}^{max}, \ j = 1, \ldots, J, \ t = 1, \ldots, T \qquad (2.4)$$

The constants p_{it}^{min}, p_{it}^{max}, s_{jt}^{max}, w_{jt}^{max} denote the minimal and maximal outputs, respectively.

Load coverage for each time interval t of the optimization horizon leads to the equations

$$\sum_{i=1}^{I} p_i^t + \sum_{j=1}^{J}(s_j^t - w_j^t) = D^t, \ t = 1, \ldots, T, \tag{2.5}$$

where D^t denotes the load to be covered in time interval t. Sudden load increases or unforeseen conditions (e.g. outage of a unit) have to be compensated on-line. Therefore, for each time interval t, some spinning reserve R^t in the termal units is required which leads to the constraints

$$\sum_{i=1}^{I}(p_{it}^{max} u_i^t - p_i^t) \geq R^t, \ t = 1, \ldots, T. \tag{2.6}$$

During the whole optimization horizon water balances in the pumped-storage plants have to be maintained. It is typical of the power system of VEAG that no additional in- or outflows arise in the upper reservoirs of the pumped-storage plants such that these operate with a constant amount of water. The possible workload for turbines and pumps is restricted by the water levels and capacities of the upper reservoirs. This is modelled by the inequalities

$$0 \leq l_j^t \leq l_j^{max}, \quad j = 1, \ldots, J, \ t = 1, \ldots, T \tag{2.7}$$

where l_j^t is connected with the variables s_j^t and w_j^t by the reservoir constraints

$$l_j^0 = l_j^{ini}, \ l_j^t = l_j^{t-1} - s_j^t + \eta_j^t w_j^t, \ t = 1, \ldots, T, \ l_j^T = l_j^{end}, \ j = 1, \ldots, J \tag{2.8}$$

Here l_j^{ini}, l_j^{max} and l_j^{end} denote the initial, maximal and terminal water levels in the upper reservoir, and η_j is the efficiency of the j-th pumped-storage plant.

Finally, there are minimal down times τ_i for the thermal units that are mainly determined by technological reasons and serve to avoid erosion of the unit by too frequent changes of thermal stress.

$$u_i^{t-1} - u_i^t \leq 1 - u_i^l, \quad l = t+1, \ldots, \min\{t + \tau_i - 1, T\}, \ i = 1, \ldots, I \tag{2.9}$$
$$t = 2, \ldots, T - \tau_i + 1.$$

3 Primal Methods

Our primal approach to the unit commitment problem relies on solving the underlying large-scale mixed-integer optimization problem by adapted branch-and-bound techniques, possibly enriched by cutting planes derived from the convex hull of feasible points. In order to apply the corresponding methodology and implementations we formulate the basic model from Section 2 as a mixed-integer linear program. More precisely, we assume that, for each thermal unit $i \in \{1, \ldots, I\}$, the fuel costs are affine-linear functions of the generated power, and the start-up costs are given by

$$SC_i(\boldsymbol{u}_i(t)) = A_i \max\{\boldsymbol{u}_i^t - \boldsymbol{u}_i^{t-1}, 0\}, \; t = 2, \ldots, T \tag{3.1}$$

where A_i is some positive constant. This piecewise linear convex function can be expressed in linear terms by introducing an additional variable and adding two linear inequalities to the constraints.

The setups for fuel as well as start-up costs are the simplest possible one could think of. On the other hand, the real-life data material from VEAG, by means of which we have validated our models, indicates that both model simplifications are tolerable. Staying within the framework of a linear model it is possible to refine the linear fuel costs by a piecewise linear setup and to take into account that startup costs vary with the preceding down-time of the unit. The tradeoff for both extensions is a growing number of variables and constraints in the programs to be solved.

It is a typical feature of the generation system of VEAG that there are thermal units which are absolutely identical. Therefore, obviously, a model reformulation is possible, treating each group of identical blocks by only one status variable $\boldsymbol{u}_i^t$ and one output variable $\boldsymbol{p}_i^t$. The status variable now is not a Boolean one but with the number of identical blocks being N we would have $\boldsymbol{u}_i^t \in \{0, \ldots, N\}$, indicating the number of blocks in on-state. The generalized lower and upper bounds (2.2) for the output $\boldsymbol{p}_i^t$ of the group of units remain valid and so does the formulation of start-up costs.

In our test runs we have tried both models. The first model corresponds to the pure Boolean setting, the second one to the modification described in the previous paragraph. Though the reduction of the number of integer variables is only by a factor of 0.3, we observe a considerable reduction of the branch-and-bound tree and the computing times. From this point of view the latter model is superior. On the other hand, the development towards a branch-and-cut algorithm on the basis of valid inequalities developed in [14] relies on a Boolean structure in the form of a knapsack problem and thus could be carried on for the first model only. Further investigation is necessary to decide whether the possible improvement of bounds by the cuts or the reduction of the number of integer variables by the model reformulation are preferable.

The test runs proved the successful applicability of the primal approach to real-life models of the type outlined above. Our program was developed on the basis of modules from the CPLEX Callable Library [4]. The details of the runs are given in Table 1. We used three parks of generating units, reflecting three different states of development. Park 1 consists of 20 thermal units and 6 pumped-storage hydro plants; park 2 has two additional coal-fired units and a modified pumped-storage unit. Park 3 comprises another three additional coal-fired units and one additional pumped-storage unit. Computations were done for three selected weeks (8 days) with 1-hour-intervals ($T = 192$) on a HP-Apollo 735/125. The figure 1 shows the load, the thermal generation and the use of pumped-storage units for park 2 in the peak load week.

variant		park 1		park 2		park 3	
week	model	time	quality	time	quality	time	quality
holiday	1	5:13	0.164 %	6:35	2.965 %	29:45	2.837 %
week	2	2:13	0.192 %	3:50	0.002 %	5:20	0.906 %
low	1	6:09	0.399 %	17:26	2.151 %	21:59	1.269 %
load	2	2:47	0.425 %	7:14	1.152 %	7:55	2.929 %
peak	1	0:01	-[1]	3:42	0.086 %	5:16	0.005 %
load	2	0:01	-[1]	1:53	0.099 %	2:48	0.269 %

[1] No feasible solution exists.

Table 1: CPU-time in minutes (HP-Apollo 735/125) and upper bound for the deviation of the objective value from the optimum

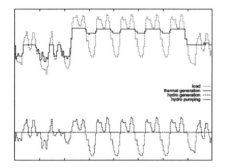 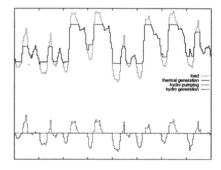

Fig. 1: Schedule of the primal approach for park 2 in the peak load week.

Fig. 2: Schedule of the dual approach for park 2 in the holiday week.

4 Dual approach

The dual approach to the unit commitment problem has been widely studied in the last 15 years (see [15]). In [12], the authors have addressed unit commitment with Lagrangian relaxation for the first time. The main idea is to incorporate the loosely coupling constraints linking operation of different units into the objective function by use of Lagrange multipliers. Then the problem decomposes into smaller subproblems. The solution of the relaxed problem provides a lower bound on the optimal solution of the original problem. The value of the lower bound is a function of the Lagrangian multipliers. This approach has found a wide application for large systems, for two reasons. It works fast due to the decomposition of the dual problem into essentially smaller subproblems. On the other hand, it has been proved in [2,3], that the duality gap, which occurs by the presence of integrality, becomes small for a large number of units.

From now on we assume that the fuel costs are a piecewise linear function of the generated power and the start-up costs are given by (3.1).
Let us associate the multipliers $\lambda, \mu \in \mathbb{R}^T$ with the demand and reserve-constraint respectively and consider the Lagrange function:

$$L(p, u, s, w; \lambda, \mu) = \sum_{t=1}^{T}\sum_{i=1}^{I} FC_i(p_i^t, u_i^t) + \sum_{t=1}^{T}\sum_{i=1}^{I} SC_i(u_i(t)) \tag{4.1}$$

$$+ \sum_{t=1}^{T}\left[\lambda^t\left(D^t - \sum_{i=1}^{I} p_i^t - \sum_{j=1}^{J}(s_j^t - w_j^t)\right) + \mu^t\left(R^t - \sum_{i=1}^{I}[u_i^t p_i^{max} - p_i^t]\right)\right].$$

which has to be minimized subject to the constraints (2.2) – (2.4) and (2.7) – (2.9). We obtain the problem:

$$P(\lambda, \mu) : \quad \min L(p, u, s, w; \lambda, \mu) \tag{4.2}$$
$$\text{subject to (2.2) – (2.4) and (2.7) – (2.9).}$$

Denoting the marginal function of $P(\lambda, \mu)$ by $d(\lambda, \mu)$ the dual problem reads:

$$\max\{d(\lambda, \mu) : \lambda, \mu \in \mathbb{R}^T, \mu \geq 0\}, \tag{4.3}$$

By the separability structure of $L(p, u, s, w; \lambda, \mu)$ and the constraints with respect to the units, we can decompose the problem $P(\lambda, \mu)$ into problems $P_i(\lambda, \mu)$ and $\tilde{P}_j(\lambda, \mu)$ as follows:

$$P_i(\lambda, \mu) : \quad \min_{u_i} \sum_{t=1}^{T}[\min_{p_i^t}\{FC_i(p_i^t, u_i^t) - \lambda^t p_i^t\} + SC_i(u_i^t) - \mu^t u_i^t p_i^{max}] \tag{4.4}$$
$$\text{subject to (2.2) and (2.9)}$$

for the coal-fired and gas-burning units ($i = 1, \ldots, I$) and

$$\tilde{P}_j(\lambda, \mu) : \quad \min_{(s_j, w_j)} \sum_{t=1}^{T}(\lambda^t + \mu^t)(w_j^t - s_j^t) \tag{4.5}$$
$$\text{subject to (2.3), (2.4), (2.7) and (2.8)}$$

for the pumped hydro storage plants ($j = 1, \ldots, J$). Denoting the marginal functions of the problems above by $d_i(\lambda, \mu)$ and $\tilde{d}_j(\lambda, \mu)$ resp., we obtain for $d(\lambda, \mu)$:

$$d(\lambda, \mu) = \sum_{i=1}^{I} d_i(\lambda, \mu) + \sum_{j=1}^{J} \tilde{d}_j(\lambda, \mu) + \sum_{t=1}^{T}[\lambda^t D^t + \mu^t R^t] \tag{4.6}$$

Useful properties of d are its separability structure, concavity and the explicit formulas for computing subgradients. Setting $d^\star := \max_{(\lambda, \mu)} d(\lambda, \mu)$ and $f_\star := \min_{(u,p)} F(u, p)$ it is known by the weak duality theorem that $d^\star \leq f_\star$. The relative duality gap $(f_\star - d^\star)/d^\star$ converges to zero as $I \longrightarrow \infty$ [2,3].
In order to solve the dual problem (4.3) efficiently, a fast non-smooth optimization method, a good initial guess for (λ, μ), efficient algorithms for solving the subproblems (4.4) and (4.5) and a proper heuristics for computing a primal feasible solution are needed. We gave our preference to the bundle method described in [7,8]. The implemented algorithm works as follows.

- Initial guess for the Lagrange multipliers $\boldsymbol{\lambda}$, $\boldsymbol{\mu}$. Based on a priority list the on/off decisions are taken to satisfy the demand constraint for each time interval. Then the relative production costs of the on-line units are used to initialize $\boldsymbol{\lambda}$. $\boldsymbol{\mu}$ is set to be zero.
- Iterative procedure of the bundle method
  - Computation of a value and of subgradients of $d(\boldsymbol{\lambda}, \boldsymbol{\mu})$ by solving $I + J$ subproblems of dimension $2T$. The first I problems are solved as follows. The minimization with respect to $\boldsymbol{p}_i^t$ is done explicitly or by one-dimensional optimization. The minimization with respect to $\boldsymbol{u}_i$ is equivalent to the search for a shortest path in the state transition graph of the unit under consideration, and it is carried out by dynamic programming. Nodes passed during the minimal down time are not included in the state transition graph (cf [16]). The next J hydro-storage subproblems are solved by the algorithm developed in [13].
  - Bundle-method iteration.
- Determination of a primal feasible solution. The algorithm works in two steps to satisfy the (possibly violated) reserve constraints. First, we try to satisfy the constraints by using the pumped-storage hydro plants in those time intervals, where the largest values of $D^t + R^t$ occur. If the reserve-constraints are still violated, we modify the schedules of the thermal units by the procedure in [17]. The idea consists in finding a period t for which the reserve constraint is violated most and then computing the smallest amount of increase $\Delta \mu^t$ to satisfy the reserve constraint in this period. This procedure is carried out recursively until the constraints are satisfied for all periods.
- Economic dispatch. In a last step we improve the feasible solution found in the previous step. We solve the primal problem keeping the integer variables fixed. The latter linear optimization problem, referred to as economic dispatch, is solved using the CPLEX Callable Library [4].

The encouraging results of the test runs proved the efficiency of the dual approach. Computations based on the same data (model 1) as in the previous section provided the numerical results reported in table 2 and figure 2.

NOA3 Optimality tolerance: 10^{-4}	park 1		park 2		park 3	
	time	gap	time	gap	time	gap
holiday week	1:36	0.74 %	0:40	0.37 %	0:35	0.93 %
low load week	0:58	1.47 %	0:36	0.23 %	0:36	0.54 %
peak load week	0:09	-[1]	3:43	0.89 %	0:39	0.52 %

[1] No primal solution exists.

Table 2: CPU-time in minutes (HP-Apollo 735/125) and upper bound of the relative duality gap

Acknowledgement. We wish to thank P. Reeh, G. Schwarzbach and J. Thomas (VEAG Vereinigte Energiewerke AG) for the outstanding collaboration which made the present paper possible.

References

[1] Aoki, K.; Itoh, M.; Satoh, T.; Nara, K.; Kanezashi, M.: Optimal Long–Term Unit Commitment in Large Scale Systems Including Fuel Constrained Thermal and Pumped–Storage Hydro. IEEE Transactions on Power Systems 4(1989), 1065 – 1073.

[2] Bertsekas, D.P.: Constrained Optimization and Lagrange Multiplier Methods. Academic Press, New York, 1982.

[3] Bertsekas, D.P.; Lauer, G.S.; Sandell, N.R.; Posbergh, T.A.: Optimal Short–Term Scheduling of Large–Scale Power Systems. IEEE Transactions on Automatic Control, AC–28(1983), 1–11.

[4] Using the CPLEX Callable Library. CPLEX Optimization, Inc. 1994.

[5] Feltenmark, S.; Kiwiel, K.C.; Lindberg, P.-O.: Solving Unit Commitment Problems in Power Production Planning. Working Paper, 1996.

[6] Guddat, J.; Römisch, W.; Schultz, R.: Some Applications of Mathematical Programming Techniques in Optimal Power Dispatch. Computing 49(1992), 193–200.

[7] Kiwiel, K.C.: Proximity Control in Bundle Methods for Convex Nondifferentiable Minimization. Mathematical Programming 46(1990), 105 – 122.

[8] Kiwiel, K.C.: User's Guide for NOA 2.0/3.0: A Fortran Package for Convex Non-differentiable Optimization. Polish Academy of Science, System Research Institute, Warsaw, 1993/1994.

[9] Lemaréchal, C.; Pellegrino, F.; Renaud, A.; Sagastizábal, C.: Bundle Methods Applied to the Unit Commitment Problem. Proceedings of the 17th IFIP–Conference on System Modelling and Optimization, Prague, July 10 – 14, 1995. (to appear)

[10] Möller, A.: Über die Lösung des Blockauswahlproblems mittels Lagrangescher Relaxation. Diplomarbeit, Humboldt–Universität Berlin, Institut für Mathematik, 1994.

[11] Möller, A.; Römisch, W.: A Dual Method for the Unit Commitment Problem. Humboldt–Universität Berlin, Institut für Mathematik, Preprint Nr. 95-1, 1995.

[12] Muckstadt, J.A.; Koenig, S.A.: An Application of Lagrangian Relaxation to Scheduling in Thermal Power–Generation Systems. Operations Research, 25(1977), 387–403.

[13] Nowak, M.: A Fast Descent Method for the Hydro Storage Subproblem in Power Generation. Working Paper, WP-96-109, IIASA, Laxenburg, 1996.

[14] van Roy, T.; Wolsey, L.A.: Valid Inequalities for Mixed 0-1 Programs. Discrete Applied Mathematics 14(1986), 199-213.

[15] Sheble, G.B.; Fahd, G.N.: Unit Commitment Literature Synopsis. IEEE Transactions on Power Systems 9(1994), 128–135.

[16] Takriti, S.; Birge, J.R.; Long, E.: A Stochastic Model for the Unit Commitment Problem. IEEE Transactions on Power Systems 11(1996), 1497–1508.

[17] Zhuang, F.; Galiana, F.D.: Towards a More Rigorous and Practical Unit Commitment by Lagrangian Relaxation. IEEE Transactions on Power Systems 3(1988), 763–773.

Optimal Trajectory & Configuration of Commuter Aircraft with Stochastic and Gradient based Methods

R. Pant[†] & C. M. Kalker-Kalkman[‡]

[†]College of Aeronautics, Cranfield University,
Bedfordshire, UK, *r.pant@cranfield.ac.uk,*
[‡]Dept. of Mechanical Engineering, Delft University of Technology,
The Netherlands, *c.m.kalker-kalkman@wbmt.tudelft.nl*

Abstract

This paper discusses the application of two stochastic optimization methods viz. Simulated Annealing (SA) and Genetic Algorithms (GA) in conceptual design of commuter aircraft. Brief description of a methodology for integrated trajectory and configuration optimization of commuter aircraft is provided, along with the three different objective functions. The features of SIMANN SA algorithm and GOOD GA program are then outlined. The results obtained with these two techniques are compared with those obtained with a proprietary gradient search procedure RQPMIN for all the three objective functions. It was seen that SIMANN & GOOD come up with similar or better configurations compared to RQPMIN. While SIMANN & GOOD required much larger number of evaluations per run compared to RQPMIN, they were found to be less prone to getting trapped in the local minima.

1 Introduction

Commuter aircraft are small, low speed aircraft aimed at the short range segment of civil air-travel. They have been found to be much more profitable than their turbofan counterparts for operations on less dense routes over short stages (up to 1000 km). Design and operation of commuter aircraft poses some interesting challenges. They are frequently required to carry out multiple short-range hops without refuelling. The flight profiles of such aircraft tend to be dominated by the climb and descent segments, with much smaller cruise segments compared to the long range aircraft. Thus, there is a very strong interconnection between the configuration and the flight trajectory, and any attempt to determine the lowest operating cost should consider the trajectory and configuration parameters at one go.

In [6], a computer code called CASTOR (Commuter Aircraft Synthesis and Trajectory Optimisation Routine) was coupled to a gradient based optimiser. Optimum configurations & profiles of typical commuter aircraft that minimize any one of the three objective functions viz. Mission Fuel Weight (M_{fuel}), Maximum Take-Off Weight (M_{TO}), & the Direct Operating Cost (DOC) were obtained for arbitrarily defined missions. Case studies for a 50 seater commuter aircraft capable of doing five hops over a range of 185 KM (without refuelling) are discussed in [6] & [12]. In the present work, CASTOR is coupled with two stochastic optimization methods, viz. Simulated Annealing (SA) and Genetic Algorithms (GA). The optimum configurations obtained by these methods that minimize the three objective functions are then compared with those reported in [6] using RQPMIN method.

2 Brief description of CASTOR & RQPMIN

CASTOR consists of 22 optimizable design variables, 10 of which are the classical aircraft design parameters, while the remaining 12 are related to the way in which the aircraft is flown. Constraints on the values of 12 parameters can be specified in CASTOR by the user, which are used to model the physical limitations and the environment within which this aircraft can be expected to operate, and to make the code non-iterative. A detailed description of the design variables, constraints and the non-optimizable constants used in CASTOR is given in [6].

RQPMIN (Recursive Quadratic Programming for MINimization) is a multivariate optimization program developed by DRA, Farnborough, UK. It is a gradient search procedure in unconstrained domain, and can be termed as a classical optimization method. Constraints are considered using a penalty function approach, and they are progressively tightened as the optimization proceeds. The step size is also changed progressively on the design surface, till the convergence criteria is met. Details of the RQPMIN method are given in [13] and its application in the area of conceptual design of combat aircraft is discussed in [14].

While tackling a multi-modal objective function, gradient based optimizers like RQPMIN are prone to getting stuck in the local minima. RQPMIN consists of a very large number (nearly 50) of intrinsic control parameters, and a considerable effort on the part of the user is required to understand their significance and assigning appropriate values. Many internal calculations are carried out during the optimization run, which at times can consume a large fraction (approx. 50%) of the total execution time for the run. It is also quite difficult to comment on the overall convergence of the optimizer, since it depends a great deal on how well the constraint scale factors have been adjusted in each case, needing several trial runs [6].

3 Stochastic Optimization techniques

Stochastic Optimization methods are the ones whose results depend on multiple evaluations of the objective function for random combinations of the design variables. They are not influenced by the non-linearity and/or discontinuity of the objective function. SA & GA have been widely reported in literature of being able to successfully and efficiently tackle complex & multi-modal objective functions ([5], [2]), including aircraft conceptual design ([3], [11]). They have been found to be quite robust and for complicated objective functions, they have been shown superior to the classical gradient based methods in arriving at the global optimum ([9], [4]).In a comparative evaluation of 5 different gradient based & stochastic optimisation methods for aircraft parametric design carried out at Lockheed Aeronautical Systems [1], the best results were obtained with modified SA & GA methods.

3.1 Optimization by Simulated Annealing

Simulated Annealing is a stochastic optimization method introduced by Metropolis et al. [10] and is based on the thermo-dynamical analogy of annealing of metals. It was first proposed by Kirkpatrick et al. [8] for optimization of combinatorial problems (in which the objective function is defined in a discrete domain) and was successfully employed for objective functions involving very large number of variables (even tens of thousands). Corana et al. [2] were among the many researchers who extended the method for objective functions involving continuous variables.

When a molten metal is allowed to anneal i.e. cool slowly, it eventually arrives at a low energy state. SA tries to minimise some analogue of energy in a manner similar to annealing to achieve the global minima. It allows excursions of the design variable away from the optimum direction once in a while, which enables it to climb out of the local minima. In the beginning, only the gross behaviour of the function is explored, but as the temperature falls, the criteria for selecting non-optimal configurations is progressively tightened. The method involves very few internal calculations. It has been found to be quite robust in tackling multi-modal problems, but it requires many more number of function evaluations compared to classical optimization methods. A few trial runs are also required to tune the optimization parameters before it can be gainfully employed.

3.2 SIMANN SA Optimizer

In the present work, the SIMANN SA algorithm developed by Goffe et al. [4], based on the methodology proposed by Corana et al. [2] for objective functions involving continuous variables was employed. This optimizer has been successfully employed for optimization of very ill-behaved objective function related to aircraft conceptual design in a previous study [11]. The algorithm starts with a high initial value of temperature T_{init} and a starting set of design variables. Trial sets are then generated using random numbers from the set [-1,1] and initial step length for each design variable vm_i. If the function value for the trial set is lower than that for the previous one, the trial set is accepted. Acceptance of a trial set yielding higher function value is random, with a probability decreasing exponentially with the temperature. After N_S steps through all design variables, their step lengths are adjusted, to ensure that roughly half of all the moves are accepted using a varying criterion c_i, in line with the approach followed by Metropolis et al. [10]. A very high acceptance rate implies that the function domain is not being fully explored, while a very low acceptance rate means that the new trial points are being generated too far away from the current optimum. After carrying the above loop N_T times, the temperature is gradually reduced employing a geometric schedule governed by the parameter R_T. The algorithm is stopped when the reduction in the function value in N_{eps} successive cycles is less than a small number eps. Further details of this algorithm are available in [2] & [4].

3.3 Parameters in SIMANN

As with all general purpose optimization methods, some control parameters in SA have to be "tuned" to suit the objective function, and to ensure that the optimizer performs efficiently. A bad choice for these parameters can make the algorithm extremely inefficient and may even result in failure to arrive at the global optimum. There are three such parameters in SIMANN viz. the initial temperature T_{int}, the temperature reduction factor R_T and the number of cycles before temperature reduction N_T. Suitable values of these three parameters were determined by several trials runs, as suggested in [4].

4 Optimization by Genetic algorithms

Genetic Algorithms (GA) are based on the Darwinian model of survival of the fittest, and are extensively described in [5]. They work by maintaining a pool of several competing designs, which are termed as the *population*. The members of the population pool are considered as carriers of good qualities that can be inherited by new generations, and are randomly combined to find improved solutions. Each population member is represented by a string of binary coding that encodes the design variables. These are analogous to the chromosomes forming the genes of the biological individual in an evolutionary chain. The probability of survival of each member to the next generation of the population depends on its fitness, which is directly related to the value of the objective function. The search proceeds towards the global optimum by iteratively selecting fitter individuals for further reproduction. A mechanism called *crossover* is employed for creating the next generation, in which the strings of two fit parents from the present generation are interchanged at a randomly selected crossover location. To avoid getting stuck in a local minimum, an operator called *mutation* is employed to each bit with a very low probability, in which the bits are flipped. As the iterations proceed, it is found that the subsequent populations are of an increasing quality, since the average fitness value of the population increases. The method is said to have converged if all (or a large fraction) of the population have similar values of fitness.

4.1 GOOD GA Optimizer

In the present work, one such methodology called GOOD (Generator Of Optimal Designs) was employed, which is a GA code coupled with a graphical interface. In GOOD, the best member of the population set is retained in the next generation, and is considered as the current optimum. The values of all design variables and the objective function of the current optimum are displayed on the screen, and subsequent such points are connected by lines. Thus, the user can get a graphical feedback of how the optimization is proceeding and the trend of change in values of the design variables. It is also possible for the user to intervene in the optimization process at any stage, and to change the parameters based on the visual feedback. For instance, it is possible to specify new search intervals for one or more design variables, change the population size or mutation rate, proceed with the present population set for a specific computational time or number of evaluations, or even to start again from scratch. The details of this method and application to other problems in previous work are given in [7].

5 Comparison of optimum configurations by the three methods

In [6], a detailed output of the minimum *DOC* configuration has been provided, hence it was possible to compare the optimum configurations obtained for this objective function by all the three methods. The geometry & the flight profiles for the main mission of the optimum aircraft obtained by the three methods for minimum *DOC* are superimposed in Figure 1. It can be seen that the optimum configurations & mission profile obtained by RQPMIN & SIMANN are quite similar. The comparison also brings out the multi-modal nature of the objective function, since even though the three objective function values differ from each other only within 0.5%, the optimum flight profile obtained by GOOD is quite different. Figure 2 shows the optimum configuration & main mission profiles obtained by three different runs of SIMANN. It can be seen that the optimum configurations & mission profiles for these three runs are quite similar, and difference in the *DOC* values are less than 0.3%. For these three cases, the seeds that were used to generate the random number series were not the same, hence the path that these three runs took to arrive at the optimum solution were totally different.

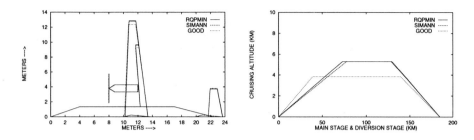

Figure 1: Minimum DOC configurations & mission profiles by the three methods

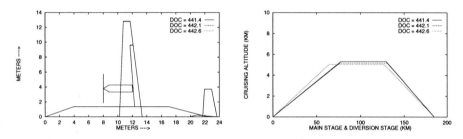

Figure 2: Minimum DOC configurations & mission profiles for 3 runs of SIMANN

For the minimum M_{TO} and minimum M_{fuel} case, the values of only four wing related design variables were listed in [6] & [12], hence only a limited comparison could be made. These parameters are the Aspect Ratio AR, gross area S_w, taper ratio τ & the thickness-to-chord ratio t/c of the wing. Table 1 compares the values of these parameters of the optimum configurations obtained by the three methods for the two objective functions, along with the number of function evaluations required for each run N_{eval}. It may be noted that in [6], N_{eval} was listed only for the minimum DOC case. It can be seen that all the optimum configurations obtained by SIMANN have better objective function values compared to those obtained by RQPMIN. For the minimum M_{fuel} case, GOOD also comes up with better configurations compared to the RQPMIN method. The number of function evaluations per run by RQPMIN were very small compared to SIMANN or GOOD, but the total number of such runs that were required by RQPMIN to reach the global minimum was not reported in [6], hence it is not possible to make a comparison of overall computational time. The computational time needed for one run of SIMANN was approx. 3 hours while running on a network of SUNSPARC workstations. For GOOD, the computational time required on a Pentium 100 MHz machine was also of the same order.

Method	DOC	M_{fuel}	M_{TO}	AR	S_w	τ	t/c	N_{eval}
Units	US$/trip	Kg	Kg		m^2			
Minimum DOC Configurations								
RQPMIN	441.9	1010	16245	12.62	52.40	0.3	0.21	2275
SIMANN	441.4	1013	16253	12.49	52.42	0.3	0.21	5232001
GOOD	443.8	1061	16157	11.54	52.92	0.3	0.21	134035
Minimum M_{fuel} Configurations								
RQPMIN	460.9	942	16413	13.00	52.86	0.3	0.18	NA
SIMANN	466.9	931	16522	13.00	53.52	0.3	0.17	4176001
GOOD	465.3	937	16342	12.92	53.41	0.3	0.19	6029680
Minimum M_{TO} Configurations								
RQPMIN	459.9	974	16030	12.28	51.84	0.3	0.21	NA
SIMANN	465.1	976	16023	11.83	52.74	0.3	0.21	3888001
GOOD	464.1	972	16104	11.86	53.18	0.3	0.21	63667

Table 1: Comparison of optimum configurations by the three methods

6 Conclusions

The present work has established the robustness of the two stochastic methods viz. Simulated Annealing (SIMANN) & Genetic Algorithms (GOOD) in arriving at the global optimum of the three multi-modal objective functions related to commuter aircraft configuration & trajectory optimization. The number of optimizer related parameters (that have to be tuned by the user) were far less in the case of SIMANN & GOOD, compared to RQPMIN. However, the number of function evaluations for one run of SIMANN & GOOD were much larger compared to the RQPMIN method.

References

[1] M. F. Bramlette and R. Cusic. A comparitive evaluation of search methods applied to parametric design of aircraft. In *Proc. of 3rd International conference on Genetic Algorithms*, 1989.

[2] A. Corona, M. Marchesi, C. Martini, and S. Ridella. Minimizing multimodal functions of continuous variables with the Simulated Annealing algorithm. *ACM Transactions on Mathematical Software*, 13(3):262–280, September 1987.

[3] Y. Crispin. Aircraft conceptual optimization using Simulated Evolution. In *Proc. of 32nd Aerospace Sciences meeting, Reno, NV, USA*, number AIAA 94-0092, January 1994.

[4] W. L. Goffe, G. D. Ferrier, and J. Rogers. Global optimization of statistical functions with Simulated Annealing. *Journal of Econometrics*, 60:65–100, 1994.

[5] D. E. Goldberg. *Genetic Algorithms in Search, Optimization and Machine Learning.* Addison-Wesley, Reading, USA, 1st edition, 1989.

[6] L. R. Jenkinson and D. Simos. The study of energy-efficient short-haul aircraft with emphasis on environmental effects. Technical Report TT87R01, Department of Transport Technology, Loughborough University of Technology, Loughborough, UK, 1987.

[7] C. M. Kalker-Kalkman and M. F. Offermans. A general design program based on Genetic Algorithms with applications. In *Proceedings of 21st ASME Design Automation Conference, Boston, USA*, September 1995.

[8] S. Kirkpatrick, C. D. Gelatt Jr, and M. P. Vecchi. Optimization by Simulated Annealing. *Science*, 220:671–680, 1983.

[9] C.-Y. Lin and P. Hajela. Genetic Algorithms in optimization problems with discrete and integer design variables. *Engineering Optimization*, 19:309–327, 1992.

[10] N. Metropolis, A. Rosenbluth, M. Rosenbluth, A. Teller, and E. Teller. Equation of state calculations by fast computing machines. *Journal of Chemical Physics*, 21:1087–1090, 1953.

[11] R. Pant. Application of stochastic optimization techniques for aircraft conceptual design optimization. In *Proceedings of the First World congress on Structural & Multidisciplinary Optimization, Goslar, Germany*, pages 827–832, May-June 1995.

[12] D. Simos and L.R. Jenkinson. Optimization of the Conceptual Design and Mission Profiles of Short-Haul aircraft. *Journal of Aircraft*, 25(7):618–624, July 1988.

[13] J. J. Skrobanski. RQPMIN version 2.0 user guide (unpublished). Technical report, MVA Consultancy, London, UK, October 1992.

[14] J. Smith and C. Lee. The RAE combat aircraft Multi-Variate Optimization method. In *Proceedings of the AIAA/AHS/ASEE Aircraft Design, Systems & Operations conference, Seattle, USA, AIAA 89-2080*, July-August 1989.

A general multi-objective optimization program for mixed continuous/integer variables based on genetic algorithms

C.M.Kalker-Kalkman
Department of Mechanical Engineering,
Delft University of Technology, NL 2611 NE Delft, The Netherlands

Abstract

A method is described for producing optimal Pareto Sets for multicriterion optimization problems. The variables are either continuous or discrete. A genetic algorithm is proposed. Cross-over and mutation methods are discussed. An implementation is described with a graphic interface that enables the user to modify the parameters of the genetic algorithm. Intervals for the variables can also be modified during runtime. As example the design of an externally pressurized bearing is treated.

1 Introduction

Many problems in engineering design can be expressed as optimization problems. The variables may have continuous or discrete values and may be grouped in tables, like physical properties of materials. Usually a number of conflicting demands has to be satisfied, thus giving rise to a multicriterion optimization problem. The equality constraints are usually physical laws and geometric identities. The inequalities can be rules-of-thumb or logical geometric equations, like inner radius < outer radius. The objective functions are costs, capacities or times, usually there is more than one objective function and they are in conflict with each other. In many problems the objective function does not exist for all combinations of variables in the search domain, or it has discontinuities. This means that gradient methods are not appropriate and stochastic methods are imperative.
In [1], [2] and [3] a computer program called GOOD (Generator Of Optimal Designs) is described that is based on Monte Carlo methods and genetic algorithms. GOOD is an interactive program that enables the user with the aid of graphic functions to determine in what regions of the search intervals the better solutions are found. The search intervals can be modified and entirely new populations can be started in the new intervals. It was found that this method gives fast convergence,

and so-called premature convergence can be avoided. Moreover the parameters of the algorithm, like population size, mutation rate, cross-over and the scaling parameters for the fitness functions can be modified during runtime. The program is suitable for both continuous and discrete values, and there is a possibility to use variables that are structured in tables.

This program has been tested for a number of optimization problems. For multi-criterion optimization a compromise function can be used, that employs weight-factors to account for the "importance" of a certain variable. In this paper we will approach the multi-criterion problem from another viewpoint.

In section 2 a multi-criterion optimization problem will be formulated and the idea of using Pareto Sets [4] will be proposed. In section 3 the use of Pareto Sets will be discussed and in section 4 the application of a genetic algorithm is treated. In section 5 the implementation is discussed and in section 6 some results are given.

2 Formulation of multi-criterion optimization problem

A multi-criterion optimization problem can be formulated as follows :

$$f_i(x_1, x_2, ..., x_n) = 0 \quad \text{for i=1, 2, ..., m; m < n.} \quad (1)$$
$$g_j(x_1, x_2, ..., x_n) > 0 \quad \text{for j=1, 2, ..., k}$$

where x_1 to x_n are either continuous or take a finite number of discrete values. For x_1 to x_n search intervals can be given. A solution of (1) is wanted that has maximum values for

$$h_i(x_1, x_2, ..., x_n) \quad \text{i=1, 2, ..., l} \quad (2)$$

where (1) is satisfied.

Throughout this paper a very simple optimization problem will be employed to illustrate the method. It is about the design of a cylindrical externally pressurized bearing, this example is also treated in [3].

The variables involved are :

β	Pressure ratio	P_s	Pump pressure
W	Stationary load	Q	Flow
h	Film thickness	N	Pumping power
S	Stiffness	T	Temperature rise
λ	Shape parameter	Re_1	Reynolds number
	(ratio of diameters)	Re_2	Reynolds number
R	Maximum outer radius		

The following variables are structured in a table for the lubricant :

ρ	density
η	viscosity
c	specific heat

Equality constraints are :

$$3(1-\beta)W = hS$$
$$Q\eta\ln(\lambda) = -0.5421h^3 P_s \beta$$

$$P_s = W\frac{\ln(\lambda)}{1.572\lambda(\lambda-1)\beta R^2}$$
$$Re_1 = \frac{\rho Q}{\pi\eta R(1+\lambda)}$$
$$N = QP_s \tag{3}$$

$$P_s = 4.1868\rho c \Delta T$$
$$Re_2 = \frac{\rho Q h}{\pi\eta R^2(1-\lambda^2)}$$

Inequality constraints are :

$$\begin{array}{llll}
0.01 < R < 0.35 & 1E6 < P_s < 3E7 & 0.2 < \beta < 0.8 & \\
1E8 < S < 1E10 & T < 30 & 0.2 < \lambda < 0.8 & \\
3.5E10 < W < 2E6 & 1E-6 < Q < 1E-4 & Re_1 < 2000 & (4)\\
1E-5 < h < 1.5E-4 & N < 200 & Re_2 < 10 &
\end{array}$$

A lubricant has to be selected from a nine row table. A bearing with a maximum load capacity and a minimum pumping power is wanted, this means that the following functions have to be maximized:

$$h_1 = W \qquad\qquad h_2 = -N \tag{5}$$

2.1 Solution of equality constraints

In most optimization programs, both the equality and inequality constraints are contained in the objective function in the form of penalty functions. This means that all variables involved are considered as independent variables, and in order to keep the information about a particular solution, all those variables have to be stored. For a genetic algorithm this means that all variables have to be coded in bitstrings and those bitstrings have to be stored for all members of a population. However, in many engineering design problems it is found that the equality constraints can be considered as a set of non-linear equations that can easily solved one by one [3]. This means that we have to do with a number of independent variables and a number of dependent variables that can be solved from the equality constraints. The inequality constraints can often be written as bounds for the dependent or independent variables.

This means that only the values for the independent variables have to be stored and this gives a reduction in computer memory storage that is important in genetic algorithms. In GOOD it is possible to give the equalities as input, if they are a solvable set of equations, a subroutine solving them is automatically written.

2.2 Reduction to one objective function

In order to formulate the problem described as an optimization problem for one objective function, a so-called compromise function can be used (see [9]):

$$f_{obj} = \sum_i w_i \frac{h_i - h_{i,bad}}{h_{i,good} - h_{i,bad}} \qquad (6)$$

It is clear that f_{obj} has to be maximized. The values of $h_{i,bad}$ and $h_{i,good}$ have to be provided, like the values of the weight factors w_i. After determination of those values the optimization procedure can be started. The results of various sets of values for w_i can be compared, in fig. 1 it is seen that a high value for $h_1(W)$ corresponds to a low value for h_2(high value for N). It is clear that the optimization program has to be run for several sets of values for w_1 and w_2. In the next sections a method will be proposed that gives rise to a set of various combinations of W and N, from which the user can make a choice.

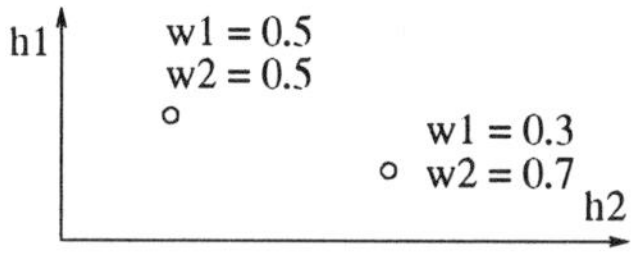

Figure 1: Result of optimization program GOOD for optimization of (5) with the aid of a compromise function.

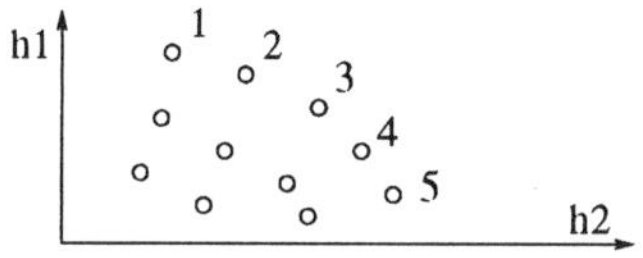

Figure 2: Non-dominated and dominated solutions.

2.3 Pareto sets

When we look qualitatively at fig. 1, we see that a designer is interested in a so-called PARETO set of solutions, that is the solutions 1 to 5 in fig. 2. The solutions 1 to 5 are said to dominate the other solutions in fig. 2. In general, a solution A dominates a solution B, when the following condition is satisfied :

$$\forall_i : h_{i,A} > h_{i,B} \qquad (7)$$

It is clear that the solution A is preferred. When A and B are members of the same Pareto Set, we can state:

$$\exists_{i,j} : i \neq j, h_{i,A} > h_{i,B}, h_{j,A} < h_{j,B} \qquad (8)$$

In this case it is not clear what solution is to be preferred. When no solutions exist that dominate one of the points 1 to 5, this set is the optimal Pareto Set of solutions. This Pareto set usually consists of an infinite number of solutions. Usually the set is bound by lower bounds for the values h_i in our case by lower bounds for W and -N.

In the following sections we will discuss a method that gives rise to such an optimal PARETO set. In the following we will talk of a Pareto set when we have a set that satisfies (7) for all possible combinations of A and B.

We talk of an Optimal Pareto Set when no solutions can be found that dominate any points of this Pareto Set. As in engineering design we often have to do with two conflicting demands, we will discuss the question of the optimal Pareto Set in more detail in section 3.

3 Optimal Pareto Set

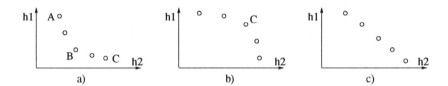

Figure 3: Various shapes for Optimal Pareto Sets.

It is clear that an Optimal Pareto Set is interesting when no weight factors can be given, or when one wants insight in the possible combinations of all objective functions. In fig. 3 various shapes for the optimal Pareto Sets for the bearing problem are given, it is clear that Fig. 3b is a most desirable shape, as point C will satisfy both demands rather well. Fig. 3a is a very undesirable shape, as point C is a very bad choice, and both point A and B are a good choice from only one viewpoint. Fig. 3c gives rise to many discussions as to what is most important: W or N. In section 6 the optimal Pareto set resulting from the program discussed is shown.

4 Genetic algorithm that uses PARETO sets as populations

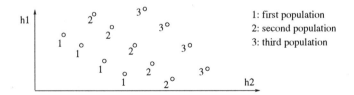

Figure 4: Successive populations for multicriterion optimization.

In [5], [6] and [7] genetic algorithms are extensively described. It is a stochastic search method based on evolution theory where the principle of survival of the fittest is employed and can be used for optimization problems where gradient methods fail. Instead of one starting value, a genetic algorithm uses a set of starting values, called a start population. From this start population new populations are created with the aid of a number of operators, called selection, cross-over and mutation. The new populations tend to have better values for the objective function(s), this function is called fitness. When we have to do with one fitness, the members of a population can be sorted according to increasing fitness. The member of a population with the best fitness is the current optimum. Pareto sets can not be sorted, as there is no criterion to sort the solutions. It was Belegundu's idea to use Pareto Sets in a population [4]. From a Pareto set another Pareto set is to be created, that dominates the previous set, see fig. 4. It is easy to generalize the idea of Pareto sets for more than two objective functions.

5 Implementation of Pareto optimization for genetic algorithm

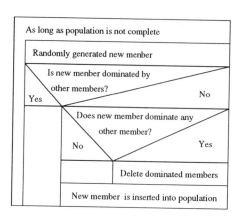

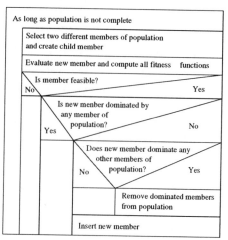

Figure 5: Program Structure Diagram for setting up start population

Figure 6: Program Structure Diagram for creation of new population.

It is clear that for implementation of the method discussed in the preceding sections a number of algorithms has to be devised. In fig. 5 a program structure diagram is given for construction of a start population. In order to create new populations, the operators selection, cross-over and mutation are used. Usually for selection the members of a population are sorted according to increasing fitness. It is clear that for a Pareto set we are not able to sort the members, as we have more than one fitness. For this reason random selection is used, where all members have an equal

1. Write last optimum to file.
2. Go on with this number
   of generations.
3. Stop.
4. Modify search interval,
   start new population.
5. Modify population size.
6. Modify scaling factors.
7. Modify mutation rate.
8. Modify number of generations.
9. Go on for specified time.

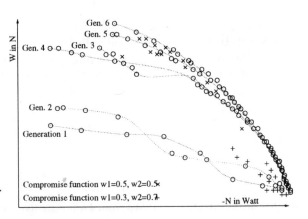

Figure 7: Main Menu of Genetic Algorithm Program

Figure 8: Results of bearing problem for Pareto optimization and compromise function.

chance to be selected. Cross-over can be realized by representing the solutions by bitstrings. Another method is heuristic cross-over, given by setting

$$x_3 = x_2 + r(x_2 - x_1), \qquad r \text{ is random number in } (0,1) \qquad (9)$$

x_1 and x_2 are the values of the variables of the parents (x_2 is better than x_1) and x_3 is the value of the child-variable. Both methods of cross-over have been implemented and it was found that the heuristic cross-over gave the fastest results in terms of computation time. It appeared that the coding and decoding of the bit strings took a relatively large time in comparison with the computation of the objective functions. For mutation a random value can be generated for a randomly chosen variable. In the case of representation by bitstrings a randomly chosen bit was modified. Mutation rate, population size, search intervals, scaling factors for fitness can all be modified during runtime, as is shown in the MAIN MENU of the program (fig. 7). A Program Structure Diagram for the creation of a new Pareto Set is shown in fig. 6.

	Compromise function				Pareto set	
	w1=0.5	w2=0.5	w1=0.7	w2=0.3	pop.size=100	
	W	N	W	N	W	N
	350084	18.84	491974	146.9	350764	18.8
	389424	51.8	448254	69.7	527100	144.1
					and many more	
Average number of computed solutions	16000		17000		815700	
Average time for one solution	0.002 sec.		0.002 sec.		0.002 sec.	

Table 1: Results compromise function and Pareto optimization.

6 Results and conclusions

The program has been run for a number of problems and gave very fast results. In fig. 8 results for the bearing problem are shown. In table 1 numerical results are shown. Various values of the weight factors have been used. It is seen that the results agree, but that multicriterion optimization gives more information per run.

References

[1] C.M.Kalker-Kalkman / M.F.Offermans: A general design program based on genetic algorithms with applications, ASME Design Engineering Conferences, Boston, U.S.A., DE-vol.82 (september 1995).

[2] C.M.Kalker-Kalkman: A design program based on the Monte Carlo method with applications, Advances in computer-aided engineering, CAD/CAM-research at Delft University of Technology, The Netherlands, Report of the VF-project CAD/CAM Delft University Press, ISBN 90-407-1017-1. (1989-1994), pp.61-70.

[3] C.M.Kalker-Kalkman: Optimal design with the aid of randomization methods, Engineering with computers 7 (1991), pp.173-183.

[4] A.D.Belegundu et al.: A general optimization strategy for sound power minimization, Structural Optimization 8, Springer-Verlag (1994), pp.113-119.

[5] Z.Michalewicz: Genetic Algorithms + Data Structures = Evolution programs, Springer-Verlag, New York (1992).

[6] David E.Goldberg: Genetic Algoriths in Search, Optimization and Machine learning, Addison-Wesley Publishing Company Inc. (1989).

[7] Lawrence Davis: Handbook of Genetic Algorithms, Van Nostrand Reinhold, New York (1991).

[8] R.Pant / C.M.Kalker-Kalkman: Optimal trajectory and configuration of commuter aircraft with stochastic and gradient based methods, Proceedings of ECMI conference, Copenhagen (1996).

[9] A.Oscyczka: Multicriterion Optimization in Engineering, Ellis Horwood Ltd, Chichester (1984).

The optimization of natural gas liquefaction processes

Gabriele Engl, Hans Schmidt
Linde AG, Process Engineering and Contracting Division
Dr.-Carl-v.-Linde-Str. 6-14
D-82049 Höllriegelskreuth near Munich, Germany
E-mail: gabriele_engl@linde-va.de

Abstract

An optimization method is presented for the cost minimization during the design of process engineering plants. The mathematical model of process plants, the general formulation of an optimization problem and the numerical method are illustrated by a natural gas liquefaction process.

Special requirements of design optimization yield discontinuous functions arising in the objective function and the constraints. The process model, a large system of nonlinear equations (flowsheet equations) with discontinuous functions and sparse Jacobian matrix, is included by the equality constraints of the optimization problem. A feasible path SQP (Sequential Quadratic Programming) method has been developed by which the special structure of the problem is taken into account. The new approach leads to a significant cost reduction which is illustrated by some computational results.

1 Introduction

Linde AG has a worldwide reputation as a process designer, as a manufacturer of industrial equipment and process plants as well as a producer of industrial gases. The Process Engineering and Contracting Division in Munich/Germany designs and builds various plants including air separation, gas processing and chemical plants.

For the steady-state and dynamic simulation and optimization of process plants, the program system OPTISIM® has been developed by the Linde AG company [1]. As well as being used for design purposes, OPTISIM has formed the basis of several on-line optimization and training simulator systems delivered to customers. While most conventional simulators are of sequential modular type, OPTISIM is an equation-oriented simulator: It automatically generates a large system of model equations (nonlinear or differential algebraic equations), and then solves all of the equations simultaneously by advanced numerical methods (inexact Newton or BDF methods). Due to the equation-oriented concept and the availability of analytical derivatives, modern optimization methods can be applied in an efficient way. Based on feasible path SLP and SQP methods and on special in-house algorithms to account for integer parameters and for the optimization of heat exchanger networks, OPTISIM has become a powerful optimization tool. See [2] for a general survey.

Recently, considerable success has been achieved in the design optimization of natural gas liquefaction processes which will be discussed in the following.

2 Natural gas liquefaction processes

Natural gas, a mixture of N_2, CO_2, CH_4 ($\sim$ 75–95 %), C_2H_6, C_3H_8, C_4H_{10} and small quantities of heavier hydrocarbons, is liquefied for two reasons: the storage for times of peak consumption (peakshaving plants) and the transportation by tankers (baseload plants). In a pretreatment step those components are removed which would freeze at low temperatures (CO_2 etc.). The purified natural gas is liquefied by cooling it down to temperatures of approximately 120 K (–153 °C).

Figure 1 shows the flowsheet of a natural gas liquefaction (LNG) process with two mixed refrigerant cycles (MRCs). Basic components of the flowsheet are process streams and units (heat exchangers etc.).

The natural gas (NG) is precooled by warming up and vaporizing the warm refrigerant mixture (C_2H_6, C_3H_8) in the heat exchangers E1 and E2. Refrigeration for the liquefaction and subcooling in E3 and E4 is produced by the cold refrigerant cycle (N_2, CH_4, C_2H_6, C_3H_8 and C_4H_{10}). The subcooled natural gas is expanded to atmospheric pressure by X1. After separating the resulting vapour in LNG tank T1, the LNG product is obtained.

The vaporized refrigerant mixtures, which leave the main heat exchangers E1–E4, are compressed in several stages (C1A, C1B, C2A, C2B, C2C), aftercooled (E5, E6) and cooled down in the main heat exchangers. The cold refrigerant cycle is split into vapour and liquid in drum D1. This yields two refrigerant mixtures which condense and vaporize at different temperature levels. Refrigeration is produced by the expansion valves V1, V2, V3 and by the expansion turbine X2 where the subcooled liquid refrigerant fractions are cooled down by Joule-Thomson effect and by Claude effect, respectively.

An optimal process design implies cost minimization: The operating costs correspond to the power of the cycle compressors and can be reduced by using more separator stages and heat exchangers or additional refrigerant cycles. On the other hand, increasing complexity of the process involves higher investment costs (e.g.

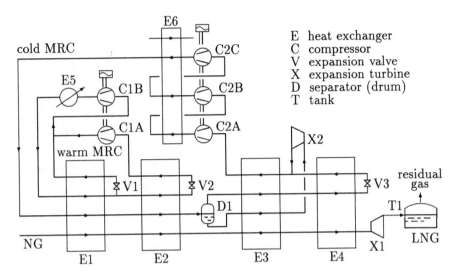

Figure 1: Flowsheet of an LNG process

heat exchanger areas).

For a given process structure, the costs depend on process parameters like refrigerant cycle compositions, outlet temperatures of heat exchangers as well as feed and discharge pressures of compressors. Among other restrictions, mean and minimum temperature differences of heat exchangers (characteristic values of the corresponding Q-T-, i.e. total enthalpy-temperature-diagrams), have to be taken into account as constraints.

See [4] for more engineering details.

3 Mathematical formulation of the optimization problem

The mathematical formulation of the optimization problem reads as follows:

$$\min_{p} \varphi(x, p, h(x, p)) \qquad (1)$$

subject to the constraints

$$p_{min} \leq p \leq p_{max}, \quad x_{min} \leq x \leq x_{max}, \quad h_{min} \leq h(x, p) \leq h_{max} \text{ and } f(x, p) = 0$$

where
- $x \in \mathbb{R}^{n_x}$, $p \in \mathbb{R}^{n_p}$ flowsheet variables and optimization parameters
- $f : \mathbb{R}^{n_x + n_p} \to \mathbb{R}^{n_x}$ function of flowsheet equations
- $h : \mathbb{R}^{n_x + n_p} \to \mathbb{R}^{n_h}$ heat exchanger function
- $\varphi : \mathbb{R}^{n_x + n_p + n_h} \to \mathbb{R}$ objective function

Flowsheet equations

A mathematical model of the process is given by the nonlinear system of flowsheet equations $f(x,p) = 0$ which appear as equality constraints to the optimization problem. The solution x includes state variables of the process streams, e.g. component flow rates, temperatures, pressures and enthalpies, and energy streams like the power of compressors. Conservation laws for mass and energy together with physical property equations (equations of state) and additional process conditions yield the mathematical model.

For example, consider a heat exchanger with ns streams and let $T_{in,i}$, $H_{in,i}$ and $T_{out,i}$, $H_{out,i}$, $i = 1, \ldots, ns$, denote the temperatures and enthalpies of the inlet and outlet streams, respectively. The unit equations determine the outlet variables for given inlet conditions and $ns{-}1$ additional specifications, e.g. outlet temperatures:

$T_{out,i} = T_i^{spec}$, $i = 1, \ldots, ns$, $i \neq i_0$ for a given i_0 , $1 \leq i_0 \leq ns$
$H_{out,i} = H_i(T_{out,i})$, $i = 1, \ldots, ns$: equations of state
$\sum_{i=1}^{ns}(H_{out,i} - H_{in,i}) = 0$: conservation of enthalpy (1st law of thermodynamics)

Phase changes (bubble and dew points) may lead to discontinuous derivatives of the nonlinear function $H = H(T)$. Since for pure streams the function itself becomes discontinuous ($T = const$ for vaporizing/condensing streams at constant pressure), the model is based on $T = T(H)$ in this case.

Heat exchanger function

Heat exchanger areas, minimum and mean temperature differences are the components h_i, $i = 1, \ldots, n_h$, of the nonlinear function h. The computation is illustrated by Figure 2.

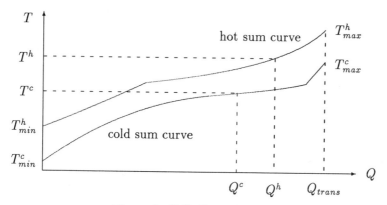

Figure 2: Q-T-diagram

A simple Q-T-diagram of a heat exchanger is shown which describes the heat transfer from hot streams to cold streams as a function of temperature, where

$$Q_{trans} = \sum_{i \in I^h}(H_{in,i} - H_{out,i}) = \sum_{i \in I^c}(H_{out,i} - H_{in,i})$$

denotes the total heat transferred (the indices $i \in I^h$ and $i \in I^c$ refer to hot and cold streams, respectively), and

$$T^h_{min} = \min_{i \in I^h}\{T_{out,i}\}, \quad T^c_{min} = \min_{i \in I^c}\{T_{in,i}\}, \quad T^h_{max} = \max_{i \in I^h}\{T_{in,i}\}, \quad T^c_{max} = \max_{i \in I^c}\{T_{out,i}\}.$$

The hot and cold sum curves $T^h = T^h(Q)$ and $T^c = T^c(Q)$, $0 \leq Q \leq Q_{trans}$, are implicitly defined as follows: By cooling down the hot streams to a given temperature T^h, the heat $Q_{trans} - Q^h$ is transferred, and similarly, for a given temperature T^c the sum of the cold stream enthalpies is increased by Q^c, where

$$Q^h = \sum_{i \in I^h} \Delta H^h_i(T^h), \quad \Delta H^h_i(T^h) = \begin{cases} 0, & T^h < T_{out,i} \\ H_i(T^h) - H_{out,i}, & T_{out,i} \leq T^h \leq T_{in,i} \\ H_{in,i} - H_{out,i}, & T^h > T_{in,i} \end{cases}$$

$$Q^c = \sum_{i \in I^c} \Delta H^c_i(T^c), \quad \Delta H^c_i(T^c) = \begin{cases} 0, & T^c < T_{in,i} \\ H_i(T^c) - H_{in,i}, & T_{in,i} \leq T^c \leq T_{out,i} \\ H_{out,i} - H_{in,i}, & T^c > T_{out,i} \end{cases}$$

Since the reduction of the compressor power implies decreasing temperature differences $\Delta T(Q) = T^h(Q) - T^c(Q)$, $0 \leq Q \leq Q_{trans}$, lower bounds on the minimum and mean temperature differences

$$\Delta T_{min} = \min_Q \Delta T(Q), \quad \Delta T_{mean} = \frac{Q_{trans}}{\int_0^{Q_{trans}} \frac{1}{\Delta T(Q)} dQ} \quad (MTD)$$

have to be imposed by the optimization constraints to enforce the 2nd law of thermodynamics. Furthermore, a measure for the heat exchanger area is given by $Q_{trans}/\Delta T_{mean}$ (Q/MTD) which might be included in the objective function.

Discontinuities in the derivatives of the function $\Delta T(Q)$ may occur for inlet and outlet temperatures (see the definitions for Q^h and Q^c) or result from phase changes.

Objective function

The objective function φ is given by a linear combination of operating costs (energy) and investment costs (heat exchanger areas).

Inserting the solution of the flowsheet equations $x = x(p)$ into (1) yields the formulation

$$\min_p \tilde{\varphi}(p) \quad \text{subject to}$$
$$p_{min} \leq p \leq p_{max} \quad , \quad \tilde{g}(p) \geq 0 \quad , \quad \tilde{g} : \mathbb{R}^{n_p} \to \mathbb{R}^{n_{\tilde{g}}} \tag{2}$$

where $\tilde{\varphi}(p) = \varphi(x(p), p, h(x(p), p))$, and $\tilde{g}(p)$ includes the constraints on $x(p)$ and $h(x(p), p)$.

4 Numerical method

The optimization problem is solved by a feasible path method which is based on formulation (2), i.e. the flowsheet equations $f(x, p) = 0$ are not treated as equality constraints of the optimization problem but solved separately. An SQP (Sequential Quadratic Programming) method is applied where the solution is iteratively computed by minimizing quadratic subproblems. Each optimization step $p^k \to p^{k+1}$ involves the solution of the flowsheet equations, an evaluation of the heat exchanger function and an SQP step.

Solution of the flowsheet equations

Due to the large dimension of the flowsheet equations and to the sparse structure of the Jacobian matrix (which is analytically computed), an inexact Newton method is applied where the resulting linear equations are solved by an iterative method and sparse matrix techniques are used. The discontinuities are treated by an in-house algorithm. The solution $x(p^k)$ and the Jacobian $\frac{\partial f}{\partial (x,p)}(x(p^k), p^k)$ are obtained.

Evaluation of the heat exchanger function

The hot and cold sum curves are approximated by piecewise linear functions which include points of discontinuities. At each discretization point the enthalpies are computed by physical property calculations. Furthermore, the in-house algorithm provides the partial derivatives of the linearized function $\Delta T(Q)$ with respect to the flowsheet and parameter variables at discrete points. Approximations to $h(x(p^k), p^k)$ and to $\frac{\partial h}{\partial (x,p)}(x(p^k), p^k)$ are obtained.

SQP step

The new iterate p^{k+1} is given by $p^{k+1} = p^k + \alpha^k d^k$, where d^k denotes the solution to the quadratic optimization problem

$$\min_{d \in \mathbb{R}^{n_p}} \{ \tilde{\varphi}(p^k) + \nabla \tilde{\varphi}(p^k) d + \frac{1}{2} d^T B^k d \}$$

$$\text{subject to} \quad p_{min} \leq p^k + d \leq p_{max} \quad \text{and}$$
$$\tilde{g}_i(p^k) + \nabla \tilde{g}_i(p^k)\, d \geq 0 \; , \; i = 1, \ldots, n_{\tilde{g}} \; ,$$

and the step length α^k is determined by an appropriate line search algorithm, see e.g. Schittkowski [3]. B^k denotes an approximation to the Hessian matrix which is computed by BFGS updates. Since the analytical derivatives of f and h are available, the gradients $\nabla \tilde{\varphi}$ and $\nabla \tilde{g}_i$, $i = 1, \ldots, n_{\tilde{g}}$, can be computed in an efficient way.

5 Computational results

Computational results are presented for the LNG process introduced in chapter 2. The process is modeled by approximately 1300 flowsheet equations with about 5300 non-zero Jacobian elements. In order to minimize the power of the compressors, 22 process parameters have to be adjusted. Starting from a manually pre-optimized solution, the optimization problem was solved by 9 SQP steps which took $\sim$ 100 CPU seconds on an IBM mainframe 9021. An energy reduction of 10.5% could be achieved. Figure 3 shows the resulting Q-T-diagram of the heat exchangers E1, E2, E3 and E4. Enforcing a minimum temperature difference of 2 K, the mean temperature difference was reduced from 6.3 K to 4.0 K.

By use of OPTISIM®, the time for an optimal process design as well as the process operating costs have been significantly reduced.

6 Conclusions

An optimization method has been presented for the design of process engineering plants. As an example, a natural gas liquefaction process was introduced.

The mathematical model of a process plant yields a large set of nonlinear equations with discontinuous functions and sparse Jacobian matrix. These equations are included as equality constraints to the optimization problem. A special problem of design optimization is given by the modelling of enthalpy-temperature distributions in heat exchangers which yields discontinuous functions arising in the objective function and the constraints.

A feasible path SQP method has been developed which is based on a decoupled solution of the flowsheet equations and approximation of the heat exchanger function in each optimization step. The algorithm includes an appropriate treatment of discontinuities and allows the fast computation of gradients, since the analytical derivatives of the flowsheet equations and heat exchanger function are available.

Implemented in the process simulator OPTISIM, the new method has been successfully used for fast and reliable optimization of process plants.

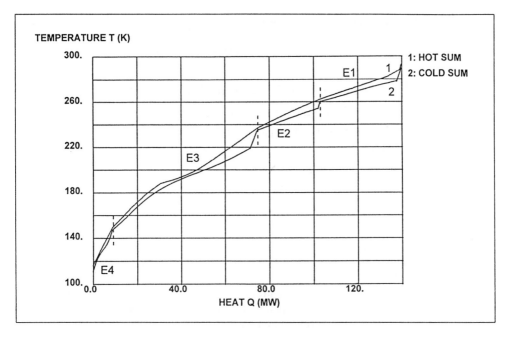

Figure 3: Q-T-diagram of the optimized LNG process

References

[1] P. Burr: The design of optimal air separation and liquefaction processes with the OPTISIM equation-oriented simulator and its application to on-line and off-line plant optimization, presented at the AIChE Spring National Meeting, Houston, Texas, April 7-11, 1991.

[2] E. Eich, P. Lory, P. Burr, A. Kröner: Stationary and dynamic flowsheeting in the chemical engineering industry, to appear in Surv. on Math. in Ind. (1997).

[3] K. Schittkowski: The nonlinear programming method of Wilson, Han and Powell with an augmented Lagrangian type line search function, Numer. Math. 38 (1981), pp. 38–127.

[4] H.-R. Zollner: Liquefaction of natural gas with the aid of refrigerant mixtures, Linde Reports on Science and Technology 32 (1981), pp. 19–28.

Genetic algorithm methodologies for scheduling electricity generation

C. J. Aldridge*, S. McKee* and J.R. McDonald[†]
*Department of Mathematics
[†]Centre for Electrical Power Engineering,
University of Strathclyde, Glasgow, UK,
email c.j.aldridge@strath.ac.uk

Abstract

Scheduling generator units in a power system to meet customer demand at minimum cost is a key activity for power utilities. Finding efficient solution methods for this problem continues to be an active area of research. This paper reviews the implementation of genetic algorithms (GAs) for the unit commitment/economic dispatch problem. In particular we focus on the solution representation, fitness evaluation and genetic operators which have been employed in recent studies.

1 Introduction

Scheduling generating units in a network to supply the daily needs of consumers as economically as possible, while satisfying a multitude of operating constraints, is a tough optimisation problem. The unit commitment/economic dispatch problem involves the solution of the commitment (on or off) and dispatched generation for each generating unit in the network at each time interval in the scheduling period. The total cost of generation depends on the operating, startup and shutdown costs of the individual generating units, summed over all units and time intervals. This is minimised subject to a variety of constraints. The total generation must meet the forecast demand; moreover there must be a certain level of reserve capacity available. Individual generating units are characterised by constraints including the minimum and maximum generation levels, ramp rate limits on the increase and decrease in generation, and the minimum up and down times. There may also be limits on the generation within local areas.

A variety of different techniques have been employed to solve the unit commitment/economic dispatch problem in the last thirty years. A recent overview of the field is given by Sheble & Fahd [13]. The principal methods which have been used include Dynamic Programming (DP) and Lagrangian Relaxation (LR), and more recently Expert Systems and Artificial Neural Networks (ANN) have been implemented for unit commitment problems. In the last few years a new technique — Genetic Algorithms (GAs) — has been applied to generation scheduling.

2 Genetic Algorithms

GAs are search and optimisation methods based on a model of evolutionary adaptation in nature [2,3,5]. A GA works with a 'population' of possible solutions, and creates successive 'generations' of the population by several simple 'genetic' operators. The objective function is absorbed into a fitness function, which may take account of constraint violations via penalty terms. In each generation, solutions are selected stochastically according to their fitness in order to contribute to the next generation. Relatively 'fit' solutions survive, 'unfit' solutions tend to be be discarded. A new generation is created by stochastic operators — typically 'crossover', which swaps parts of binary-encoded solution strings, and 'mutation', which changes random bits in the strings. Successive generations yield fitter solutions which approach the optimal solution to the problem. GAs are inherently simple, naturally parallelisable, and can generate a set of near-optimal solutions for evaluation.

A GA is thus characterised by three main elements: (i) solution representation, (ii) fitness evaluation, and (iii) genetic operators. The recent studies reviewed below have used a variety of different choices for these elements in order to improve the accuracy and efficiency of the GA solution.

3 Problem Specification

GAs have been used to solve unit commitment only [1] as well as calculate both commitment and dispatch [4,6,7-12,14]. All the test problems considered have constraints of system demand, system reserve, unit minimum up and down times, and unit minimum and maximum generation. Other constraints considered include ramp rates [11,14], unit inflexibilities [4], maximum number of shifts [12], and transmission losses [7,10].

Operating costs are specified by fuel costs, startup costs and shutdown costs. Fuel cost is generally taken to be a quadratic function of dispatched generation. Since only the commitments are calculated in [1], the fuel cost in this case is taken as the (constant) average full load cost (AFLC) for each unit. Start-up cost is typically formulated as an exponential function of the unit down time; shut down cost is included in some studies as a constant. GAs have been applied to test problems with up to 110 thermal units and typically 24 (hourly) time periods.

4 Solution Representation

The commitment and dispatch variables for each time interval in the scheduling period are coupled by dynamic constraints of minimum up and down times and ramp rate limits. Hence most studies [4,6,7,8,9,12,14] have solved the problem using a *single* GA for the entire scheduling period. In this case the solution may be represented in alternative ways. Firstly, an explicit representation may be used in which both the commitment and the dispatch variables are encoded in the solution string [9]. Secondly, the commitments only may be represented in the solution string, and the dispatch subsequently calculated in the fitness function. In this case the obvious representation is simply a binary array A of the commitments,

$$(A)_{it} = \alpha_i^t, \tag{1}$$

where $\alpha_i^t = 1$ if unit i is committed in time interval t, otherwise $\alpha_i^t = 0$, and there are n units and T time intervals. This representation has proved the most common choice in recent studies [4,6,7,8,14].

In order to apply the standard mutation and crossover operators used in a basic GA, the commitments must be listed in a one-dimensional string, by ordering the elements of A column-wise [4] or row-wise [6]. Ma et al. [7] found that the column-wise ordering

$$(\alpha_1^1, \ldots, \alpha_n^1; \ldots \ldots; \alpha_1^T, \ldots, \alpha_n^T) \tag{2}$$

gave quicker convergence to a lower cost solution.

The representation in (1) admits a multitude of solution strings which violate minimum up and down time constraints, and therefore much of the search space of the GA will be infeasible. Hence Saitoh et al. [12] introduced a representation in which the minimum up and down times are automatically respected. Suppose each unit i is allowed to change status a maximum of m_i times, and let t_{ik}, $k = 1, \ldots, m_i$ denote the number of time intervals in which unit i remains up (or down) beyond the minimum up (or down) time. Then the commitments of the units may be represented by the string

$$(t_{11}, \ldots, t_{1m_1}, \ldots \ldots, t_{n1}, \ldots, t_{nm_n}). \tag{3}$$

Instead of using a single GA for the entire period, the solution may be calculated *sequentially* [1,10,11]. This approach involves using a GA for each individual time interval in turn, taking account of the effect of the variables set at the previous scheduling points. The representation in (1) gives a relatively large string of length nT, and a large search space of 2^{nT} possible strings. As is noted in [1], this leads to a large population size, and brings associated memory and computational time requirements. Moreover the problem is highly 'epistatic' (e.g. see [1]), with the commitments at one time interval being highly dependent on those at previous time intervals, and this can lead to problems in convergence. These problems are avoided by decomposing the problem with respect to the time intervals. In this case the solution string of the GA for time interval t is simply $(\alpha_1^t, \ldots, \alpha_n^t)$.

5 Fitness Evaluation

The quality of each solution created within the GA is measured using a fitness function. For a constrained minimisation problem such as unit commitment/economic dispatch, low cost feasible solutions are assigned high fitness values, while high cost infeasible solutions take low values of fitness. The fitness is therefore typically calculated as follows. Using the commitments given by the solution string, the economic dispatch sub-problem is solved by a standard (non-evolutionary) algorithm. For a GA solving the problem over the entire scheduling period, for example, the operating cost is given by

$$\text{cost} = \sum_{t=1}^{T} \sum_{i=1}^{n} \text{start up cost} + \text{shut down cost} + \text{fuel cost}, \qquad (4)$$

where the summation is over all units $i = 1, \ldots, n$ and time intervals $t = 1, \ldots, T$. This cost is then penalised by

$$\text{evaluation} = \text{cost} + \sum_{t=1}^{T} w_1 \theta_1^t + w_2 \theta_2^t + \sum_{i=1}^{n} w_3 \theta_{3i}, \qquad (5)$$

where w_1, w_2 and w_3 are penalty weights, θ_1^t and θ_2^t are the violations of the demand and reserve constraints, and θ_{3i} is the violation of the minimum up and down constraints of unit i.

Alternatively in [8,14] the minimum up and down time constraints have been incorporated in the fitness function as follows. A temporary matrix is created from A by changing the commitments which violate minimum up and down times. The operating cost is then calculated using this temporary matrix, with the modified commitments incurring banking costs. In [1] a utility factor is introduced to measure the efficiency of the chosen commitments, defined as the ratio (load − reserve requirement)/total committed output, and then incorporated in the fitness evaluation using an extra penalty function.

The weights associated with the penalty terms in the fitness function must be chosen sufficiently large enough to discourage evolution of solutions in infeasible search space. It is simplest to keep weights fixed throughout the GA run. Kazarlis et al. [6] set the weights as linearly increasing with generation number. Hence the GA is allowed more freedom in the initial generations and more pressure is applied as the GA progresses to penalise infeasible strings.

The fitness is then given by a monotonically decreasing function of the evaluation value, for example [10,11]

$$\text{fitness} = \left(\max_{\text{population}} \text{evaluation} \right) - \text{evaluation} \qquad (6)$$

Finally, fitness scaling can be done by applying a nonlinear transformation to the fitness values in each generation, in order to emphasise the differences between near-optimal solutions [6,12].

6 Genetic Operators and Results for Single GAs

In a basic binary GA, strings selected from the previous generation are used to form the current generation of strings by applying standard mutation and crossover operators with some chosen probabilities. In one-point crossover, two strings are randomly chosen, split at some randomly chosen bit position, and two new strings produced by exchanging the left and right parts. The standard mutation operator simply changes randomly chosen bits. In general, crossover acts to exploit good solutions, while mutation explores the solution space. The standard operators are problem independent, and the performance of the GA can be enhanced by employing problem specific mechanisms within the operators. A number of such operators have been introduced in studies using single GAs for the entire scheduling period.

Hassoun & Watta [4] employed two different GAs. The first GA uses a randomly chosen initial population, and standard one-point crossover and mutation operators. In the second GA, the initial population is chosen to satisfy minimum up and down time but otherwise is generated randomly, and the mutation and crossover operators are modified to act unit-wise. The modified crossover selects a unit and exchanges the commitments for that unit only. The modified mutation operator selects a unit and chooses a new random commitment, making it satisfy the minimum up and down time specified. Both GAs were applied to a test problem of 10 units and 24 time intervals. The best solution found by the first and second GA cost respectively 4% and 1.2% more than the best solution found by Lagrangian relaxation. However the computational time for the GA was several hours, compared with around one minute for LR.

Four new operators were introduced by Kazarlis et al. [6]. All are unary in nature in that they operate on elements of a single chromosome, and are thus variants on mutation. The first two are applied throughout the population, the remaining pair act only on the best string in each generation. The combination of these four operators and linearly increasing penalty weighting enabled the optimum solution to be found by the GA for a test problem of 10 units and 24 hours. Solving 60–100 unit problems, all GA runs achieved lower operating costs than LR. However the GA run time on a workstation for 100 units was nearly four-and-a-half hours.

Ma et al. [7] demonstrated that quicker convergence to near-optimal solutions was obtained using two-point rather than one-point standard crossover. Moreover a new unary operator 'forced mutation' was introduced, and applied after selection, crossover and standard mutation. This ensures that the reserve and demand constraints may be met by full generation of the committed units, but that the units are not over-committed with the demand and reserve less than the sum of the minimum outputs of the generators. The repair is achieved by forcing a unit, randomly chosen from the down (or up) units, to be on (or off) for the whole time period.

In the GA used by Maifeld & Sheble [8], the commitment matrix A was initialised by randomly choosing one of the 10 cheapest economic dispatch solutions for each hour. Crossover was done by exchanging the first m columns of two commitment matrices. In addition to standard mutation, three extra 'mutation' operators were introduced to act on the commitments for a randomly chosen unit i. The GA was applied to test problems with 9 generators and 24 or 48 hours and results compared with LR. The GA solutions were on average approximately 0.1% to 3% lower cost than those found by LR, with both methods taking around one minute of computing time on a workstation.

Sheble & Maifeld [14] used a GA with column-wise crossover and mutation for a test problem with 6 units and 24 time intervals. The best GA solutions were on average about 0.5% higher cost than that found by Lagrangian relaxation. The GA was then modified, so that each solution created is checked to establish whether units marked as 'banking' during the fitness evaluation should be turned on in order to reduce the operating cost. Thus modified, the GA achieved up to 3% lower cost solutions than LR. However, with a lower demand (so that all units need not be committed) only 1 out of 5 GA runs achieved lower cost solutions than LR.

Oliveira [9] addressed a test problem of two thermal units plus one pumped storage and one hydro plant. Because of the different operating constraints, distinct genetic operators were designed for each unit. All local constraints (such as minimum up and down time for thermal units) were respected by these operators. The best solutions produced by the GA were within 2% of the optimum.

7 Results for Sequential GAs

The sequential GA approach was used by Dasgupta & McGregor [1] to calculate multiple strategies. In each time period, a GA is used for each strategy, in which the initial population is taken as a combination of solutions from the final population at the previous stage of that strategy, and randomly generated strings. The best solution is adopted as the next stage of the strategy. The GA, which uses standard mutation and crossover, was applied to a test problem with 10 units over 24 hours. The least cost solution found using 5 strategies was 7% cheaper than using a single strategy.

Orero & Irving [10] used a sequential GA incorporating a priority list heuristic algorithm. This was applied at each time interval, and the heuristic solution included in the initial population of the GA, together with the GA solution at the previous time interval. The 'hybrid' GA gave quicker convergence than the GA without the priority list algorithm. For the 10 unit problem considered in [4], both GAs found a solution 4% cheaper than the simple priority list solution, using about 1 second of computing time on a workstation. The GA solution was 0.2% higher cost than that obtained using Lagrangian relaxation. Applying the GAs to a larger system of 110 units, the hybrid GA took 17 minutes to produce a solution 0.5% cheaper than the priority list, compared to 3 hours for the normal GA.

The same authors [11] compared a sequential GA and a single GA for a test network of 26 units and 24 time intervals. The computational time is much reduced for the sequential GA, which produces solutions with a modest improvement in cost on the single GA solutions. Moreover, for a problem with a relatively low demand and ramping rate constraints on the units, the GA solutions were some 5% lower cost than those found by a method combining ANN, DP and heuristics.

8 Discussion and Conclusions

A variety of methods have been used in applying GAs to the unit commitment/ economic dispatch problem, involving different solution representations, fitness evaluations and genetic operators. The sequential approach has been found to require less computational time than for a single GA unaided by problem specific operators. Including heuristic solutions in the starting population has been shown to give quicker convergence, and lower cost solutions have been obtained by constructing multiple strategies over the scheduling period. Most studies using a single GA have introduced problem specific operators, in particular new mutation operators. It has been demonstrated that using such operators leads to lower cost solutions than with standard genetic operators. A variety of new operators have been proposed and it is difficult to make comparisons between performance for different test problems and on different computing environments.

The results reviewed above have demonstrated that GAs can find solutions comparable or lower in cost to rival methods, in particular Lagrangian relaxation. However the computational time required by a GA can be typically much greater than for LR and this remains a chief drawback, though parallelisation of the GA offers a potential speed up of the method.

It is clear that GAs augmented by knowledge-based systems show much promise as solution methods for unit commitment/economic dispatch. There are several distinct sources of problem specific knowledge with which to augment the GA. Problem specific knowledge may be drawn from operating archives and elicited from experts. Moreover successive computations will also build up a store of knowledge about good solutions to the problem. Work in progress at the University of Strathclyde is exploring these issues in order to develop a GA underpinned by a knowledge-based system for generation scheduling.

Acknowledgements

This research is supported by the Engineering and Physical Sciences Research Council and The National Grid Company plc.

References

[1] Dasguptar, D. & McGregor, D.R., Thermal unit commitment using genetic algorithms, *IEE Proceedings C — Generation, Transmission and Distribution* **141**(5):459–465, 1994.

[2] Davis, L., *Handbook of Genetic Algorithms*, Van Nostrand Reinhold, 1991.

[3] Goldberg, D.E., *Genetic Algorithms in Search, Optimisation, and Machine Learning*, Addison-Wesley, 1989.

[4] Hassoun, M.H. & Watta, P., Optimization of the unit commitment problem by a coupled gradient network and by a genetic algorithm, report no. TR-103697, Electric Power Research Institute, 1994.

[5] Holland, J.H., *Adaptation in Natural and Artificial Systems*, University of Michigan Press, 1975.

[6] Kazarlis, S.A., Bakirtzis, A.G. & Petridis, V., A genetic algorithm solution to the unit commitment problem, *IEEE Transactions on Power Systems* **11**(1):83–90, 1996.

[7] Ma, X., El-Keib, A.A., Smith, R.E. & Ma, H., A genetic algorithm based approach to thermal unit commitment of electrical power systems, *Electrical Power Systems Research* **34**(1): 29–36, 1995.

[8] Maifeld, T.T. & Sheble, G.B., Genetic-based unit commitment algorithm, in: *Proceedings of North American Power Symposium*, 1995.

[9] Oliveira, P., *Optimal Scheduling of hydro-thermal power generation systems*, Ph.D. thesis, University of Strathclyde, 1992.

[10] Orero, S.O. & Irving, M.R., Scheduling of generators with a hybrid genetic algorithm, in: *Proceedings of First International Conference on Genetic Algorithms in Engineering Systems: Innovations and Applications*, IEE Conference Publication no. 414, 200–206, 1995.

[11] Orero, S.O. & Irving, M.R., A genetic algorithm for generator scheduling in power system, *International Journal of Electrical Power and Energy Systems* **18**(1):19–26, 1996.

[12] Saitoh, H., Inoue, K. & Toyoda, J., Genetic algorithm approach to unit commitment, in: Hertz, A., Holen, A.T. & Rault, J.C. (eds), *Proceedings of the International Conference on Intelligent System Application to Power Systems*, 583–589, 1994.

[13] Sheble, G.B. & Fahd, G.N., Unit commitment literature synopsis, *IEEE Transactions on Power Systems* **9**(1):128–135, 1994.

[14] Sheble, G.B. & Maifeld, T.T., Unit commitment by genetic algorithm and expert system, *Electrical Power Systems Research* **30**(2):115–121, 1994.

Target Zone Models with Price Inertia: A Numerical Solution Method

Isa Scheunpflug   Norbert Köckler

Abstract

For examining the logic of exchange rate regimes we analyze a political based approach in an interdependent macroeconomic framework. The stochastic exchange rate model yields an ordinary differential equation (ODE) with the interval limits as unknown parameters. This problem is solved numerically by a special method which merges the ODE solution process with the parameter identification. This method is fast and black box–like because it yields results to a prescribed precision in all cases considered.

1 Introduction

The research of the functioning of exchange rate regimes has a long tradition in the field of international monetary economics. The traditional theoretical work–horses of the area are only able to distinguish flexible and fixed exchange rate regimes in more deterministic frameworks. But empirically, absolutely and permanently fixed or flexible exchange rate regimes never exist. The "fixed" exchange rate regimes of Bretton–Woods or the European Monetary System are defined by an explicitly announced target zone around a national central parity – exchange rate movements are possible and allowed for the given range. The regimes which are normally treated as "flexible" like the DM/$, $/Yen or DM/Yen exchange rate are controlled to a "managed–floating" and/or temporarily informal target zone commitments like the Plaza–Agreement or the Louvre–Accord – these are "quiet" target zones. Despite these explicit examples, the main argument of constructing an exchange rate theory by analysing target zone regimes is political and institutional based: The target zone approach takes into account that political authorities are trying to reach special domestic targets like fighting unemployment, inflation etc. with their monetary and fiscal instruments, independent from an explicitly announced exchange rate regime. Political authorities normally react when well–defined trigger points, indicating the maximum allowed divergence in both directions from the main target, are reached. Every defended target leads via the reaction function of the authorities to a special restriction on the foreign currency market – nothing else than a special exchange rate target zone regime, which works independently from

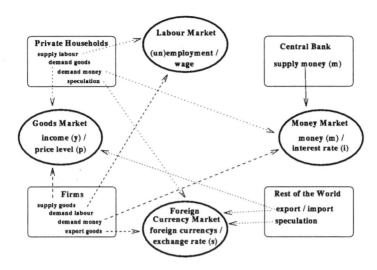

Figure 1: The interdependencies of the markets

international agreements. If the implicit created restriction of the foreign exchange market is actively taken into account by the authorities, an explicit exchange rate target zone regime is established. A target zone system can be seen as the general case of exchange rate regime, where fixed and flexible exchange rates represent only the extremes.

Here we are only concerned with explaining the functioning of an explicitly announced exchange rate target zone regime in an abstract theoretical stochastic economic model; that means evaluating the impacts for the remaining economic variables under an exchange rate target zone regime, as well as stability conditions and dynamical behavior of the solutions and, first of all, finding the solution itself. Essentially, the economic problem of constructing an exchange rate target zone model involves controlling a stochastic process. In our special model this leads to a two point boundary value problem with unknown parameters which we cannot solve in closed form. So we use a numerical method for solving the ODE combined with a special root finding method for the parameter identification problem.

2 The Model

The model used is a stochastic exchange rate model of a small open economy with price inertia and rational expectations, which is capable to explain the resulting macroeconomic adjustment processes in terms of interest rate i, price level p, output y and employment, see [1] for details. The interdependent market system can be formalized as followed[1]:

$$m - p = \phi y - \lambda i \tag{1}$$

$$E_t[ds(t)] = (i - i^*)\,dt \tag{2}$$

$$y = -\gamma\,[i - E_t(dp)] + \eta\,(s - p + p^*) \tag{3}$$

$$dp = \varphi\,(y - \bar{y})\,dt + \sigma\,dw \tag{4}$$

The variables of the model are linked by several feed–back rules, see Fig. 1 or [9] for a detailed explanation. The fundamental forcing variable of the model, the price level p, follows therefore an autoregressive process whose trend is endogenously determined. This non–recursive structure is the general problem of solving the stochastic model under a target zone regime.[2]

3 A Nominal Currency Band

Since the markets and therefore the variables of the model are interdependently connected, a given restriction on the domain or range of a special target variable will have great impacts to the domains or ranges of other variables. If the evaluations of the exchange rate are restricted to a prespecified interval like a target zone, other variables will be restricted as well. The most obvious case of a target zone regime is a nominal currency band, i.e. restricting the evolution of the nominal exchange rate s to a symmetrical zone[3] $b_u = \bar{s}$, $b_l = \underline{s}$ around a central parity s_0, like the European Monetary System.

The target zone technique looks for a mapping between the variables s and p, m of the form $s(t) = G(p(t), m(t))$.

The desired solution $G(\cdots)$ has to satisfy several conditions:

(1) $G(\cdots)$ is a continuously twice differentiable function of the arguments p and m. Anticipated profits cannot occur to the rational agents by this no–jump condition.

(2) $G(p(\cdot), m(\cdot))$ depends only on realizations at time t, since the Brownian motion fulfills the Markov property.

The first step of evaluating the unknown function $G(p, m)$ is reducing the model. Subsituting the equations (1) to (4), the model can be formulated in the forcing

---

[1] All variables are logarithms despite the interest rate; Greek letters indicate constant parameters; starred variables indicate variables of the rest of the world.

[2] For a treatment of the qualitative solution see [5] – [7]

[3] b_u for upper band and b_l for lower band

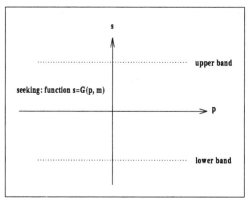

Figure 2: Range of the desired solution

variable p, the control variable m and the target variable s:

$$\begin{bmatrix} dp \\ E(ds) \end{bmatrix} = A \begin{bmatrix} p \\ s \end{bmatrix} dt + b\, dt + \begin{bmatrix} \sigma \\ 0 \end{bmatrix} dw \qquad (5)$$

$$\text{with} \quad A = \begin{bmatrix} a_{11} & a_{12} \\ a_{21} & a_{22} \end{bmatrix} = \frac{1}{\Lambda} \begin{bmatrix} -\varphi(\lambda+\gamma\eta) & \varphi\gamma\eta \\ 1-\phi\eta-\varphi\lambda & \phi\eta \end{bmatrix}$$

$$b = \begin{bmatrix} b_1 \\ b_2 \end{bmatrix} = \frac{1}{\Lambda} \begin{bmatrix} \varphi\lambda \\ (\varphi\lambda-1) \end{bmatrix} m$$

$$\Lambda = \phi\lambda + \gamma - \varphi\lambda\gamma$$

and $i^* = 0$, $p^* = 0$, $\bar{y} = 0$ for simplicity. The only variables we consider are the relationship between the price level p and the exchange rate s.

Since the determinant of $\mathbf{A}$ is negative[4], the system (5) is saddlepoint stable. Of economic reasons, we will assume that the slope of the (overall) stable (uncontrolled) saddlepath is negative, this requires $a_{21} > 0$.

Since the model is stochastic, we are faced with a problem known as Jensen's inequality: $E[G(p,m)] \neq G[E(p), E(m)]$, if $G(p,m)$ is a nonlinear function. Applying the rules of stochastic calculus to $s = G(p,m)$, we overcome the stochastic problem by using Ito's Lemma (see [2]):

$$ds = G_p(p,m)\, dp + G_m(p,m)\, dm + \frac{\sigma^2}{2} G_{pp}(p,m)\, dt. \qquad (6)$$

Taking expection yields

$$E(ds) = G_p(p,m)\, E(dp) + \frac{\sigma^2}{2} G_{pp}(p,m)\, dt. \qquad (7)$$

and substituting the system (5) into (7) we end up with the following ordinary differential equation for $s = G(p,m)$:

$$G_{pp} = \frac{2}{\sigma^2} \{a_{21}\, p + a_{22}\, G + b_2 - (a_{11}\, p + a_{12}\, G)\, G_p\} \qquad (8)$$

---

[4]$\det \mathbf{A} = -\frac{\varphi\eta}{\Lambda} < 0$ is only possible if $\Lambda > 0$, or equivalently, if $\frac{\omega}{\gamma} > \varphi - \frac{1}{\omega}$.

Because of the internal feed-back rule in p, equation (8) has no closed-form solutions. So we use a numerical method for solving the saddlepaths.[5]

4 The Numerical Method

Given a restriction interval b_u, b_l for the exchange rate, the target zone regime itself is modeled by choosing appropriate boundary value conditions on $G(p,m)$ which are consistent with the process of a reflected Brownian motion, i.e.:
Find an interval $[p_l, p_u]$ such that the ODE (8) for the function $G(p,m)$ fulfills the boundary value conditions (see [2]):

$$G(p_l, m) = b_u, \qquad G_p(p_l, m) = 0, \tag{9}$$
$$G(p_u, m) = b_l, \qquad G_p(p_u, m) = 0. \tag{10}$$

We have developed a numerical procedure for finding these values. It consists of a combination of a find algorithm, a modification of Newton's method and – just in case – the bisection method. In each step of this root finding method an initial value problem (IVP) is solved with the ODE (8) and initial values (9). The upper limit $p = p_u$ has to be determined during the solution process so, that the solution takes on both the boundary value conditions (10). For the moment we assume that we can always find a point p_u such that at least one of the conditions (10) is fulfilled. Then we search a root of the function

$$f(p_l) := \begin{cases} f_1(p_l), & \text{if } f_2(b_l) = 0, \\ f_2(p_l), & \text{if } f_1(b_l) = 0 \end{cases} \tag{11}$$

with $\quad f_1(p_l) := G(p_u, m) - b_l \quad$ and $\quad f_2(p_l) := G_p(p_u, m)$. $\qquad(12)$

Let $\bar{p}$ be the value for p_l looked for, then the infinite region $[-\infty, p_0]$ is divided into

region 1 $:= [-\infty, \bar{p}] := \{p \,|\, f_1(p_l) = 0$ and $f_2(p_l) < 0\}$ and

region 2 $:= [\bar{p}, p_0] := \{p \,|\, f_1(p_l) > 0$ and $f_2(p_l) = 0\}$

with $p_0 := (-b_u\, a_{22} - b_2)/a_{21}$. Values $p > p_0$ need not be considered, because we are only looking for a decreasing solution ($G_p < 0$). Starting from $p = p_0$ and moving to the left, a find algorithm, see [3], is designed resulting in an inclusion interval $[p_1, p_2]$ for $\bar{p}$. We can now apply Newton's method to the function f from (11) with some starting value $p^{(0)} \in [p_1, p_2]$.

---
[5]A power-series solution method is developed by A. Sutherland, see [10]

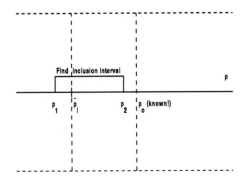
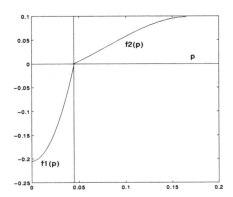

Figure 3: The Find Algorithm   Figure 4: The Split Function

We have to take into account, that f is a split function (see Fig. 4) with a derivative f_p, which is discontinuous at $\bar{p}$, and that the form of the functions f_1 and f_2 may lead to divergence. Additionally, one function evaluation of f means solving the IVP from p_l to p_u, where p_u has to be determined during the solution process such that at least one of the functions f_1 or f_2 attains the value zero. We modified Newton's method accordingly. The convergence behavior of this special Newton method is sensible to the selection of its parameters. The easiest, but slowest way to find $\bar{p}$ after having included $\bar{p}$ with the find method, is bisection. This will automatically be applied, when Newton's method fails.

5 Economic Implications

The resulting saddlepaths show some interesting features (see Fig. 5): first, they are non–linear; under a nominal target zone system, the expectation of exchange rate changes in the foreign exchange market is always (except at the origin) not equal to zero. Suppose that the Brownian motion creates cumulated positive realizations of the price level in the interior of the band. The stochastic inflation process lowers the real money supply. This induces an increase in the nominal and real interest rate and the exchange rate. The goods demand reacts with lower national investments and exports. The demand falls under the potential output, unemployment rises and induces a deflationary tendency. The inflation process consists now of two components: the realizations of the stochastic process (which can be positive) and the deflationary tendency. Therefore the agents expect a deflation on the right side of the stable saddlepoint, since a higher price level $p > 0$ is connected with unemployment.[6] Since the probability, that the realizations of the Brownian motion

---

[6]On the left side of the origin, the agents expect an inflationary process, since the goods demand exceeds the potential output.

reaches the lower boundary over a given time interval is quite small, it seems reasonable, that the deflationary tendency dominates the price adjustment. The system is therefore self–stabilizing, it converges back to the stable saddlepoint.

If the cumulated realizations of the Brownian motion reach the lower boundary at b_l, the agents expect a deflationary tendency and an intervention of the central bank. At this point, the interest rate differential reaches its maximum (see Fig. 6). The central bank changes the money supply to prevent a further increase in the interest rate. Because the domestic interest rate must be lowered for the exchange rate stabilization at the boundary, the central bank has to expand the money supply. The intervention shifts the $\mathcal{S}$-curve to the right. As the money supply changes the $\mathcal{S}$-curves become asymmetric: they do not reflect a new long–run equilibrium with an higher money supply but an overall mean–reverting tendency to the equilibrium point at the origin. The agents expect in addition to the deflation of the goods market effect conditional decreasing realizations of the stochastic process. This intensifies the deflationary process and the domestic exchange rate is likely to hit the upper boundary.

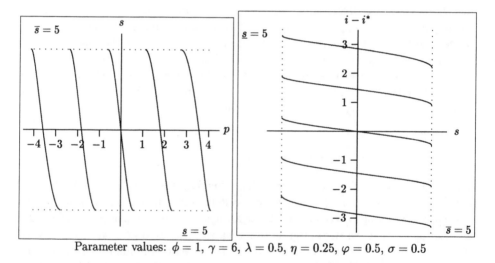

Parameter values: $\phi = 1, \gamma = 6, \lambda = 0.5, \eta = 0.25, \varphi = 0.5, \sigma = 0.5$

Figure 5: A nominal target zone with interventions

Figure 6: Interest rate differential of a nominal target zone

The central bank reverses the intervention and lowers the money supply at the upper boundary. The intervention schemes goes on until the original saddlepath through the origin is reached. In a nominal target zone regime, the reduced exchange rate variability is shifted to an higher variability of domestic interest rates. Empirically, the argument for a stabilization of exchange rates in form of target zone commitments is trade based. A reduced variability in the foreign currency market should reduce transaction costs and risks for export and import industries.

But the variability trade-off in the interest rates affects not only the international industries but the entire economy including the private sector. The adjustment mechanism to reach the long-run equilibrium works through the channel of higher unemployment and/or prices, whereas the maximum of both variables for a given money supply is bounded. A generalized version of our algorithm should be able to translate every exchange rate restriction into the resulting saddlepath dynamics. The parameters b_u and b_l could be treated as (non-)linear functions which are determined by special political targets like stabilization of growth rates. A next stage of investigation will be an extension for the analysis of implicit exchange rate regimes: given a political reaction function, the resulting exchange rate restriction should be identified.

References

[1] Dornbusch, R.: 'Expectations and Exchange Rate Dynamics', Journal of Political Economy 84 (1976), pp. 1116-1176

[2] Harrison, J.M.: 'Brownian Motion and Stochastic Flow Systems', John Wiley and Sons, New York, 1985

[3] Köckler, N.: Numerical Methods and Scientific Computing, Oxford University Press, Oxford, 1994

[4] MacDonald, R.:, 'Long-Run Purchasing Power Parity: It is for Real?', The Review of Economics and Statistics 75, 4 (1994), pp. 690-695

[5] Miller, M./Weller, P.: 'Solving Stochastic Saddlepoint Systems: A Qualitative Treatment with Economic Applications', CEPR Discussion Paper No. 308, 1989

[6] Miller, M./Weller, P.: 'Exchange Rate Bands with Price Inertia', The Economic Journal 101, 409 (1991), pp. 1380-1399

[7] Miller, M./Weller, P.: 'Stochastic Saddlepoint Systems Stabilization Policy and the Stock Market', Journal of Economic Dynamics and Control 19, 1/2 (1995), pp. 279-302

[8] NAG Fortran Library Manual: The Numerical Algorithm Group Ltd. Wilkinson House, Jordan Hill Road, Oxford OX2 8DR, U.K.

[9] Scheunpflug, I./Köckler, N.: 'Target Zone Models with Price Inertia: Nominal and Real Exchange Rate Bands', forthcoming

[10] Sutherland, A.: 'Target Zone Models with Price Inertia: Solutions and Testable Implications', The Economic Journal, 104 (1994), pp. 96-112

Dynamics of machinery

Model Reduction of Random Vibration Systems

Ralf Wunderlich, Jörg Gruner and Jürgen vom Scheidt

TU Chemnitz-Zwickau, Department of Mathematics
09107 Chemnitz, Germany
E-Mail: wunderlich@mathematik.tu-chemnitz.de

Abstract

The paper considers the computation of statistical characteristics of the response of a high-order linear vibration system to a random excitation. Since standard methods fail because of tremedous computational problems model reduction techniques are applied. We find approximations of the desired response characteristics of the high-order system by solving a "suitable" low-order system. Numerical results for the torsional random vibrations of a generator shaft are presented.

1 Problem

Mathematical modeling of real-world vibration systems (e. g. rotating generator shafts excited by random fluctuations of the generator torque, vehicles moving on a rough guideway) results in a system of ordinary differential equations (ODE) containing random parameters. For a linear system with n degrees of freedom (DOF) and random excitation we get

$$\mathbf{J}\,\ddot{\mathbf{p}}(t) + \mathbf{D}\,\dot{\mathbf{p}}(t) + \mathbf{K}\,\mathbf{p}(t) = \mathbf{C}\,\mathbf{f}(t,\omega) \tag{1}$$

with the initial conditions $\mathbf{p}(0) = \mathbf{p}_0$ and $\dot{\mathbf{p}}(0) = \mathbf{p}_1$. Thereby the vector $\mathbf{p} = (p_1,\ldots,p_n)^\tau$ contains the response variables, $\mathbf{J},\mathbf{D},\mathbf{K},\mathbf{C}$ are $n \times n$ matrices and $f(.,\omega)$ is a n-dimensional random vector function representing the random excitation terms.

For the case of a discretized model of a generator shaft (see Fig. 1) $p_i(t)$, $\dot{p}_i(t)$ and $\ddot{p}_i(t)$, $i = 1,\ldots,n$, denote the angular deflection, velocity and acceleration of the i-th mass. The matrices $\mathbf{J},\mathbf{D}$ and $\mathbf{K}$ contain the moments of inertia, damping and stiffness coefficients, respectively, while $f_i(t,\omega)$ is the random excitation of the i-th mass by which the fluctuating torques on the steam turbines or the electric generator can be modeled.

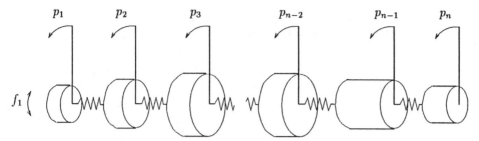

Figure 1: Discretized model of a generator shaft

Since the right hand side of (1) contains random terms (which are explained in Section 2) the response p is a random function, too. In this paper we consider the problem of computation of statistical characteristics (such as mean, variance, correlation and spectral density function) of the random response $\mathbf{p} = \mathbf{p}(t,\omega)$ to given characteristics of the random excitation term $\mathbf{f} = \mathbf{f}(t,\omega)$.

For the present case of a linear system there exist a number of methods for solving this problem (see e.g. [2],[3]). Without loss of generality we can assume that $\mathbf{f}(t,\omega)$ is centered, since in the case of linear systems a non-zero mean of $\mathbf{f}$ only affects the mean of the response $\mathbf{p}$, which can be obtained by the solution of a non-random system of ODE. In Section 3 our method which is based on the theory of the weakly correlated random functions is sketched. In practice these methods only can be applied for moderate numbers of DOF. Tremedous computational problems arise if the number of DOF is very high, which is the typical case in the discretization of continuous vibration systems as in the mentioned example above. Therefore we consider in Section 4 model reduction techniques to find approximations of the statistical characteristics of the response of a high-order system by solving a "suitable" low-order system, which still can be handled numerically. Some numerical results for the random vibration of a generator shaft are presented in Section 5.

2 Random Excitation

The random excitation term is described by the vector of dimension n

$$\mathbf{f}(t,\omega) = (f_1(t,\omega),\ldots,f_r(t,\omega),0,\ldots,0)^\tau, \quad 0 < r \leq n,$$

where r is the number of excitations. The random functions $f_p(t,\omega)$, $p = 1,\ldots,r$, are expressed by linear functionals

$$f_p(t,\omega) = \int_{-\infty}^{t} Q_p(t-s) h_\varepsilon(s,\omega)\, ds \qquad (2)$$

of a weakly correlated, wide-sense stationary and centered random function $h_\varepsilon(t,\omega)$ (see [1]) with the correlation length ε.
The kernel functions $Q_p(.) \in \mathbf{L}_1(\mathbb{R}_+) \cap \mathbf{L}_2(\mathbb{R}_+)$ are non-random.

The essential property of a weakly correlated function $h_\varepsilon(s,\omega)$ is that the influence of the values of the function does not reach far. Values of the function at two points do not correlate (are independent) if the distance of the two points exceeds the correlation length ε. The quantity ε is assumed to be sufficiently small to apply expansions as to ε for approximations of the distribution of linear functionals as (2). For the correlation function we get the expansion

$$\begin{aligned}
R_{f_p f_q}(t,s) &= \mathbf{E}\{f_p(t) f_q(s)\} \\
&= \int_{-\infty}^{t} \int_{-\infty}^{s} Q_p(t-u) Q_q(s-v) \mathbf{E}\{h_\varepsilon(u) h_\varepsilon(v)\}\, du\, dv \\
&= \varepsilon \cdot a \int_{-\infty}^{\min(t,s)} Q_p(t-u) Q_q(s-u)\, du + o(\varepsilon)
\end{aligned} \quad (3)$$

Here, $\mathbf{E}\{.\}$ denotes the expectation and a is the so-called intensity defined by

$$a = \lim_{\varepsilon \downarrow 0} \frac{1}{\varepsilon} \int_{-\varepsilon}^{\varepsilon} \mathbf{E}\{h_\varepsilon(t) h_\varepsilon(t+\tau)\}\, d\tau.$$

If the non-random kernel functions in (2) are choosen as

$$Q_p(t) = \exp(-\gamma_p t), \; \gamma_p > 0,$$

we get an excitation which is centered and wide-sense stationary with an autocorrelation function for which the expansion (3) gives

$$R_{f_p f_p}(\tau) = \mathbf{E}\{f_p(t) f_p(t+\tau)\} = \varepsilon \cdot \frac{a}{2\gamma_p} \exp(-\gamma_p |\tau|) + o(\varepsilon). \quad (4)$$

This approximation of the correlation function corresponds to an in applications widely used model $R(\tau) = \sigma^2 \exp(-\gamma |\tau|)$, $\sigma^2, \gamma > 0$.

As can be seen in the next Section, the response variables can be represented as sums of linear functionals of weakly correlated functions with appropriate kernel functions. Hence, the correlation function (as well as other characteristics) of the response can be expanded as to ε in a similar way.

3 Linear Random Vibration Systems

Forming the state vector $\mathbf{z} = (\dot{\mathbf{p}}, \mathbf{p})^\tau \in \mathbb{R}^{2n}$ the system (1) is transformed into a system of first-order ODE of dimension $2n$

$$\dot{\mathbf{z}}(t) = \mathbf{A}\, \mathbf{z}(t) + \mathbf{B} \begin{pmatrix} \mathbf{f}(t,\omega) \\ 0 \end{pmatrix}, \quad \mathbf{z}(0) = \mathbf{z}_0, \quad (5)$$

with $2n \times 2n$ matrices

$$\mathbf{A} = -\begin{pmatrix} \mathbf{J} & \mathbf{O} \\ \mathbf{O} & \mathbf{I}_n \end{pmatrix}^{-1} \begin{pmatrix} \mathbf{D} & \mathbf{K} \\ -\mathbf{I}_n & \mathbf{O} \end{pmatrix} \text{ and } \mathbf{B} = \begin{pmatrix} \mathbf{J} & \mathbf{O} \\ \mathbf{O} & \mathbf{I}_n \end{pmatrix}^{-1} \begin{pmatrix} \mathbf{C} & \mathbf{O} \\ \mathbf{O} & \mathbf{O} \end{pmatrix},$$

where $\mathbf{J}$ is assumed to be regular, $\mathbf{I}_n$ denotes the $n \times n$ identity matrix and $\mathbf{z}_0 = (\mathbf{p}_1, \mathbf{p}_0)^\tau$.

We assume stability of the system (5), that is the eigenvalues of $\mathbf{A}$ have negative real parts only. Furthermore, the eigenvalues are assumed to be simple. Then there exists the decomposition

$$\mathbf{A} = \mathbf{V}\boldsymbol{\Lambda}\mathbf{V}^{-1} \text{ with } \boldsymbol{\Lambda} = \operatorname{diag}(\lambda_1, \ldots, \lambda_{2n}), \; \operatorname{Re}\lambda_i < 0, \; \lambda_i \neq \lambda_j,$$

for $i, j = 1, \ldots, 2n$, $i \neq j$, and a $2n \times 2n$-matrix $\mathbf{V}$ containing the eigenvectors of $\mathbf{A}$. By the substitution $\mathbf{z} = \mathbf{V}\mathbf{x}$ the system (5) is transformed into a decoupled complex system of first-order ODE

$$\dot{\mathbf{x}}(t) = \boldsymbol{\Lambda}\mathbf{x}(t) + \mathbf{g}(t,\omega), \quad \mathbf{x}(0) = \mathbf{x}_0 = \mathbf{V}^{-1}\mathbf{z}_0, \tag{6}$$

with $\mathbf{g}(t,\omega) = \mathbf{V}^{-1}\mathbf{B} \begin{pmatrix} \mathbf{f}(t,\omega) \\ \mathbf{0} \end{pmatrix}$.

The stationary solution of (6), that is the solution after a certain transient time, is found as

$$\mathbf{x}(t,\omega) = \int_{-\infty}^{t} \exp(\boldsymbol{\Lambda}(t-s))\mathbf{g}(s,\omega)\, ds. \tag{7}$$

Equation (7) can be used for the stochastic analysis of the response. For the response mean we get $\mathbf{E}\{\mathbf{x}(t)\} = 0$. It is equal to the solution of the so-called averaged problem, which is obtained from (6) by replacing the random input term $\mathbf{g}(t,\omega)$ by its (non-random) expectation $\mathbf{E}\{\mathbf{g}(t,\omega)\} = 0$.

For the evaluation of the variances and correlation functions we have to consider the fluctuation $\mathbf{x}(t,\omega) - \mathbf{E}\{\mathbf{x}(t)\} = \mathbf{x}(t,\omega)$. Using our excitation model (2) we find for the i-th component $x_i(t,\omega)$, $i = 1, \ldots 2n$,

$$x_i(t,\omega) = \sum_{p=1}^{r} \int_{-\infty}^{t} G_{ip}(t-u)h_e(u,\omega)\, du, \tag{8}$$

with

$$G_{ip}(s) = \int_{0}^{s} \exp(\lambda_i(s-v))[\mathbf{V}^{-1}\mathbf{B}]_{ip} Q_p(v)\, dv,$$

where $[.]_{ip}$ denotes the (i,p)-element of a matrix. The right hand side of (8) is a sum of linear functionals of $h_e(s,\omega)$ for which we can apply the expansion (3) to find the following approximation of the correlation function

$$R_{x_i x_j}(\tau) = \mathbf{E}\left\{x_i(t) \cdot x_j^*(t+\tau)\right\}$$
$$\underset{\tau \geq 0}{=} \varepsilon \cdot a \sum_{p,q=1}^{r} \int_0^\infty G_{ip}(u) G_{jq}^*(\tau+u) du + o(\varepsilon), \quad \text{for } \tau \geq 0,$$

$i,j = 1,\ldots,2n$, where $(.)^*$ denotes complex conjugate. The corresponding correlation function of the original system (5) can be found by applying the inverse transformation. Using matrix notation we get

$$R_{\mathbf{zz}}(\tau) = \mathbf{E}\left\{\mathbf{z}(t)\mathbf{z}^\tau(t+\tau)\right\} \underset{\mathbf{z}=\mathbf{V}\mathbf{x}}{=} \mathbf{V} R_{\mathbf{xx}}(\tau) \mathbf{V}^*.$$

$R_{\mathbf{zz}}(\tau)$ contains the correlation functions for the response variables $p_i(t,\omega)$ and its first derivatives $\dot{p}_i(t,\omega)$. Spectral density functions can be computed applying the Fourier tranformation. The characteristics of the second derivatives $\ddot{p}_i(t,\omega)$ (response accelerations) can be deduced in a similar way (see [2]).

4 Model Reduction

In the previous Section we have described a method for the computation of statistical characteristics of a linear random vibration system. The computer implementation of this method works for systems with some hundreds of DOF. If the dimension is higher we try to find a "suitable" low-order system which can be used to compute approximations of the desired statistical characteristics of the high-order system. For that purpose we again consider the decoupled system (6) which consists of $2n$ scalar complex first-order ODE

$$\dot{x}_i(t) = \lambda_i x_i(t) + g_i(t,\omega); \quad x_i(0) = x_{0i}; \quad i = 1,\ldots,2n.$$

The idea of reduction is to neglect "small" components x_i. Since the mean $\mathbf{E}\{x_i(t)\}$ is zero we use as the criteria of "smallness" the variance $\mathbf{E}\{|x_i(t)|^2\}$. We can obtain a reduced-order model of dimension $m < 2n$ by substituting the $n - m$ components with the smallest variances by zero and removing the corresponding equations in (6).

For this procedure we have to determine the order of the variances of the components x_i. Since we are not able to find these variances without solving the high-order system exactly, upper bounds of the variances which are easy to compute are used.

Applying the representation (7) of the stationary solution we derive for the covariance matrix of $\mathbf{x}$

$$|\mathbf{E}\{\mathbf{x}(t)\mathbf{x}^*(t)\}| \leq \mathbf{dd}^\tau$$
$$\text{with} \quad \mathbf{d} = \left|(\text{Re}\Lambda)^{-1}\right|\left|\mathbf{V}^{-1}B\right|\left(\sqrt{\mathbf{E}f_1^2},\ldots,\sqrt{\mathbf{E}f_r^2},0,\ldots,0\right)^\tau$$

where for a matrix M we denote by $|M|$ the matrix $\{|m_{ij}|\}$. For the standard deviations of f_p, $p = 1, \ldots, r$, follows from (4)

$$\sqrt{\mathbf{E} f_p^2} = \sqrt{R_{f_p f_p}(0)} = \sqrt{\varepsilon \frac{a}{2\gamma_p} + o(\varepsilon)}.$$

Especially, we have the following upper bounds for the variances of x_i

$$\mathbf{E}\left\{|x_i(t)|^2\right\} \leq d_i^2.$$

Let $\mathbf{T}$ be the $2n \times 2n$ permutation matrix (with $\mathbf{T}^{-1} = \mathbf{T}^\tau$) which transforms the vector $\mathbf{d}$ of upper bounds into $\hat{\mathbf{d}} = \mathbf{T}\mathbf{d}$, where $\hat{d}_1 \geq \hat{d}_2 \geq \ldots \geq \hat{d}_{2n}$. Further we define a $m \times 2n$ "projection matrix" $\mathbf{R} = (\mathbf{I}_m, \mathbf{O})$ and a $2n \times m$ "expansion matrix" $\mathbf{S} = \mathbf{R}^\tau = \begin{pmatrix} \mathbf{I}_m \\ \mathbf{O} \end{pmatrix}$.

Then the reduced-order system of dimension $m < 2n$ for the state vector $\mathbf{y} = \mathbf{RT}\mathbf{x}$ can be written as

$$\dot{\mathbf{y}}(t) = \mathbf{RT}\,\mathbf{\Lambda}\,\mathbf{T}^\tau \mathbf{R}^\tau\,\mathbf{y}(t) + \mathbf{RT}\,\mathbf{g}(t,\omega), \quad \mathbf{y}(0) = \mathbf{RT}\,\mathbf{x}_0. \tag{9}$$

An approximation $\tilde{\mathbf{x}}$ of the solution of the high-order system we obtain by expanding the solution $\mathbf{y}$ of (9) and the inverse permutation

$$\tilde{\mathbf{x}} = \mathbf{T}^\tau \mathbf{S}\,\mathbf{y} = \mathbf{T}^\tau \begin{pmatrix} \mathbf{y} \\ \mathbf{0} \end{pmatrix}. \tag{10}$$

From (10) follows the approximation of the correlation function

$$R_{\tilde{\mathbf{x}}\tilde{\mathbf{x}}}(\tau) = \mathbf{T}^\tau \mathbf{S}\,R_{\mathbf{yy}}(\tau)\,\mathbf{S}^\tau \mathbf{T}$$

Inverse transformation gives an approximation of the correlation function of the original system (5)

$$R_{\tilde{\mathbf{z}}\tilde{\mathbf{z}}}(\tau) \underset{\tilde{\mathbf{z}} = \mathbf{V}\tilde{\mathbf{x}}}{=} \mathbf{V}\,R_{\tilde{\mathbf{x}}\tilde{\mathbf{x}}}(\tau)\,\mathbf{V}^*.$$

For the approximation error we get the estimate

$$|R_{\mathbf{zz}}(\tau) - R_{\tilde{\mathbf{z}}\tilde{\mathbf{z}}}(\tau)| \leq |\mathbf{V}\mathbf{T}^\tau|\,\mathbf{D}\,|(\mathbf{V}\mathbf{T}^\tau)^*|,$$

where

$$D_{ij} = \begin{cases} 0 & 1 \leq i,j \leq m, \\ d_i d_j & \text{else}. \end{cases}$$

5 Numerical Results

In this Section we consider as an example torsional random vibrations of a generator shaft excited by a random fluctuating load in the electric generator. The shaft vibration is modeled by a discrete vibration system (see Fig. 1) with $n = 200$ DOF. As input data we have for each mass moments of inertia J_i, $i = 1, \ldots, n$, for each spring-damper coupling stiffness and damping coefficients k_i and d_i, $i = 1, \ldots, n-1$, and coefficients d_i^e, $i = 1, \ldots, n$, for the external frictional damping of the masses (e. g. in the bearings or steam turbines). Further we have a centered, wide-sense stationary random excitation $f_1(t, \omega)$ at the one end of the shaft, which represents the fluctuations of the torque on the electric generator around its mean. The parameters of our excitation model (2) are $\varepsilon = 0.01$, $a = 1$ and $\gamma = 100$.

From these data we have to calculate the variances, correlation and spectral density functions of

- the angular deflections p_i, velocities $\dot{p}_i$ and accelerations $\ddot{p}_i$ of the n masses,
- the so-called torsional torques

$$M_i(t,\omega) = k_i(p_i - p_{i+1}) + d_i(\dot{p}_i - \dot{p}_{i+1}), \quad i = 1, \ldots, n-1,$$

which are important characteristics of the shaft torsion.

For the given number of DOF $n = 200$ we are able to compute the desired characteristics without using a reduced-order model. It takes some hours on a high-end workstation. These results are used to assess the goodness of the corresponding approximations derived from a reduced-order model, where the original system (5) of dimension $2n = 400$ is replaced by a system of dimension $m = 10$ and $m = 6$. The computation of these approximations only needs some seconds.

The results show that for dimensions $m \geq 10$ of the reduced-order system the approximation error can be neglected for practical purposes. This effect can be observed in Figure 2, where the variances $\mathbf{E}\{\dot{p}_i^2\}$ of the speeds of rotation for the 200 masses and in Figure 3, where the correlation function $R_{M_{100}M_{100}}(\tau)$ and the spectral density function $S_{M_{100}M_{100}}(\alpha)$ (α denotes the circular frequency) of the torsional torque M_{100}, which acts at the interface between mass 100 and 101, are plotted. The corresponding curves of the original system and the reduced-order system of dimension $m = 10$ are nearly undistinguishable while for $m = 6$ significant differences can be observed.

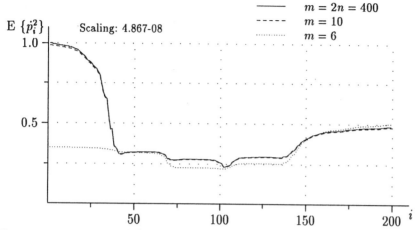

Figure 2: Variances $\mathbf{E}\{\dot{p}_i^2\}$ of the speeds of rotation

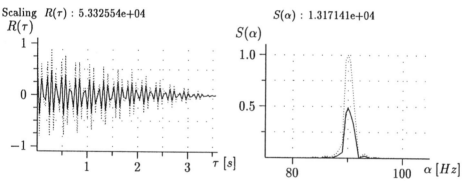

Figure 3: Correlation function $R_{M_{100}M_{100}}(\tau)$ and spectral density function $S_{M_{100}M_{100}}(\alpha)$ of the torsional torque M_{100}

References

[1] VOM SCHEIDT, J.: Stochastic Equations of Mathematical Physics. Akademie-Verlag, Berlin 1990.

[2] VOM SCHEIDT, J.; FELLENBERG, B.; WÖHRL, U.: Analyse und Simulation stochastischer Schwingungssysteme. B. G. Teubner, Stuttgart 1994.

[3] SOONG, T.T.; GRIGORIU, M.: Random vibration of mechanical and structural systems, Prentice Hall, Englewood Cliffs NJ 1993

On Compensation of Vibration of High-Speed Press Automatic Machine

Alexandra **Rodkina**

Voronezh State Academy of Construction and Architecture,
20-letiya Oktyabrya Str. 84, Voronezh 394006, Russia,
rod@vgasa.voronezh.su

Valery **Nosov**

Moscow State Institute of Electronics and Mathematics,
Trehsvetitelskii Str. 3/12, Moscow 109028, Russia,
root@avt.isrir.msk.su

Abstract

To compensate the vibration of the high-speed press automatic machine the authors construct the crank of executive mechanism in special way and introduce the feed-back by means of electromagnetic vibro-stimulant and proportional and proportional-differential regulators. The authors perform computer simulation for some actual high-speed blanking sheet-stamping press automatic machine which is produced in Russia. Various nonlinear random noises of the control systems are simulated. The computer simulation allows to decrease the control parameters values, to lower watts consumed, to evaluate the level of random noise values for which press vibration magnitude is satisfactory.

1 Introduction

The vibration protection problem arises in all fields of contemporary industry. Its solutions essentially lean upon the specific character and dynamics of the system. Stroke frequency of the executive mechanisms of some forge and press machines may be very high: for cold drawing automatic machine it reaches 200-300 r.p.m (rotation per minute) for high-speed sheet-stamping press-automatic machine — 2000 r.p.m. It leads to the increase of the vibration level of the machine itself and adjacent equipment, of the noise and of the kinematic couples wearing. In order to avoid it the different compensation systems can be applied. The foundation of the theory of compensation of the mechanisms was laid down in the works of I. I. Artobolevsky and N. G. Bruevich [1] and was developed in the works of V. A. Shepetilnikov [2], E. N. Lanskoi, M. D. Cerliuk [3] *etc.* At present the theory of the static compensation with the aid of the correcting masses and the theory of the partial dynamic compensation with the aid of the connection to the special compensation equipment were worked out sufficiently complete. But compensation equipment has some shortcomings: it is expensive, complicated, short-lived, increases noise level *etc.*

In this paper we propose to compensate partially the rotating masses of the crank connecting rod mechanism. With this purpose we form the crank in such a way that the centre of gravity of the whole mechanism moves along the slide block leading axis. Then we introduce the feed-back which is realized by means of electromagnetic vibro-stimulant and proportional (P) and proportional-differential (PD) regulators.

For the analytical description of the press vibration we use some mathematical model. We consider actual object: the high-speed blanking sheet-stamping press-automatic machine AA6324. We carry out the theoretical calculation of the control parameters values, which provide the vibration magnitude $\epsilon \leq 5\cdot 10^{-8}$ m for various values of angular velocity ω (ω =80, 150 radian per second). We also present the computer simulation of the gravity centre oscillation and watts consumed by the vibrator. The computer simulation does not only illustrate the theoretical results, but it is the original instruments of research.

2 Arrangement Description and Control Parameter Calculation

It was shown in [2] that the crank can be constructed in such a way that the centre of gravity S of all press automatic machine moves along the slide block leading axis. In this case the action of unstable force in the direction perpendicular to the slide block leading axis is compensated. In the parallel to OX direction (see fig.1) unstable force $P(S) = -l\omega^2 m(\cos\omega t + A_2 \cos 2\omega t + A_4 \cos 4\omega t + A_6 \cos 6\omega t + \cdots)$ acts to the frame of mechanism. Here $\omega = \frac{\pi n}{60}$ is an angular velocity respecting to stroke frequency n, l is crank radius, m is mass adjustment to point E of slide-block. Coefficients A_{2n} are defined by the ratio λ of crank radius to the connecting-rod length (see [2]).

Let high-speed press automatic machine be rigidly established on the supporting plate A (see fig.1) and the crank have such form, that force $P(S)$ acts upon the press frame and the supporting plate. This force induces the vibration of aggregate. We install a vibration sensor C on a supporting plate. We set up a vibro-stimulant B on the plate symmetrically to the slide axis. It is connected by

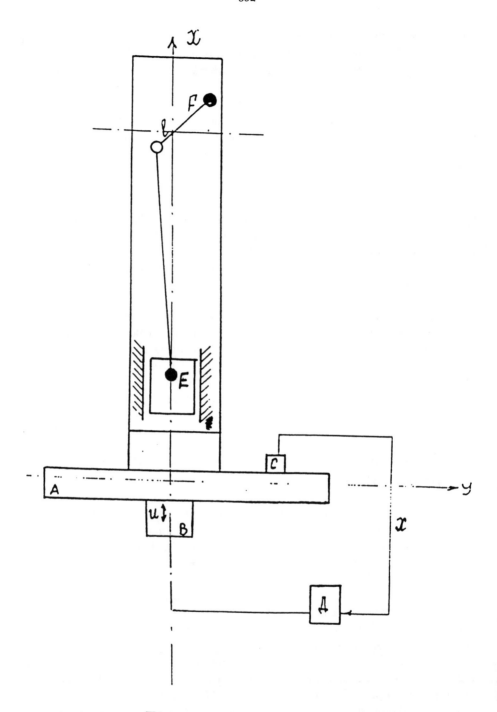

Figure 1

its input with the output of shaping electromagnetic device D. The last one is connected by its input with the output of vibration sensor C. Amplified sensor signal comes to the shaping electromagnetic device D which together with the force transformator B realizes the dependence $u = -a_1\dot{x} - b_1 x$, where x is a deviation of centre of gravity S from the equilibrium, $\dot{x}$ — its speed, u is a force, which create a counteraction for the vibration extinguishing.

For the analytical description of press we model it as elastic body with rigidity $k > 0$ and viscosity $\alpha > 0$. Then the equation of the motion of gravity centre has the form

$$M\ddot{x} + \alpha\dot{x} + kx = P + u, \qquad (1)$$

where M is aggregate mass and control u is constructed with the help of standard regulator in the form $u = -a_1\dot{x} - b_1 x$, where a_1 and b_1 are control parameters. After substituting the meanings of u, P and ωt instead of φ in (1) we obtain the following linear differential equation

$$\ddot{x} + a\dot{x} + bx = K\omega^2(\cos\omega t + A_2\cos 2\omega t + \cdots), \qquad (2)$$

where $a = \frac{\alpha+a_1}{M}$, $b = \frac{k+b_1}{M}$, $K = -\frac{lm}{M}$. For the simplicity we assume that $4b - a^2 = 4\beta^2 \geq 1$, $b > 1$, $a < 2b$ and $x(0) = 0$. Then the general solution of the equation (2) has the form

$$x(t) = e^{-\frac{at}{2}}(H_1\cos\beta t + H_2\sin\beta t) +$$
$$\bar{A}_1\cos\omega t + \bar{B}_1\sin\omega t + \sum_{n=1}^{\infty}\left(\bar{A}_{2n}\cos 2n\omega t + \bar{B}_{2n}\sin 2n\omega t\right). \qquad (3)$$

It is clear that this control can not give an ability to amortise the oscillation of an aggregate. But we can choose a_1, b_1 in such a way that the amplitude of the oscillation becomes sufficiently small from some moment T_0. Suppose that from practice we know T_0 and admissible level of vibration D_0. Choose parameter b_1 in such a way that for every $d > 0$ $\bar{A}_1 + \bar{B}_1 + \sum_{n=1}^{3}\left(\bar{A}_{2n} + \bar{B}_{2n}\right) \leq d$ and parameter a_1 such that $e^{-\frac{(\alpha+a_1)T_0}{2M}}(H_1 + H_2) < d$. Omitting the cumbersome calculation we obtain that for

$$b_1 = 2^5 lm\omega^2 d^{-1}, \quad a_1 = 2M(1+\omega^{1/2})T_0^{-2} \qquad (4)$$

the vibration value is not greater than $D_0 = 7d$ for $t > T_0$.

3 Examples and Simulation

We examine the high-speed press-automatic machine AA6324 with a capacity 250 kN. We have: mass of machine $m_1 = 4350$ kg, mass adjustment to connecting-rod $m = 400$ kg, crank radius $l = 0,008$ m, $A_2 = 0,2918$, $A_4 = 0,0062$, $A_6 = 0,000128$, angular velocity ω of crank shaft means 80 and 150 (rad· sec^{-1}), press vertical rigidity $k = 4,46 \cdot 10^8$ (N·m^{-1}), corresponding viscosity coefficient $\alpha = 7,1 \cdot 10^3$ (N·sec·m^{-1}). Using experimental information we can consider that the press vibration magnitude having the order $10^{-5} - 5 \cdot 10^{-6}$(m) is satisfactory. Carrying out the calculation for $d = 7 \cdot 10^{-7}$(m), $T_0 = 10^{-2}$(sec), by formulas (4) we obtain the following values of control parameters: $a_1 = 1,7 \cdot 10^8$(N·sec·m^{-1}), $b_1 = 1,1 \cdot 10^{12}$(N· m^{-1}) for $\omega = 80$(rad · sec^{-1}), $b_1 = 7,4 \cdot 10^{12}$(N· m^{-1}) for $\omega = 150$(rad · sec^{-1}). Right calculations show that to realise such control it needs to consume the power value in 100 time greater than the main electric drive power. For finding the real control parameters values we make a computer simulation. If we take into account the several random noises which caused by the nonlinearities and hysteresis phenomenon of the executive mechanism ets , then the real control law has the form $u = -a_1\dot{x} - b_1 x + \eta_1 + \eta_2 + \eta_3$. We simulate the following noises η_1, η_2, η_3:

$$\eta_1(t) = \int_0^t e^{-10(t-s)}(s+1)^{-1}l_1(|\varphi(s)|+1)ln(|\varphi(s)|+2)ds;$$

$$\eta_2(t) = \begin{cases} l_2, & \varphi(t) > l_3 \\ -l_2, & \varphi(t) < l_3 \end{cases} \dot{\varphi}(t) > 0; \\ \begin{cases} l_2, & \varphi(t) > -l_3 \\ -l_2, & \varphi(t) < -l_3 \end{cases} \dot{\varphi}(t) < 0; \qquad \eta_3(t) = l_3\varphi(t)\dot{w}_t.$$

The results of computer simulation for AA6324 and watts consumed by the vibro-stimulant are presented in Figures 2–3 . Every figure has 3 graphs. Upper one (a) shows the time dependences of oscillation x_1 (mm) of centre of gravity of aggregate relative to equilibrium state, the middle (b) – the oscillation of its speed x_2 (mm·sec^{-1}), the lower (c) – the oscillation of power consumption P . Three types of noises defined above with $l_1, l_2 \le 100$, $l_3 \le 300$ are presented. For Figure 2 the values of control parameters $a_1 = 10^6$ (N·sec· m^{-1}), $b_1 = 5 \cdot 10^8$ (N·m^{-1}) with $\omega = 80$ (rad · sec^{-1}). Here $|x_1(t)|_{max} \simeq 0,018$ (mm) and $|P|_a \simeq 1$(kWt); where P_a is an average value of power and it is essentially less than the main electric drive power (which is equal to 8 kWt). For the realization of a such control it is sufficient to use only one vibrator of middle size. For Figure 3 $a_1 = 8,7 \cdot 10^7$(N·sec·m^{-1}), $b_1 = 3,48 \cdot 10^{10}$(N·m^{-1}), $\omega = 150$(rad · sec^{-1}), $|x_1(t)|_{max} \simeq 0,018$(mm) and $|P|_a \simeq 7$(kWt).

4 Conclusions

The proposed arrangement can be constructed of familiar blocks, in particular P and PD regulators, electromagnetic vibrator of the middle size and *etc*. Computer simulations allow to decrease the control parameters values, to watch for watts consumed and to evaluate the level of values of various random noises presented in real control.

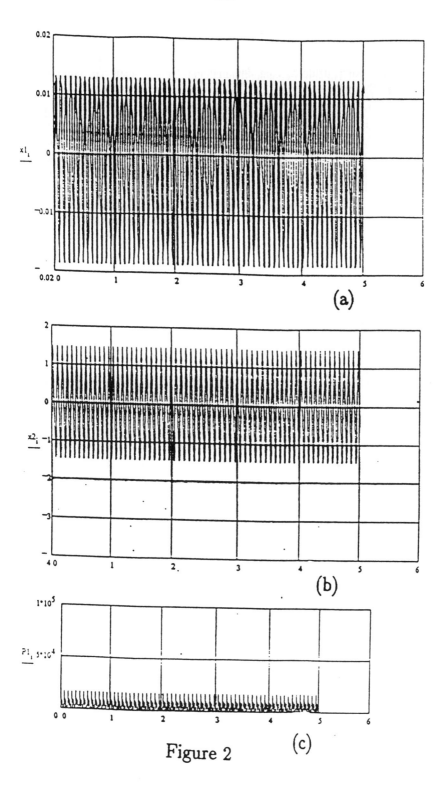

Figure 2

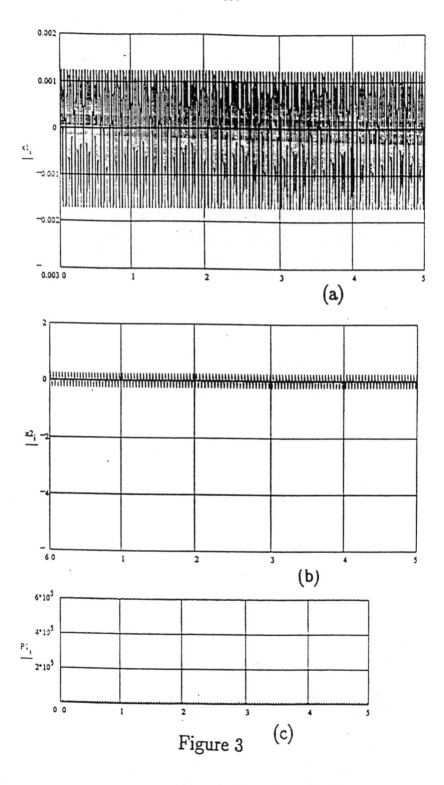

Figure 3

References

[1] Artobolevsky, I.I.: Theory of Machines and Mechanisms, Moscva, Nauka, 1975.

[2] Shepetilnikov, V.A.: Equilibration of Mechanisms, Moscva, Machinostroenie, 1982.

[3] Lanskoi, E.N., Cerliuk, M.D.: Choice of Rational System of Dynamic Equilibration for Forge and Press Automatic Machine, Kuznechno-Pressovoe Proizvodstvo, 2 (1987), pp.34-45.

Identification of external actions on dynamic systems as the method of technical diagnostics

Yuri Menshikov

Faculty of Mech. and Mathem. Dniepropetrovsk Univ. Nauchny line 13,Dniepropetrovsk, 320626,Ukraine,E-mail: mensh@ftf.dsu.dp.ua

Abstract

The problem of technical diagnostics of the real object is reduced to the problem of identification additional external action on some subsystem of initial system which is modeling the behaviour of real object.This additional external action is related to the mathematical description of control parameter of real object. Tikhonov's regularization method is used for solution of ill-posed (instable) problem of action identification. Here the method of choise of the optimal model was used for improvement of accuracy of regularized solution. The paper deals with two practical problems: the operative evaluation of rotor unbalance characteristics and the Krilov's problem .

1 Introduction

In this paper the method of technical diagnostics of real object is suggested which can attribute to the methods capable to avoid tests (see[1]).

Let the behaviour of real object in normal condition is described by some mathematical model. Let the part of variables of state $X_j(t)$ ($1 \leq j \leq n$) of studied mahematical model permits the direct experimental measuring .The function $X_j(t)$ obtained experimentally can be interpreted as two external actions on initial system $C_j X_j(t)$ and $-C_j X_j(t)$, C_j-const. Then the initial system is reduced to the equivalent more simple system (or systems).Such transformation will be called the "j-section".The subsystem of the initial system with all known external actions and one the khown variable of state $x_1(t)$ are being picked out with the help of the series of "sections" of such kind.This subsystem has to contain the control physical parameter the true value of which is taked into consideration within the mathematical description.

The additional external action $Z(t)$ is introduced into the mathematical description of the control parameter.Then we are going to consider the problem of external action identification $Z(t)$ that is the problem of value and character evaluation of change $Z(t)$ by the use of experimental measurings of external actions on subsystem and a certain response $X_1(t)$. If the control parameter has a true value then $Z(t) \equiv 0$. The change of control parameter in process of work will lead to the change of additional external action in a way, that the process within a real object continues to be described by the old incorrect mathematical model. The deviation inside of the control parameter might be estimated on the basis of change character and value of $Z(t)$.

The following might be considered as the main advantages of such approach:
i) there is no need of obtaining the testing signals;
ii) the possibility of the continious observation of the control parameter;
iii) the stability of diagnostic results with respect to the small change of measuring apparatus parameter.

The suggested approach is illustrated by the solution of the problem of rotor unbalance identification and Krilov's problem.

2 The evaluation of the unbalance characteristics of a rotor

Current methods of unbalance definition of machine rotor in its own bearings are not effective in case when the unbalance arises in the machine work process because they demand the special conditions of work or installation of trial plummets (see[2,3]). Besides, these methods do not give the complete information about the position of unbalance if the rotor has a large size along the axis of rotation. The suggusted algorithm of unbalance evaluation uses the experimental data about accelerations of rotor supports in two mutually perpendicular directions during the work for few rotor rotations as the initial information. This algorithm doesn't demand the special conditions of work or installation of trial plummets.

Let us consider a deformable rotor rotating on two non-rigid supports [4]. We introduce rectangular right-hand coordinate system $0\xi\eta\zeta$. The axis 0ζ coincides with axis of the rotating shaft of rotor. The axis 0ξ belongs to the plane of rotor in horizontal position. The axis 0η has vertical direction.
The equations of motion have the form in this system

$$m\ddot{\xi} = P_1 + z_1(t), \quad m\ddot{\eta} = P_2 + z_2(t),$$

$$T_1\ddot{\gamma} + T_1^z(\dot{\varphi}\dot{\psi} + \ddot{\varphi}\psi) = M_1^0, \quad T_1\ddot{\psi} - T_1^z(\dot{\varphi}\dot{\gamma} + \ddot{\varphi}\gamma) = M_2 + z_3(t),$$

$$T_1^z\ddot{\varphi} = M_3, \quad m_a\ddot{\xi}_a + b_a^1\dot{\xi}_a + c_a^1\xi_a = l_1(bP_1 + M_2), \quad (1)$$

$$m_a\ddot{\eta}_a + b_a^2\dot{\eta}_a + c_a^2\eta_a = l_1(bP_2 - M_1), \quad m_b\ddot{\xi}_b + b_b^1\dot{\xi}_b + c_b^1\xi_b = l_1(aP_1 - M_2),$$

$$m_b\ddot{\eta}_b + b_B^2\dot{\eta}_b + c_b^2\eta_b = l_1(aP_2 + M_1),$$

where $z(t) = m_g r \dot\varphi^2 \sin(\vartheta + \varphi)$, $P_1 = -c_1\tilde\xi + c_2\tilde\psi$, $M_1 = c_1\tilde\xi + c_3\tilde\psi$,

$$P_2 = -c_1\tilde\eta + c_2\tilde\gamma + mg, \quad z_2(t) = -m_g r \dot\varphi^2 \cos(\vartheta + \varphi),$$

$$M_1^0 = c_2\tilde\eta - c_3\tilde\gamma + hz_2(t), \quad M_2^0 = -c_2\tilde\xi - c_3\tilde\psi + z_3(t),$$

$$z_3 = -hz_1(t), \quad \tilde\xi = \xi - \xi_0, \quad \tilde\eta = \eta - \eta_0, \quad \tilde\gamma = \gamma - \gamma_0,$$

$$\tilde\psi = \psi - \psi_0, \quad \xi_0 = l_1(a\xi_b + b\xi_a), \quad \eta_0 = l_1(a\eta_b - b\eta_a), \quad \psi_0 = l_1(\xi_b - \xi_a),$$

$$\gamma_0 = l_1(\eta_a - \eta_b), \quad l_1 = (a+b)^{-1};$$

c_1, c_2, c_3 are constants, m_g is the mass of unbalance, φ is the angular velocity of rotation around axis $O\zeta$, h is unbalance arm, ϑ is angular deviation of the factor of unbalance with respect to correction plane, m is the mass of rotor, T_1 is the moment of rotors inertia, m_a is the mass of the left support, m_b is the mass of the right support , c^1, c^2 is the stiffness of support with respect to the horizontal and vertical direction, a is the distance of gravity centre of rotor to the left support, l is the shaft length of rotor. As the additional external actions in mathematical model of motion are choosen the projections of forse $z_1(t), z_2(t)$ and unbalance moment $z_3(t)$. If the unbalance is absent then the functions z_1, z_2, z_3 will be equal to zero. As an example , we consider the equation for the unknown function $z_1(t)$.

We suppose that with the help of acceleration transducers the function $\ddot\xi_a(t)$ and $\ddot\xi_b(t)$ have been recorded. The functions P_1 and M_2 may be expressed in terms of functions $\xi_a(t)$ and $\xi_b(t)$. At the same time the function $\xi(t)$ is expressed through P_1, M_2 and the transference of supports

$$\xi = \tilde\xi_0 + \xi = (c_3 P_1 - c_2 M_2)\Delta + (a\xi_b + b\xi_a)l_1,$$

where $\Delta = (c_1 c_3 - c_2^2)^{-1}$. The solution of the first equation of system (1) is presented in form

$$\int_0^t (t-\tau) z_1(\tau) d\tau = \tilde u_1(t). \tag{2}$$

The equations for the unknown functions $z_2(t), z_3(t)$ are analogous to (2).

The unknown function $z_1(t)$ must be continuous according to the physical sense. Therefore we suppose $z_1(t) \in C[0,T]$, ($[0,T]$ is the interval of time where the function $z_1(t)$ is studied). The function $\tilde u_1(t)$ in (2) is obtained using the experimental data $\ddot\xi_a(t), \ddot\xi_b(t)$ where the noise is present. Therefore it is convenient to suppose that $\tilde u_1(t) \in L_2[0,T]$. The deviation of function $\tilde u_1(t)$ from $u_1^T(t)$ is known

$$\| \tilde u_1 - u_1^T \|_{L_2[0,T]} \leq \delta,$$

where $(.)^T$ denotes the exact data.

Let us represent (2) as follows

$$Az_1 = \tilde u_1, \tag{3}$$

where A is a linear integral operator $(A : Z \to U)$.
In this case the inverse operator A^{-1} is not defined for all $u_1 \in U$ and is not continuous on U, see, for example, [5].
Consider the vector of parameters of the mathematical model (for function $z_1(t)$)

$$\vec{p} = (E, J, m, a, b, m_a, m_b, c_a^1, c_b^1, b_a^1, b_b^1)^*,$$

where $(.)^*$ is symbol of transposition.
The subjective factors affect the definition of parameters of MM system "rotor-supports" and therefore the parameters are supposed to have their values in the certain limits :

$$0 < E^0 \leq E \leq \hat{E}, \quad 0 < J^0 \leq J \leq \hat{J}, \quad 0 < m^0 \leq m \leq \hat{m}, \quad 0 < a^0 \leq a \leq \hat{a},$$

$$0 < b^0 \leq b \leq \hat{b}, \quad 0 < m_a^0 \leq m_a \leq \hat{m}_a, \quad 0 < m_b^0 \leq m_b \leq \hat{m}_b, \quad 0 < c_a^{01} \leq c_a^1 \leq \hat{c}_a^1,$$

$$0 < c_b^{01} \leq c_b^1 \leq \hat{c}_b^1, \quad 0 \leq b_a^{01} \leq b_a^1 \leq \hat{b}_a^1, \quad 0 < b_b^{01} \leq b_b^1 \leq \hat{b}_b^1.$$

In this way the vector $\vec{p}$ can be changed within the known closed region $\bar{D}; \vec{p} \in \bar{D} \subset R^{11}$.

Function $\tilde{u}_1$ might be written in the form

$$\tilde{u}_1 = B_{\vec{p}} \tilde{f}, \qquad (4)$$

where

$$\tilde{f}(t) = (\ddot{\xi}_a(t), \ddot{\xi}_b(t), \xi_a(0), \xi_b(0), \dot{\xi}_a(0), \dot{\xi}_b(0), \xi_a^{(3)}(0), \xi_b^{(3)}(0))^*$$

is vector function within the function space F, $B_{\vec{p}}$ is a linear unreversible operator. The component of a vector function $\tilde{f}(t)$ is defined with error.

The operator $B_{\vec{p}}$ in (4) depends on the given parameters of mathematical model "rotor - supports", i.e. on $\vec{p}$, that is why $B_{\vec{p}}$ defined approximately. Therefore

$$\| B_{\vec{p}} - B_{\vec{p}}^T \|_{F \to U} \leq h.$$

The value δ is

$$\delta = \| B_0 \| \delta_1 + h \| \tilde{f} \|_F,$$

where

$$\| f^T - \tilde{f} \|_F \leq \delta_1, \quad \| B_0 \| = \sup_{\vec{p} \in \bar{D}} \| B_{\vec{p}} \|_{F \to U}.$$

Let us suppose that such the exact solution of equation (3) z_1^T $(A z_1^T = u_1^T = B_{\vec{p}}^T f^T)$ belongs to the function space $W_2^1[0, T]$ [5].

The set of possible solutions $Q_{h,\delta}$ of equation (3) have a form

$$Q_{h,\delta} = \{z : \| Az - \tilde{u}_1 \|_U \leq \delta, \quad z \in W_2^1[0, T], \tilde{u}_1 = B_{\vec{p}} \tilde{f}\}.$$

The operator A in (3) is exact.

The Tikhonov's method of regularization is used for solution of unsteady problem (3). One of the possible ways of construction of regularized solution is the search of element $z_{h,\delta}$ on which the greatest lower bound of stabilizing functional $\Omega[z]$ was reached on the set of possible solutions $Q_{h,\delta}$ of equation (3),[5]. The functional $\Omega[z]$ is choosen as follows

$$\Omega[z] = \int_0^T (q_0 z^2 + q_1 \dot{z}^2) dt,$$

where $q_0 \geq 0, q_1 > 0$.

In this paper the modification of regularization method described in [6] was used. It permits to raise the "quality" of regularized solution at the expense of the decrease of an error value and to exclude the use of parameter h.

Let us consider the sets :

$$Q_{\vec{p},\delta} = \{z : \|Az - B_{\vec{p}}\tilde{f}\|_U \leq \delta_{\vec{p}}, z \in W_2^1[0,T]\},$$

$$\hat{Q} = \bigcup_{\vec{p} \in \bar{D}} Q_{\vec{p},\delta} \quad (\bigcup \text{ is the union}),$$

where $\delta_{\vec{p}} = \|B_{\vec{p}}\| \delta$.

It is evident that $\hat{Q} \subset Q_{h,\delta}$ for any $h > 0$ and any $\delta > 0$. Therefore the use of the set $\hat{Q}$ instead of $Q_{h,\delta}$ allows to obtain the more detailed solution. The proposed approach is based on [6].

For the realization of such approach it is necessary to choose within the vectors $\vec{p} \in \bar{D}$ some vector $\vec{p}_0 \in \bar{D}$ such that from equalities

$$\Omega[z_1] = \Omega[A^{-1}B_{\vec{p}}f], \quad \Omega[z_2] = \Omega[A^{-1}B_{\vec{p}_0}f]$$

follows the inequality

$$\Omega[z_1] = \|z_1\|_W^2 \geq \Omega[z_2] = \|z_2\|_W^2$$

for all possible $f \in \bar{F}$ and all $\vec{p} \in \bar{D}$. The operator $B_{\vec{p}_0}$ with parameter $\vec{p}_0 \in \bar{D}$ will be called the optimal operator.

T h e o r e m . The optimal operator $B_{\vec{p}_0}$ in (4) for function $z_1(t)$ exists, is unique and corresponds to vector

$$\vec{p}_0 = (E^0, J^0, \hat{m}, \hat{a}, \hat{b}, \hat{m}_a, \hat{m}_b, c_a^{01}, c_b^{01}, b_a^{01}, b_b^{01})^*.$$

If it is known a priori that the unbalance characteristics are insignificant, then we propose a parametrization of this problem. In such case explicit algebraic expressions are obtained for the unbalance characteristics m_g, ϑ, h:

$$\vartheta = \arctan(a_1 b_2 - a_2 b_1)(a_1 b_1 + a_2 b_2)^{-1},$$

$$m_g = b_1(a_1 \cos\vartheta - a_2 \sin\vartheta)^{-1} T^{-1} \omega^{-2}, \quad h = b_3 b_2^{-1},$$

where $a_1 = [\alpha(1+\omega^2) + T^2/2\omega^2], a_2 = T/\omega(T^2/3 - 1/\omega^2), \alpha$ is the parameter of regularization, b_1, b_2, b_3 are constants. The suggested method permits to evaluate operatively all characterictics of unbalance on working machinery in real time. It can be used for technical diagnostics of unbalance and for balancing of rotors in their own bearings . The method can be adapted to measuring of velocity or displacement of supports.

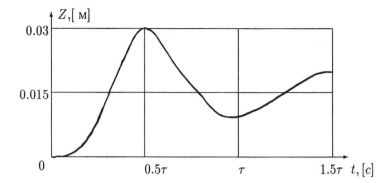

Figure 1: The curve of piston displacement of compressor cylinder according the Viccers's indicator

3 The problem of A.N.Krilov

In 1914 during the testing of the guns of warships it was discovered that the pressure of the powder gas exceed the permissible value almost three times according to the reading of Viccers's indicator of pressure (see[7]). The curve of pressure inside the compressor cylinder has been shown on fig.1. This diagram shows that the greatest pressure reaches 45 MPa. But the tests of compressor cylinders must be executed under the double pressure 90MPa according the contract with the factory. Meanwhile all 48 compressors have been ready and they had been designed on normal pressure 25MPa (the trial pressure is 50MPa). The compressor substitution inquired the expenses about 2.5 million of the gold roubles and would shift the terms of warships readiness.

The research conducted by A.N.Krilov had shown that the records of Viccerss's indicator was abble to differ from the real change of studied value if the period of natural oscillations of indicator is insufficiencly small as compared to the duration of rise of measured value.

The following inverse problem has been examined by A.N.Krilov for the consideration of character of real pressure change:the curve of pressure has been substituted for analitical curves: then the pressure value has been calculated by means of the differentiation from the motion equations.Here the solution of this problem is given (see[7]).

The equation of compressor piston motion has been selected in form

$$\ddot{X}(t) + n^2 X(t) = G(t)/M, \tag{5}$$

where $G(t)$ is the pressure force to piston, M is the reduced mass of piston, $X(t)$ is the piston motion during test, n is the frequency of natural oscillations of piston.

The curve of piston motion has been substituted for following relations:

$$X_1(t) = (1.5(1 - \cos 2\pi t/\tau) - 0.25(1 - \cos 4\pi t/\tau))10^{-2} \mathrm{M} t \in [0, 0.5\tau];$$

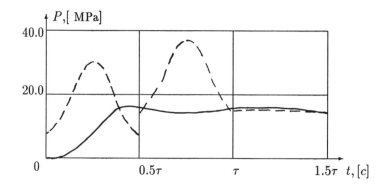

Figure 2: The functions $P_k(t)$ (dotted line) and the regularization results (unbroken line).

$$X_2(t) = (2.0+1.0\cos 2\pi(t-0.5\tau)/\tau - 0.25(1-\cos 4\pi(t-0.5\tau)/\tau))10^{-2}\text{M}, t \in [0.5\tau, \tau];$$

$$X_3(t) = (1.5 - 0.5\cos 2\pi(t-\tau)/\tau)10^{-2}\text{M}, t \in [\tau, 1.5\tau];$$

where τ is the period of natural oscillations of piston on the spring: $\tau = 0.00357c$. The difference between the values of the real motion $X(t)$ and the functions $X_k(t)$ is small and it is not exceeded as a rule the recording thickness line.

The piston motion equation for the obtaining of pressure force value on the piston has a form

$$M(\ddot{X} + n^2 X) = G(t) = SP(t), \qquad (6)$$

where S is the area of piston surface, $P(t)$ is the pressure on the piston, M=0.48kg, n^2=3.110$^6 c^{-2}$. We have from the equation (6) that :

for $t \in [0, 0.5\tau]$ $\quad P_1(t) = 11.16(1.667 - \cos 3520t)MPa$;
for $t \in [0.5\tau, \tau)$ one have $\quad P_2(t) = 11.16(2.33 - \cos 3520(t-0.5\tau))MPa$;
for $t \in [\tau, 1.5\tau]$ one have $\quad P_3(t) = 14.88 MPa$.

The graph of functions $P_k(t)(k=1,2,3)$ is shown on fig.2 by dotted line. This results did not corresponded the physical sence and so A.N.Krilov had to give up the such approach.

Let us take the solution of real pressure problem using the primary version of A.N.Krilov. Here the whole value $G(t)$ is chosen as the additional external action on model (6) as the real value of pressure $G(t)$ is unknown.

The equation (6) can be represented in the form

$$\int_0^t \sin n(t-\tau)P(\tau)d\tau = A_h P = \tilde{u}(t), \qquad (7)$$

where $\tilde{u}(t) = nM/S[X(t) - X(0)\cos nt - \dot{X}(0)/n \sin nt]$ is given function, A_h is the linear integral operator. This operator is given approximately as the parameters of mathematical model (mass M and frequency n) are defined with error.

In complete analogy with foregoing problem we define the optimal operator A_0 with parameters: $M_{\min} = 0.42kg, n_{\max} = 1800c^{-1}$. According to the regularization method the initial problem of equation (7) solution is reduced to the problem of function definition which ensures the minimum of functional $M^\alpha[P, A_h, \tilde{u}]$ [5].The parameter of regularization is defined by disripancy method.

The function which is illustrated by unbroken line on fig.2 is finally obtained as the solution of equation (7) by regularization method. The results of identification confirm the conclusion of Krilov that the pressure into the cylinder do not exceed 30 MPa.

4 Conclusions

In the article the stable algorithm of external actions identification on dynamic system was given which can be used as the technical diagnostics method of wide class of real objects.Two problems were considered here as an example : evaluation of the unbalance characteristics of a rotor and the Krilov's problem.

References

[1] Dmitriev A.N./Egunov N.D.:The identification of external actions of linear unstationary systems on the basis of transmission function generalization, Proceedings of USSR Academy of Science.Technical cybernetics/ 1975,1,pp.212-218.

[2] Brazhco A./Levit M.: To balance of the flexible rotors with the help of computers, Balancing rotors and mechanismes/Moscow, Mechanical engineering, 1978, pp.183-186.

[3] Darlow Mark S.: Balancing of high-speed machinery:theory, methods and experimental results, Mech.Syst.and Signal Process. /1987, 1, n^0 1, pp.103-134.

[4] Dondoshansky B.: The computation of vibrations of elactic systems/ Moscow, Mechanical engineering, 1965.

[5] Tikhonov A.N./ Arsenin V.Y.: The methods of solution of the incorrectly formulated problems/ Moscow, 1979.

[6] Menshikov Y.: The choice of the optimal mathematical model in the problems of the external actions, Differential equations and their applications in Physics/The University of Dnepropetrovsk, 1989,pp.34-43.

[7] Krilov A.N.: Vibrations of ships/ ONTI,L., 1936.

Fluids in industry

Capillary effects in thin films

S.B.G.O'Brien, Department of Mathematics,
University of Limerick, Ireland.

Abstract

We consider two problems associated with the well-known Laplace Young capillary equation modelling the shape of a free liquid-gas interface. We consider a simplified model for the line pinning of a contact line on a single (physical) spherical heterogeneity and predict that the contact line pins, distorts and finally slides across the defect in agreement with experimental results. Secondly we consider the unstable equilibrium shape of a small hole in a thin liquid film. If a hole occurs in a film which is smaller than the equilbrium shape, it tends to heal over: if a hole occurs which is larger, it tends to open out.

1 Introduction

Many industrial processes involve the coating of substrates with liquids. For example, television screens require an anti-reflective coating of some sort in order to prevent glare for the viewer [16], [7]. Such coatings might typically consist of small submicron sized particles which on adhering to the glass of the screen disperse the incident light and inhibit glare. In order to get a **uniform** coating of such particles onto a screen (otherwise the picture may be distorted), one possibility is to suspend the particles in some inert liquid forming a uniform suspension and then to coat the screen with this liquid by spinning or dip-coating. Finally the excess liquid is allowed to evaporate off the screen (or encouraged to do so by the introduction of heating elements) and if the suspension is evenly distributed at this point then so too will be the layer of coating particles left behind on the screen. In the initial stages of such processes wetting phenomena play a important role as the liquid spreads over the initially dry screen. In particular the occurrence of heterogeneities on the coating surface can have an effect on the quality of the final coating. Here we concentrate on the influence of small (physical roughness) defects on the uncoated substrate and in particular the so-called line-pinning effect [2] which such defects exert on the free surface as it passes over them. In [8], the microscopic pinning of a contact line on a single heterogeneity [2] was investigated in a series of carefully conducted experiments. A variety of phenomena was observed including sticking, stretching and jumping of the contact line. Some of the results were compared with approximate models [5], [6] based on the assumption of small contact angle and point heterogeneity. We also consider the shape of a hole occurring in a thin liquid film as such holes are detrimental to the final coating uniformity.

Both of our models will express a static or quasi-static equilibrium and in exploiting the existence of a small Bond number ($\epsilon \equiv (\rho g L^2/\sigma)^{1/2}$), where $\rho(\cong 10^3 \text{Kgm}^{-3})$ is liquid density, $g(\cong 9.81 ms^{-2})$ is acceleration due to gravity, $\sigma(\cong 0.072 Nm^{-1})$ is surface tension and L is a characteristic particle length scale, we can obtain an asymptotic solutions to both problems in the limit as $\epsilon \to 0$.

2 Line-pinning

2.1 The basic model

In the experimental paper [8], a defect was created on a silicon wafer (spin-coated with a thin film of polysterene) by moulding a small circular polymethylmethacrylate bead (of radius $\cong 50\mu$m.) to it and this bead was then forced to pass through a water/air phase boundary. The phase boundary was observed and photographed during this process and phenomena such as sticking, stretching and jumping of the free surface were observed as the defect passed through it. The speed of the contact line relative to the bead was of order 1μm./s so quasi-static models can be used to analyse the process [5], [6], [12] since the capillary number $\mu V/\sigma << 1$ (where μ is dynamic viscosity and V is velocity of immersion).

If we nondimensionalise all lengths using the capillary length $a \equiv (\sigma/\rho g)^{1/2}$ as reference length we obtain the well known Laplace Young capillary equation (in cylindrical coordinates (r, z)):

$$\frac{z''}{(1+z'^2)^{3/2}} + \frac{z'}{r(1+z'^2)^{1/2}} - z = 0 \tag{1}$$

subject to the boundary conditions (see fig.1):

$$z(r = \epsilon \cos\phi) = \epsilon \sin\phi; \quad \frac{dz}{dr}(r = \epsilon \cos\phi) = -\cot(\theta + \phi) \tag{2}$$

with z and its derivatives finite at $r = \infty$.

2.2 Perturbation solution

The details of what follows are to be found in [11] where it is shown that the solution to (1) can be found using matched asymptotics. Inner and outer solutions can be found; in the present instance the significant physical quantity is the height of the undisturbed meniscus level H_∞ above the centre of the sphere (fig.1). This is found to be:

$$\epsilon f(\theta, \phi)(\ln \epsilon + \ln(\cos\phi + \sqrt{(\cos^2\phi - \cos^2\phi \cos^2(\theta+\phi)}) - \ln 4 + \gamma_c) - \epsilon \sin\phi \tag{3}$$

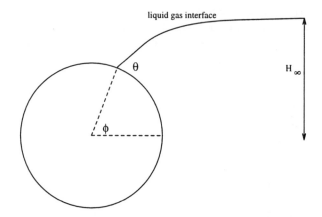

Figure 1: Schematic of the immersion of a small sphere.

where γ_c is Euler's constant, θ is the contact angle, $f(\theta, \phi) = \cos\theta \cos(\theta + \phi)$ and ϕ is the angle subtended at the centre of the sphere by the line of contact.

2.3 Interpretation of solutions

We consider the case where $\theta = 90^\circ$, $\epsilon = 0.01$ corresponding to the situation in [8]. In fig.2 we show the effect which the sphere in the bottom left hand corner has on the liquid free surface as the sphere (moving downwards) passes through the air/water boundary. We graph the solutions for a number of different cases.

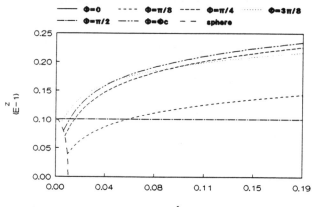

Figure 2: Meniscus height for the case $\theta = \pi/2$, $B = 0.01$.

Consider first the situation where the sphere is half-immersed in the liquid (see fig.2). As the contact angle is assumed here to be 90°, the equilibrium position clearly consists of a straight line meniscus running along the r axis. As the liquid level rises (or the sphere is immersed), we observe a gradual movement of the contact line (more correctly the contact circle) along the profile of the sphere as a series of new quasi-static equilibrium positions becomes possible. Note also that as the contact point slides along the sphere, a critical point is reached where the location of the meniscus at ∞ reverses. The position of this critical point as defined by the value of the angle ϕ at which it occurs depends on the Bond number ϵ. Consider now H_∞ which gives the relationship between the centre of the sphere and the height of the undisturbed meniscus at ∞. This has a turning point when its derivative with respect to ϕ is zero. In the case when $\theta = \pi/2$, a simple computation shows that the turning point occurs when (approximately):

$$\phi_c = \pi/4 - 1/(\sqrt{2}\ln\epsilon) - 0.2311/(\ln\epsilon) \qquad (4)$$

This indicates that beyond this point no static equilibrium is possible as suggested in the experiments in [8].

The existence of such a turning point is interesting as it essentially represents a bifurcation point of the equation: $H_\infty = H$ (where H is allowed to vary continuously). The significance of H_∞ having a turning point is that this corresponds to a point at which the implicit function theorem does not hold and bifurcations can occur. Ths situation is represented schematically in fig.3. During submersion of the sphere, the sequence of events is A,B,C. At this point the position of maximum distortion has been attained and on further submerging the sphere, no new equilibrium is possible and the meniscus slides along the sphere and breaks free so position D is never attained (See [8]).

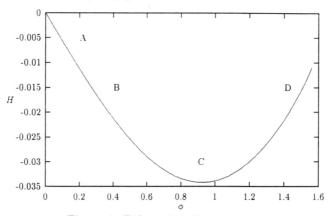

Figure 3: Bifurcation diagram.

3 Holes in thin films

Basically a very thin film (typically of the order of microns) is very sensitive to outside disturbances and holes may form in the film. There are many ways in which this can occur; an instructive example occurs when the film has an alcohol/water base. Small fluctuations in the (local) alcohol concentration may occur (due to differential evaporation) and this can give rise to surface tension gradients and Marangoni type effects. If these give rise to an outgoing radial type flow, a hole may start to open (and this will be further exacerbated by the effects of disjoining pressure (van der Waals forces) which ultimately may lead to complete film break-up and hole formation. An analysis of the stability of pinholes is necessary [15]. In this seminal paper, it was shown that for the case of an infinite expanse of liquid, corresponding to each undisturbed film thickness there exists one critical static equilibrium solution which is unstable and some numerical solutions are calculated. The present paper calculates an asymptotic expansion for this unstable equilibrium solution. (The problem has already been solved numerically in ([15]). The emphasis here will be on posing a mathematical problem and obtaining solutions to the problem using matched asymptotics.

3.1 Mathematical model

The basic problem [13] is represented schematically in fig.4 and shows an infinite film configuration after a hole has opened up. The governing equation is again the Laplace Young capillary equation though one difficulty which arises in the present instance is the double-valuedness of the height for some values of the radial coordinate. This can be resolved by breaking the problems into two sub-problems. (An alternative would be to reformulate the problem and look for $r = r(z)$ as in [15]).

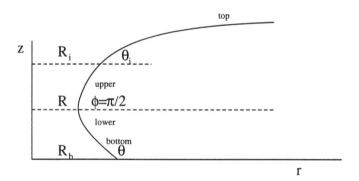

Figure 4: Schematic of hole in thin film.

Following previous work on capillary problems ([3], [4], [9], [10]), we again first non-dimensionalise using the capillary length a

$$\frac{z''}{(1+z'^2)^{\frac{3}{2}}} + \frac{z'}{r(1+z'^2)^{\frac{1}{2}}} \pm z = 0 \tag{5}$$

(5) is a second under ODE and requires two boundary conditions. We pose these in the following way: At

$$\text{As } r \to \infty, z, z', z'' \to 0; \qquad \text{As } r \to \epsilon, z' \to \pm\infty \qquad (6)$$

where the $\pm$ sign depends on whether we are considering the upper or lower branch and $\epsilon = R/a$ is the dimensionless half width of the profile at its narrowest point (and also a Bond number). The parameter ϵ is assumed to be small.

Before proceeding we outline the basic form of the solution (fig.4): we integrate inwards starting from $r = \infty$ using a small slope approximation to develop a solution which we refer to as the top solution. This is matched into an inner solution referred to as the upper solution. The integration stops at the point where the profile become vertical and is then restarted in the form of a solution referred to as lower. Finally this is matched to a small slope solution referred to as the bottom solution. In general we will assume the contact angle on the substrate is π radians. This is the most general possibility. If the actual contact angle is smaller than this, the profile can be cut off at the point where it attains the required inclination.

3.2 The top and upper solutions

The top solution is obtained by making the (well known) assumption of small slope and rescaling as follows: $x \equiv r$, $y = \frac{1}{\epsilon}z$. The scaled "outer" equation a modified Bessel equation with solution (after matching and applying the boundary condition at $r = \infty$) $y_0 = -K_0(x)$.

The upper solution is obtained by rescaling as follows: $r = \epsilon\xi$; $z = \epsilon\zeta$ (which essentially brings the curvature terms into a balance in a thin boundary layer region while neglecting gravity effects to leading order) and is:

$$\zeta = \ln\left(\xi + \sqrt{\xi^2 - 1}\right) - \ln 4 + \gamma_c + \ln\epsilon \qquad (7)$$

where the boundary condition at $\xi=1$ has been satisfied by choosing the positive root and switchback terms [1] have been added during the matching as required. We note that in the case where the contact angle $\theta < 90^0$, the upper and top solutions take the form:

$$\zeta = \sin\theta(\ln\xi + \sqrt{\xi^2 - \sin\theta^2}) - 2\ln 2 + \gamma + \ln\epsilon) \qquad (8)$$

$$y = -\cos\theta K_0(X) \qquad (9)$$

3.3 The bottom and lower solution

The development for the bottom and lower solutions is quite similar to that for the top and upper solutions in that the basic balances (and hence scales) are the same. However referring to fig.4 we note that the governing equation for the lower part of the meniscus must incorporate the + sign of eq.(1) rather that the - sign once the profile has passed through the vertical (singular point).

We will use the same notation for the lower and bottom solution as was used for the top/upper solutions i.e. (ζ, ξ) and (y, x) respectively and we will demand continuity at the crossover point where $\xi = 1$.

Using the rescaling $r = \epsilon \xi$, $z = \epsilon \zeta$, the lower solution to leading order is found to be:
$$\zeta = -\ln(\xi + \sqrt{\xi^2 - 1}) - \ln 4 + \gamma_c + \ln \epsilon \qquad (10)$$
which agrees with the upper solution at $\xi = 1$ (continuity). The bottom solution with the rescaling $x \equiv r, y = \frac{z}{\epsilon}$ is found to be:
$$y = 2\ln \epsilon J_0(x) + J_0(x)(2\gamma - 4\ln 2) - \frac{\pi}{2} Y_0(x) \qquad (11)$$

Composite solutions can be formed in the usual way ([11]).

3.4 Discussion of zero order solutions

As a test of the accuracy of the solutions, we compare our results with numerical computations. Referring to fig.4, we estimate the value of R_b (in fact its dimensionless equivalent). We consider the case where $\epsilon = 0.0739397$ and the contact angle $\theta = \pi$. Using the asymptotic solutions we find the value of r at the base to be $r = 0.6039$ using the bottom solutions (and $r = 0.60606$ using the composite solution [13]). The "exact" numerical answer from [3] is $r = 0.608160$.

As an illustration of how the solutions may be used we consider the case where the contact angle is less than $\pi/2$. Then, we can estimate the critical pinhole radius given the film thickness or vice versa. If the dimensional film thickness is H, the capillary length a is assumed known so the dimensionless film thickness $h = H/a$. In this case R is the radius of the dry spot so the dimensionless radius is $\epsilon = R/a$ and we need to solve the following non-linear algebraic equation for ϵ: $-H = \epsilon \sin\theta(\ln(1 + \cos\theta) - 2\ln 2 + \gamma + \ln \epsilon)$ For very thin films, we clearly have $H \sim -\sin\theta \, \epsilon \ln \epsilon$. Note that the critical pinhole size is smaller than the thickness of the layer by a factor of the logarithm of ϵ and in practical applications, ϵ is about 0.01.

4 Discussion

The line-pinning model introduced here is an improvement on previous models [5], [6] in that it does not make the assumption of small surface slope. Essentially we have analysed the asymptotic solutions appropriate to the situation where a small spherical object is passed through a liquid/gas interface. An investigation of the consequences of the model presented here shows that it, at least qualitatively, captures the characteristics of the experiments described in [8] i.e. sticking, stretching and sliding of the contact line and in addition predicts hysteresis in the immersion/withdrawal cycle. This should aid fully numerical approaches to this problem [14].

The asymptotic solutions for holes in thin films are in good agreement with numerical computations: in addition they illustrate the basic balances in the Young Laplace equation and pave the way to better higher order approximations.

Useful discussions with C. Jordan are acknowledged.

References

[1] Van Dyke M., (1964), *Perturbation methods in fluid mechanics*. Academic Press.

[2] de Gennes P.G., Rev. Mod. Phys., **57**, (1985), 827.

[3] Hartland S., Hartley R.W., (1976), *Axisymmetric fluid-liquid interfaces*, Elsevier, Amsterdam.

[4] James D.F., (1974), J. Fluid Mech., **63**, 659-669.

[5] Joanny J.F. and de Gennes P.G., J. Chem. Phys., **81** (1984), 352.

[6] Joanny J.F. and Robbins M.O., J. Chem. Phys., **92** (1990), 3206.

[7] Lammers J.H., O'Brien S.B.G., Proceedings of first European Coating Symposium, Leeds (1995).

[8] Nadkarni G.D. and Garoff S., Europhys. Lett., **20** (1992), 523.

[9] O'Brien S.B.G., J. Fluid Mech., **233**, 519-539, (1991).

[10] O'Brien S.B.G., Quarterly of Applied Mathematics, **LII**, 43-48, (1994).

[11] O'Brien S.B.G., J. Coll. Int. Sci., **182**, (1996).

[12] O'Brien S.B.G. and v.d Brule B.H.A.A., J. Coll. Int. Sci., **144** (1991), 210.

[13] O'Brien S.B.G, (1996) submitted to Q. App. Math.

[14] Slikkerveer P.J, van Lohuizen E.P., O'Brien S.B.G., (1996), Int. J.Num. Methods in Fluids, **22**, 851-865.

[15] Taylor G.I., Michael D.H., J. Fluid Mech., **58**, 625-639, (1973).

[16] Wimmers O.J., Beerens R.N. and O'Brien S.B.G., A.I.Ch.E. meeting April 1994, Atlanta.

The linear stability of channel flow of fluid with temperature-dependent viscosity

D. P. Wall   S. K. Wilson
Department of Mathematics, University of Strathclyde,
Livingstone Tower, 26 Richmond Street,
Glasgow G1 1XH, United Kingdom.

Abstract

The linear stability of flow of fluid with temperature-dependent viscosity through a channel with heated walls is considered. The resulting sixth-order eigenvalue problem is solved numerically using high-order finite-difference methods for four different viscosity models. For all the viscosity models considered a non-uniform increase of the viscosity in the channel always stabilises the flow whereas a non-uniform decrease of the viscosity in the channel may either destabilise the flow or, more unexpectedly, stabilise the flow. We discuss our results in terms of three physical effects, namely bulk effects, velocity-profile shape effects and thin-layer effects.

1 Introduction

This work was motivated by a practical problem associated with the uranium enrichment by laser isotope separation process currently being developed by British Nuclear Fuels Ltd. The problem is concerned with flow through a dye laser cell, a device used for the amplification of the laser beams used in the isotope separation process. In a dye laser cell an organic dye solution flows through a channel and is illuminated by light from a 'pump' laser beam through side windows along a section of the channel. The incidence of the pump beam causes an excitation of the dye molecules in the solution and some of this energy is then radiated in the form of a stimulated emission of light due to the incidence of a second, highly-tuned, weaker 'dye' laser beam which is amplified by this process. In practice only up to 30% of the pump beam's energy goes into this amplification process with the remainder going principally into heating the channel walls and the direct heating of the layers of fluid immediately adjacent to these walls. It is of interest to determine the effect of this heating upon the stability of flow through the cell. As a first model we represent the laser dye cell by a parallel-sided channel whose walls are maintained at different constant temperatures. The effect of heating upon the linear stability of flow through such a channel is considered by allowing the dynamic viscosity of the fluid, $\mu = \mu(T)$, to vary with temperature, T. Linear stability analysis of the corresponding isothermal problem gives rise to the classical Orr-Sommerfeld equation which, together with the appropriate no-slip boundary conditions forms a fourth-order eigenvalue problem. Solutions to this problem identify a finite critical Reynolds number, $R_c^0 \simeq 5772$, such that plane Poiseuille flow is linearly unstable for $R > R_c^0$ and is linearly stable for $R < R_c^0$.

The present work generalises the classical Orr-Sommerfeld problem to include the effects of a temperature-dependent fluid viscosity and heating of the channel walls. This problem was first addressed by Potter & Graber [1] using a particular viscosity model relevant to water who, however, neglected any disturbance to the basic state temperature distribution and so obtained a modified fourth-order Orr-Sommerfeld equation. Recently Schäfer & Herwig [3] derived asymptotic equations for the present problem in the limit of a small non-dimensional viscosity gradient with respect to temperature which they solved numerically. At leading order they recover the isothermal Orr-Sommerfeld equation and thermal effects enter at first order.

In the present work, in contrast with Potter & Graber [1], a disturbance is permitted to the basic state temperature distribution and, in contrast to Schäfer & Herwig [3], the stability equations are solved non-asymptotically. Results for four different viscosity models are presented and discussed. It might be expected that heating which causes viscosity to decrease throughout the channel would always destabilize the flow; perhaps surprisingly this is not always found to be the case. Further details of this work may be found in Wall [4] and Wall & Wilson [5].

2 Problem Formulation

Adopting a cartesian coordinate system whose origin is on the centreline of the channel with $x_2 = y$ the cross-channel ordinate and $x_1 = x$ parallel to the channel walls, we introduce the non-dimensional variables

$$\mathbf{x} = \frac{\mathbf{x}_*}{L}, \quad \mathbf{u} = \frac{\mathbf{u}_*}{V}, \quad p = \frac{p_*}{\rho V^2}, \quad t = t_* \frac{V}{L}, \quad \mu = \frac{\mu_*}{M}, \quad T = 2\left(\frac{T_* - T_l}{T_u - T_l}\right). \quad (2.1)$$

where L, ρ, T_u, T_l, M denote half the channel width, density, temperature at $y_* = L$, temperature at $y_* = -L$, and viscosity at $y_* = -L$ respectively; $V = JL^2/2M$ where $-J$ ($J > 0$) is the constant imposed pressure gradient along the channel in the positive x-direction and a star ($*$) denotes a dimensional quantity. The variables p, T, $\mu = \mu(T)$, t, $\mathbf{u} = (u_1, u_2, u_3) = (u, v, w)$ represent pressure, temperature, viscosity, time and velocity respectively. In the isothermal case V is the maximum velocity (or 3/2 times the average velocity) in the channel. $R = LV\rho/M$ is the Reynolds number and $P_e = RP_r$ is the Peclet number where $P_r = M/\rho\kappa$ is the Prandtl number.

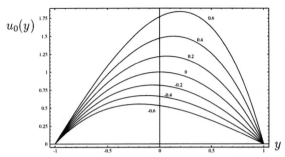

Figure 1: Basic state velocity profiles $u_0(y)$ corresponding to the viscosity $\mu(T) = e^{-K_1 T}$ for the values of K_1 indicated.

We consider the stability of basic states in the form
$$\mathbf{u}(x,y,z,t) = (u_0(y), 0, 0), \quad p(x,y,z,t) = p_0(x), \quad T(x,y,z,t) = T_0(y). \tag{2.2}$$
Upon substituting into the governing Navier-Stokes and heat equations and boundary conditions representing no slip and fixed temperature at the channel walls it is easily shown that $T_0(y) = 1 + y$, and we are left to solve
$$R\frac{dp_0}{dx} = \frac{d}{dy}\left(\mu(T_0)\frac{du_0}{dy}\right) = -2. \tag{2.3}$$
Any $\mu(T_0)$ and $u_0(y)$ which satisfy this equation subject to the boundary conditions may be investigated, in the present study we consider four monotonic viscosity / temperature relationships; $\mu(T) = e^{-K_1 T}$, $\mu(T) = 1 - K_2 T$, $\mu(T) = 1 + b(1 - e^{K_3 T})$ where unless otherwise stated $b > 0$ and $\mu(T) = Ce^{(K_4 T + F)^{-1}}$ where we fix the value $F = 0.16393$ which in turn fixes $1/C = 445.9$ in order to permit comparison with the results of Potter & Graber [1]. For viscosity models 2 and 3 we further require $K_2 < 1/2$ and $K_3 < 1/2 \log(1 + 1/b) = \hat{K}_3$ in order for the viscosity to remain positive everywhere in the channel. Given these assumptions, all these viscosity models describe a monotonic increasing relationship of viscosity with temperature when $K_i < 0$, $i = 1, \ldots, 4$, a monotonic decreasing relationship of viscosity with temperature when $K_i > 0$, $i = 1, \ldots, 4$ and the isothermal constant viscosity $\mu \equiv 1$ when $K_i = 0$, $i = 1, \ldots, 4$. Velocity profiles corresponding to viscosity model 1 for a number of different values of K_1 are plotted in figure 1. It may be noted that these profiles skew towards the hot wall $(y = 1)$ when $K_1 > 0$ and vice versa which is in agreement with expectations since when $K_1 > 0$ viscosity monotonically decreases across the channel and vice versa. The velocity profiles corresponding to the other viscosity models are qualitatively similar.

3 Linear Stability Analysis

In accordance with classical linear stability theory we seek solutions in the form $u = u_0(y) + u_1(x, y, z, t)$ with similar expressions for v, w, p and T, and neglect second and higher order products of the perturbation. We consider two-dimensional normal temporal mode disturbances and so it is convenient to introduce a stream function for the perturbation velocity,

$$\Psi(x,y,t) = \hat{\psi}(y)e^{i\alpha(x-\sigma t)} \tag{3.1}$$

defined in the usual way. Without loss of generality we take $\alpha \geq 0$ and so a given mode is unstable if $\sigma_i > 0$, stable if $\sigma_i < 0$ and neutrally stable if $\sigma_i = 0$. After some manipulation of the perturbation equations we obtain the thermal Orr-Sommerfeld equations

$$\frac{i}{R}\left[\mu(T_0)(\psi^{(4)} - 2\alpha^2\psi'' + \alpha^4\psi) + \mu'(T_0)(D_1\psi + E_1 T) + \mu''(T_0)(D_2\psi + E_2 T)\right.$$
$$\left. + \mu'''(T_0)u_0'T_0'^2 T\right] = \alpha(\sigma - u_0)(\psi'' - \alpha^2\psi) + \alpha\psi u_0'', \tag{3.2}$$

$$i\alpha(u_0 - \sigma)T - i\alpha T_0'\psi = \frac{1}{P_e}(T'' - \alpha^2 T), \tag{3.3}$$

subject to the boundary conditions

$$\psi(\pm 1) = 0, \ \psi'(\pm 1) = 0, \ T(\pm 1) = 0, \tag{3.4}$$

where we have introduced the operators

$$D_1 = 2T_0'\left(\frac{d^3}{dy^3} - \alpha^2 \frac{d}{dy}\right) + T_0''\left(\frac{d^2}{dy^2} + \alpha^2\right), \quad D_2 = T_0'^2\left(\frac{d^2}{dy^2} + \alpha^2\right),$$

$$E_1 = u_0'\frac{d^2}{dy^2} + 2u_0''\frac{d}{dy} + (\alpha^2 u_0' + u_0'''), \quad E_2 = 2u_0'T_0'\frac{d}{dy} + (2T_0' u_0'' + u_0' T_0''),$$

and have dropped the hats for clarity. The equations (3.2) and (3.3) together with the boundary conditions (3.4) form a sixth-order eigenvalue problem for σ as a function of α, R and P_e. Note that in the special case $T \equiv 0$, $\mu \equiv 1$ we recover the familiar fourth-order isothermal Orr-Sommerfeld problem.

We solve the linear differential eigenvalue problem numerically using a finite-difference technique together with the QZ algorithm implemented using NAG routine F02GJF.

4 Results

For a given temperature/viscosity relationship and value of P_e we can calculate the (discrete) eigenvalue spectrum for σ at each point in the (R,α)-plane. Ordering the spectrum of eigenvalues according to

$$\sigma_i^{(1)} > \sigma_i^{(2)} > \sigma_i^{(3)} > \ldots$$

we define the point (R,α) to be stable if $\sigma_i^{(1)} < 0$, unstable if $\sigma_i^{(1)} > 0$ and neutrally stable if $\sigma_i^{(1)} = 0$. A marginal stability curve is a curve in the (R,α)-plane on which $\sigma_i^{(1)} = 0$ which separates stable from unstable regions. On each marginal stability curve we may identify the critical Reynolds number, R_c, which is the global minimum of the marginal stability curve with respect to R. The value

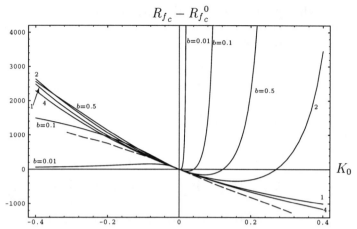

Figure 2: $R_{f_c} - R_{f_c}^0$ plotted as a function of K_0 when $P_e = 1$ for the four different viscosity models. Schäfer & Herwig's [3] asymptotic results are shown by the dashed line.

of R_c is of prime importance since any flow with $R < R_c$ is linearly stable to all disturbances whilst for $R > R_c$ there exist disturbances to which the basic state is unstable. For all four viscosity models considered here the stability characteristics were found to be only weakly dependent on the value of P_e.

In order to compare the critical point behaviour of all the viscosity models considered here we choose K_1, K_2, K_3, and K_4 so that $\mu'(0)$ agrees for the same values of the scaled parameters. Accordingly in figure 2 we plot $R_{f_c} - R_{f_c}^0$ against $K_0 = K_1 = K_2 = bK_3 = K_4/F^2$, where R_f is a Reynolds number based on flux, that is a Reynolds number based on the velocity scale $V_f = (V/2)\int_{-1}^{1} u(y)dy$ in order to facilitate comparison with Schäfer & Herwig [3] and $R_{f_c}^0 = 2/3R_c^0$ denotes the isothermal critical Reynolds number.

If we consider the results for viscosity model 1, it can be seen that R_{f_c} decreases monotonically with K_0 and we may thus conclude that the effect of heating the channel walls is always stabilising if $K_0 < 0$, and is always destabilising if $K_0 > 0$. The results obtained for model 4 are qualitatively similar. The stability behaviour associated with model 2 is clearly somewhat more complicated. Evidently R_{f_c} increases monotonically as K_0 decreases from zero as for models 1 and 4. However, as K_0 increases from zero, R_{f_c} initially decreases, before reaching a global minimum and subsequently increasing monotonically, with our results suggesting that $R_{f_c} \to \infty$ as $K_0 \to 1/2$. Thus for this viscosity model the flow may be arbitrarily stabilised not only by making the magnitude of $K_0 < 0$ sufficiently large, but also by allowing K_0 to approach sufficiently close to $1/2$. The behaviour of R_{f_c} with K_0 obtained for model 3 is also qualitatively distinct from that of models 1 and 4. When $K_0 < 0$, while initially R_{f_c} increases monotonically as K_0 decreases from zero, a global maximum is reached and subsequently $R_{f_c} \to \check{R}_{fc}(b)$ as $K_0 \to -\infty$

where $\check{R}_{fc}(b) > R_{fc}^0$. When $K_0 > 0$ the behaviour of the critical point for this viscosity model is similar to that of viscosity model 2 when $K_0 > 0$, with the flow arbitrarily stabilised as $K_0 \to b\hat{K}_3$. We note that for a given $K_0 > 0$ the viscosity profiles for the various models are quite different, in particular for both viscosity models 1 and 4 as K_1 and K_4 respectively increase viscosity falls sharply near the cool wall ($y = -1$) and declines slowly throughout the rest of the channel creating a relatively viscous boundary layer near this wall relative to the rest of the channel. Contrastingly, with viscosity models 2 and 3 as $K_2 \to 1/2$ and $K_3 \to \hat{K}_3$ respectively a relatively less viscous layer of fluid forms near the hot wall ($y = 1$).

5 Comparison with Previous Studies

Potter & Graber [1] simplified the eigenvalue problem posed by the differential equations (3.2) and (3.3) subject to (3.4) by omitting any perturbation to the basic state temperature. This modified fourth-order Orr-Sommerfeld eigenvalue problem is recovered by setting $T \equiv 0$ in equation (3.2) and the boundary conditions (3.4). It has apparently not been previously noted that Potter & Graber's [1] approximation is exact in the case $P_e = 0$, since in this case T is identically zero, but not when $P_e \neq 0$. Potter and Graber [1] found that R_c was a monotonically decreasing function of K_4 when $K_4 > 0$. In comparing our results for R_c when $P_e = 0$ with Potter & Graber's [1] we find that the results differ by up to 2%, a discrepancy which we must therefore attribute to Potter & Graber's [1] numerical calculations being less accurate than those of the present study. The accuracy of Potter & Graber's [1] *approximation*, as opposed to the accuracy of their numerical technique, may only be measured by comparing the approximate results thus obtained with solutions to the full problem. Accordingly we calculate the quantity

$$\frac{R_c(K_4, P_e) - R_c(K_4, P_e = 0)}{R_c(K_4, P_e = 0)} \tag{5.1}$$

for various values of P_e. The results of this investigation for viscosity model 4 are plotted in figure 3. Clearly the neglect of temperature perturbations is a good approximation for non-zero values of P_e and this is due to the relatively weak dependence of σ on P_e.

In their study Schäfer & Herwig [3] considered equations (3.2) and (3.3) in the limit $K_0 \to 0$ and solved the resulting leading and first order problems numerically. In figure 2 we have also plotted the results of Schäfer & Herwig [3] taken from figure 7 of their paper and the corresponding results of the present calculations when $P_e = 1$. As expected all the models agree with Schäfer & Herwig's [3] asymptotic results for sufficiently small values of $|K_0|$. Schäfer & Herwig [3] concluded that R_c decreased monotonically with K_0 and this conclusion is confirmed by our non-asymptotic results provided that $|K_0|$ is sufficiently small.

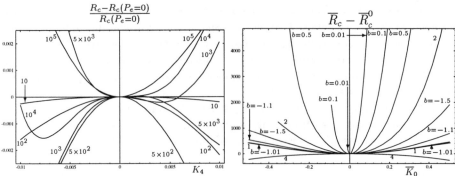

Figure 3: $(R_c - R_c(P_e = 0))/R_c(P_e = 0)$ plotted as a function of K_4 for the values of P_e indicated.

Figure 4: $\overline{R}_c$ plotted as a function of $\overline{K}_0$ when $P_e = 1$ for the four different viscosity models.

6 Discussion

The present study's findings contain several unexpected results. For instance, we have a situation where one flow whose viscosity monotonically decreases across the channel is arbitrarily *destabilised* by heating whereas another flow whose viscosity monotonically decreases across the channel is found to be arbitrarily *stabilised* by heating. In the remainder of this paper we discuss our results in terms of three physical effects; namely bulk effects, velocity-profile shape effects and thin-layer effects.

Bulk effects concern the destabilisation or stabilisation that occurs in a flow when fluid viscosity is uniformly decreased or increased respectively. This destabilisation or stabilisation occurs since the Reynolds number of the flow is increased or decreased by a uniform decrease or increase in viscosity respectively. In the present problem when K_0 is positive or negative viscosity throughout the channel is non-uniformly decreased or increased respectively and so there is a bulk effect included within this change. Bulk effects alone would therefore imply a monotonic decreasing relationship of R_c with K_0 and so any exceptions to such a relationship must be due to other effects. The bulk effect is easily filtered out of our results by the introduction of a new Reynolds number, $\overline{R}_c = R_c / \left(1/2 \int_0^2 \mu(T) dT\right)^2$, based on the average viscosity in the channel. In figure 4 we plot $\overline{R}_c - \overline{R}_c^0$ against $\overline{K}_0 (= K_1)$ where values of $\overline{K}_0(K_i)$, $i = 2, \ldots, 4$ for each value of K_i, $i = 2, \ldots, 4$ are found such that $\int_0^2 \mu(T, \overline{K}_0(K_i)) dT = \int_0^2 e^{-\overline{K}_0 T} dT$, that is for a given value of $\overline{K}_0$ the average viscosity of all models is the same. Evidently all the $\overline{R}_c$ curves are flat in the neighbourhood of $\overline{K}_0 = 0$ which suggests that bulk effects predominate in the limit $\overline{K}_0 \to 0$. A second effect relevant to the present study is the stabilisation which occurs when a symmetric basic state velocity profile becomes skewed. One study of such a phenomenon is that by Potter & Smith [2] who found skewed profiles to be more stable than the symmetric Poiseuille profile in the isothermal

Orr-Sommerfeld problem. This shape effect is thus stabilising for both for positive and negative K_0. A third effect relevant to the present study is that related to the formation of thin layers of fluid near a channel wall of differing viscosity relative to the fluid in the rest of the channel. The present results suggest that the formation of a thin layer of less viscous fluid adjacent to a channel wall stabilises the flow for the present problem and *vice versa*.

To illustrate the above effects in relation to the present problem we consider the behaviour of R_c and $\overline{R}_c$ with K_0 and $\overline{K_0}$ respectively for viscosity model 3. If we consider the region $K_3 > 0$, the bulk effect is destabilising while the shape effect is stabilising. Figure 2 demonstrates that, as for all our results, the bulk effect dominates for sufficiently small K_0. However, in the limit $K_0 \to b\hat{K}_3$ a thin layer of fluid forms next to the hot wall which is relatively less viscous compared to the fluid in the rest of the channel. We tentatively attribute the unexpected stabilisation that occurs in this limit to the formation of this relatively less viscous thin layer of fluid. We note that as $K_0 \to b\hat{K}_3$ the viscosity in this layer becomes increasingly small which, we suggest, causes the stabilisation due to the presence of the thin layer to become arbitrarily large. In contrast, the destabilisation caused by the bulk effect is bounded since $1 + b - [\log(1 + 1/b)]^{-1} < \overline{\mu} < 1$ when $b > 0$ and $0 < K_0 < b\hat{K}_3$ (or when $b < -1$ and $b\hat{K}_3 < K_0 < 0$), which may account for the overall arbitrary stabilisation even when bulk effects are included. In contrast, the bulk effect for viscosity models 1 and 4 (and viscosity model 3 when $-1 < b < 0$) becomes arbitrarily destabilising as $K_0 \to \infty$.

7 Acknowledgements

DPW would like to acknowledge the financial support of British Nuclear Fuels Ltd via a studentship.

References

[1] M. C. Potter and E. Graber. Stability of plane Poiseuille flow with heat transfer. *Phys. Fluids*, 15:387–391, 1972.

[2] M. C. Potter and M. C. Smith. Stability of an unsymmetrical plane flow. *Phys. Fluids*, 11:2763–2764, 1968.

[3] P. Schäfer and H. Herwig. Stability of plane Poiseuille flow with temperature dependent viscosity. *Int. J. Heat Mass Transfer*, 36:2441–2448, 1993.

[4] D. P. Wall. *Thermal Effects in Fluid Flow*. PhD thesis, University of Strathclyde, Glasgow, Scotland, U.K., 1996.

[5] D. P. Wall and S. K. Wilson. The linear stability of channel flow of fluid with temperature-dependent viscosity. *J. Fluid Mech.*, 323:107–132, 1996.

ADAPTIVE METHODS IN INTERNAL AND EXTERNAL FLOW COMPUTATIONS*

Jiří Felcman, Vít Dolejší

Charles University Prague
Malostranské nám. 25, 118 00 Prague
Czech Republic
e-mail: felcman@karlin.mff.cuni.cz

Abstract

To illustrate the application of mathematics to industrial problems we will discuss a mathematical model for the two-dimensional inviscid as well as viscous transonic gas flow in channels, past airfoils and cascades of profiles.

1 Physical problem

We consider gas flow in a space-time cylinder $Q_T = \Omega \times (0, T)$, where $\Omega \subset \mathbb{R}^2$ is a bounded domain representing the region occupied by the fluid and $T > 0$.

The complete system of viscous compressible flow consists of the continuity equation, Navier–Stokes equations and energy equation:

$$\frac{\partial \varrho}{\partial t} + \operatorname{div}(\varrho \boldsymbol{v}) = 0, \tag{1}$$

$$\frac{\partial}{\partial t}(\varrho v_1) + \operatorname{div}(\varrho \boldsymbol{v} v_1) = -\frac{\partial p}{\partial x_1} + \frac{\partial}{\partial x_1}\tau_{11} + \frac{\partial}{\partial x_2}\tau_{21},$$

$$\frac{\partial}{\partial t}(\varrho v_2) + \operatorname{div}(\varrho \boldsymbol{v} v_2) = -\frac{\partial p}{\partial x_2} + \frac{\partial}{\partial x_1}\tau_{12} + \frac{\partial}{\partial x_2}\tau_{22},$$

$$\frac{\partial e}{\partial t} + \operatorname{div}(e\boldsymbol{v}) = -\operatorname{div}(p\boldsymbol{v}) + \sum_{s=1}^{2}\frac{\partial}{\partial x_s}(\tau_{s1}v_1 + \tau_{s2}v_2) + \operatorname{div}(k\nabla\theta).$$

---

*This work was supported under the grant No. 201/96/0313 of the Czech Grant Agency

Here $\tau_{sl} = \lambda \operatorname{div} \boldsymbol{v}\, \delta_{sl} + \mu(\frac{\partial v_s}{\partial x_l} + \frac{\partial v_l}{\partial x_s})$, $s,l = 1,2$. From thermodynamics we have
$$p = (\gamma - 1)(e - \varrho|\boldsymbol{v}|^2/2), \quad e = \varrho(c_v \theta + |\boldsymbol{v}|^2/2). \tag{2}$$
We use the standard notation: t – time, x_1, x_2 – Cartesian coordinates in $\mathbb{R}^2$, ϱ – density, $\boldsymbol{v} = (v_1, v_2)$ – velocity vector with components v_s in the directions x_s, $s = 1, 2$, p – pressure, θ – absolute temperature, e – total energy, τ_{sl} – components of the viscous part of the stress tensor, δ_{sl} – Kronecker delta, $\gamma > 1$ – Poisson adiabatic constant, c_v – specific heat at constant volume, k – heat conduction, λ, μ – viscosity coefficients. We assume that c_v, k, μ are positive constants and $\lambda = -\frac{2}{3}\mu$. We neglect outer volume force and heat sources.

System (1), (2) is equipped with the initial conditions (which means that at time $t = 0$ we prescribe, e.g., ϱ, v_1, v_2 and θ) and boundary conditions: The boundary $\partial\Omega$ is divided into several disjoint parts. By Γ_I, Γ_O and Γ_W we denote inlet, outlet and impermeable walls, respectively, and assume that

(i) $\quad \varrho = \varrho^*, \quad v_s = v_s^*, \quad s = 1, 2, \qquad \dfrac{\partial \theta}{\partial n} = 0 \quad \text{on } \Gamma_I,$ \hfill (3)

(ii) $\quad v_s = 0, \quad s = 1, 2, \qquad\qquad\qquad\quad \dfrac{\partial \theta}{\partial n} = 0 \quad \text{on } \Gamma_W,$

(iii) $\quad -pn_l + \sum_{s=1}^{2} \tau_{sl} n_s = 0, \quad l = 1, 2, \quad \dfrac{\partial \theta}{\partial n} = 0 \quad \text{on } \Gamma_O.$

Here $\partial/\partial n$ denotes the derivative in the direction of unit outer normal $\boldsymbol{n} = (n_1, n_2)$ to $\partial\Omega$; ϱ^* and v_s^* are given functions.

In the solution of a cascade flow problem, Ω is one period of the cascade with $\partial\Omega$ formed by the inlet Γ_I, outlet Γ_O, impermeable profile Γ_W and two piecewise linear artificial cuts Γ^- and Γ^+ such that
$$\Gamma^+ = \{(x_1, x_2 + \tau); (x_1, x_2) \in \Gamma^-\}, \tag{4}$$
where $\tau > 0$ is the width of one period of the cascade in the x_2 direction. On $\Gamma^{\pm}$ we consider the periodicity condition
$$\varrho(x_1, x_2 + \tau) = \varrho(x_1, x_2), \quad (x_1, x_2) \in \Gamma^- \tag{5}$$
and similarly for other quantities.

2 Computational method

Based on the system (1) the conserved quantities ϱ, ϱv_1, ϱv_2 and e will be computed. For this we rewrite (1) in the form
$$\frac{\partial \boldsymbol{w}}{\partial t} + \sum_{s=1}^{2} \frac{\partial \boldsymbol{f}_s(\boldsymbol{w})}{\partial x_s} = \sum_{s=1}^{2} \frac{\partial \boldsymbol{R}_s(\boldsymbol{w}, \nabla \boldsymbol{w})}{\partial x_s} \quad \text{in } Q_T, \tag{6}$$

where

$$w = (w_1, w_2, w_3, w_4)^T = (\varrho, \varrho v_1, \varrho v_2, e)^T, \qquad (7)$$
$$w = w(x, t), \quad x \in \Omega, \ t \in (0, T),$$
$$f_s(w) = \left(f_s^1(w), f_s^2(w), f_s^3(w), f_s^4(w)\right)^T$$
$$= (\varrho v_s, \varrho v_s v_1 + \delta_{s1} p, \varrho v_s v_2 + \delta_{s2} p, (e+p) v_s)^T,$$
$$R_s(w, \nabla w) = (0, \tau_{s1}, \tau_{s2}, \tau_{s1} v_1 + \tau_{s2} v_2 + k \partial \theta / \partial x_s)^T.$$

The functions f_s are called inviscid (Euler) fluxes and are defined in the set $D = \{(w_1, \ldots, w_4) \in \mathbb{R}^4; w_1 > 0, w_4 > (w_2^2 + w_3^2)/2 w_1\}$. The viscous terms R_s are obviously defined in $D \times \mathbb{R}^8$.

We suppose that the viscosity and heat conduction are so small that the viscous terms R_s can be considered as a perturbation in the equation (6). Neglecting the terms R_s in (6) we obtain the system of inviscid Euler equations

$$\frac{\partial w}{\partial t} + \sum_{s=1}^{2} \frac{\partial f_s(w)}{\partial x_s} = 0. \qquad (8)$$

We add to this system the purely viscous system

$$\frac{\partial w}{\partial t} = \sum_{s=1}^{2} \frac{\partial R_s(w, \nabla w)}{\partial x_s} \qquad (9)$$

and discretize (8) and (9) separately.

The complete system (6) is then discretized via operator inviscid – viscous splitting. One time step is divided into two fractional steps:

Step I The inviscid system (8) is discretized by the explicit cell – centred finite volume (FV) method on a mesh $\mathcal{D}_h = \{D_i\}_{i \in J}$. Here the finite volumes D_i are triangles, J is a suitable index set. The upwind flux vector splitting scheme from [7] is used.

Step II The purely viscous system (9) is descretized by the explicit conforming linear finite elements (FE) on a triangulation $\mathcal{T}_h$.

The novelty of the resulting schemes lies in compatibility of $\mathcal{T}_h$ with $\mathcal{D}_h$ in such sense that the set of all vertices P_i of the triangles $T \in \mathcal{T}_h$ consists of the barycentres of all $D_i \in \mathcal{D}_h$ and vertices of $D_i \in \mathcal{D}_h$ lying on $\partial \Omega$. The details can be found in [7, 2].

3 Adaptive mesh refinement

The unified conception of the mesh refinement was worked out with the intention to improve the quality of the approximate solution of the Euler system (8) in a transonic regime and consequently of the complete system (6). This conception is independent of the numerical method for solving (8) and supposes only, that the computational mesh is triangular and that the previously computed approximate solution is piecewise constant on it.

Three types of refinement indicators are defined:

1. Shock indicator based on a divided difference approach that checks the density jumps up taking into account the flow direction.

2. Residual based indicator that uses the norm of the residual of (8) for the refinement process.

3. Residual based indicator that employs the weak formulation of (8).

A theoretical justification of a combined finite volume - finite element method is given for the model scalar convection–diffusion equation in [2]. The tests of accuracy order for a model problem can be found in [4].

Given w_h produced by a triangular finite volume (TFV) procedure on the mesh $\mathcal{D}_h$, we compute for each $D_i \in \mathcal{D}_h$ the local refinement indicator $\eta(i)$. Then the triangles on which the value of the indicator is grater then a prescribed tolerance TOL are refined. The TFV procedure continues on the refined mesh and gives back the solution w_h^{ref}. The whole process is repeated until a suitable stopping criterion is reached.

3.1 Shock Indicator

In [6] the original shock indicator

$$\eta(i) = \max_{j \in s(i)}[-(\varrho_i - \varrho_j)\,v_i \cdot n_{ij}]^+/h_i, \quad i \in J \tag{10}$$

was devised for the inviscid transonic flow problem (8). Here h_i is the length of the longest side of the triangle $D_i \in \mathcal{D}_h$, ϱ_i is the density and v_i is the velocity vector computed by TFV procedure on D_i. n_{ij} denotes the unit normal to the common side of two neighbouring finite volumes D_i and D_j pointed from D_i to D_j. The maximum is taken over the index set $s(i)$ of neighbouring volumes to D_i. The indicator (10) differs from those proposed in [13, 14] and its use in the computations gives the results that are in a good agreement with experiments. The indicator (10) was then successfully applied also in the numerical solution procedure for the viscous flow (6) in ([5]). We illustrate its use on the standard NACA 0012 test case (inlet Mach number M = 0.85, inlet angle of attack $\alpha = 1.25°$) on the Figure 1.

3.2 L^2–norm Residual Indicator

In [11] the L^2–norm of the residual was examined for an *ad hoc* scaled error indicator for finite volume approximations of compressible flow problems. This type of refinement criterion has been used also in [15] and [16]. In the present work this kind of criterion is applied using the L^2-norm of the residual. The concept of the residual differs from that of [11] because we use another finite volume approximation.

The residual of (8) cannot be simply computed by substituting the piecewise constant approximate solution into (8). Therefore we suppose additionaly that the numerical solution procedure produces moreover piecewise linear approximations of fluxes $f_1(w)$ and $f_2(w)$ in (8).

Let the t_k be the time level, when the time marching procedure stops. For this time level we compute in our code from constant values $f_1(w_i^k)$ on triangles D_i the values in the vertices. The weighted averages of values on triangles having the vertex in common are used. The weights are the areas of the corresponding volumes. From values in vertices we construct the piecewise linear approximation of the flux f_1. We denote its restriction on the triangle D_i as $(f_1)_i^k$ and do the same for the flux f_2. We introduce the residual as the piecewise constant vector-valued function r_h with values r_i^k on the triangle D_i computed from the formula

$$r_i^k = \frac{w_i^k - w_i^{k-1}}{\tau_k} + \frac{\partial}{\partial x_1}(f_1)_i^k + \frac{\partial}{\partial x_2}(f_2)_i^k, \qquad (11)$$

where the space derivatives are expressed analyticaly. Then we define the local error indicator

$$\eta(i) = \|r_i^k\|_{L^2(D_i)}. \qquad (12)$$

Because the statement that L^2-norm of the residual is of order $O(h^{\frac{1}{2}})$ is valid for 1D case only, we have abandoned the local scaling factor proposed in [11]. On Fig. 2 the application of the error indicator (12) is compared with the shock indicator (10) for a turbine cascade flow described in [3].

3.3 H^{-1}-norm Residual Indicator

The error estimates from [9] for compressible potential flows as well as the analysis in [10] have motivated the the new error indicator

$$\eta(i) = \left[\sum_{\ell=1}^{4} \left(\eta^\ell(i)\right)^2\right]^{1/2}, \qquad (13)$$

where

$$\eta^\ell(i) = \max_{j \in s(i)} \frac{h_{ij}}{2|\varphi_{ij}|_{H_0^1(\Omega_h)}} \left|\sum_{s=1}^{2} n_{ij}^s \left(f_s^\ell(w_i) - f_s^\ell(w_j)\right)\right|, i \in J, \ell = 1, \ldots, 4. \qquad (14)$$

The details can be found in [8]. In order to verify the indicator (13), a two-dimensional inviscid flow through the GAMM channel (10% circular arc in the channel of width 1 m) for air, i.e. $\gamma = 1.4$, and inlet Mach number $M = 0.67$ was solved. In Figure 3 the mesh refined with the aid of the shock indicator (10) and Mach number isolines are plotted. Figure 4 shows the mesh refinement based on the residual indicator (13) and corresponding Mach number isolines.

Acknowledgements. The results presented here were obtained in cooperation of the authors with M. istauer and A. Kliková.

References

[1] M. Feistauer: *Mathematical Methods in Fluid Dynamics*, Pitman Monographs and Surveys in Pure and Applied Mathematics 67, Longman Scientific & Technical, Harlow, 1993.

[2] M. Feistauer/ J. Felcman: Theory and applications of numerical schemes for nonlinear convection–diffusion problems and compressible viscous flow. Proc. of the MAFELAP96 Conf., Wiley (to appear).

[3] J. Felcman/ V. Dolejší/ M. Feistauer: Adaptive Finite Volume Method for the Numerical Solution of the Compressible Euler Equations, Computational Fluid Dynamics 94 (Wagner, S., Hirschel, E. H., Périaux, J., Piva, R. - Eds.), Wiley, 1994, pp. 894–901.

[4] J. Felcman: On properties of numerical schemes for transonic flow simulation, Proc. of the Seminar Euler and Navier–Stokes Equations, Institute of Thermomechanics of the Czech Academy of Sciences, Praha, 3. - 5. April 1996.

[5] M. Feistauer/ J. Felcman/ V. Dolejší: Numerical simulation of compressible viscous flow through cascades of profiles, **ZAMM 76** (1996) S4, pp. 297–300.

[6] J. Felcman/ V. Dolejší: Adaptive methods for the solution of the Euler equations in elements of blade machines; **ZAMM 76** (1996) S4, pp. 301–304.

[7] M. Feistauer/ J. Felcman/ M. Lukáčová: Combined finite element – finite volume solution of compressible flow. Journal of Comput. and Appl. Math. **63** (1995), pp. 179–199.

[8] J. Felcman/ G. Warnecke: Adaptive computational methods for gas flow, Proc. of Prague Mathematical Conference 1996, July 8 – 12, 1996 (to appear).

[9] U. Göhner/ G. Warnecke: A second order finite difference error indicator for adaptive transonic flow computations, Numer. Math. 70, 1995, pp. 129-161.

[10] J. Mackenzie/ E. Süli/ G. Warnecke: A posteriori analysis for Petrov-Galerkin approximations of Friedrichs systems, Oxford Univ. Comp. Lab. report 19/01, 1995.

[11] Th. Sonar: Strong and weak norm refinement indicators based on the finite element residual for compressible flow computation. Impact of Computing on Sci. and Engrg, 5, 1993, pp. 111-127.

[12] Th. Sonar/ G. Warnecke: On finite difference error indication for adaptive approximations of conservation laws, Manuscript, 1996.

[13] R. Hentschel/ E. H. Hirschel: Self adaptive flow computations on structured grids, Computational Fluid Dynamics 94 (Wagner, S., Hirschel, E. H., Périaux, J., Piva, R. - Eds.), Wiley, 1994, pp. 242–249.

[14] K. Riemslagh/ E. Dick: Mixed discretization multigrid methods for TVD-schemes on unstructured grids; Computational Fluid Dynamics 94 (Wagner, S., Hirschel, E. H., Périaux, J., Piva, R. - Eds.), Wiley, 1994, pp. 317–324.

[15] H. Ritzdorf/ K. Stüben: Adaptive Multigrid on Distributed Memory Computers, Proc. 4th European Multigrid Conf., Amsterdam, July 6-9, 1993, in Book: Multigrid Methods IV, eds. P. W. Hemker, P. Wesseling, Birkhäuser Verlag, 1993.

[16] J. Wu: Adaptive Parallel Multigrid Solution of Navier-Stokes Equations: Adaptive Criteria and Parallelization Strategy, Workshop Adaptive Finite Element Methods, IWR, Heidelberg, 1995.

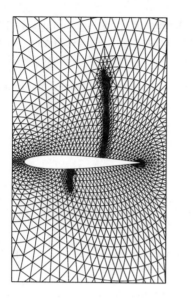 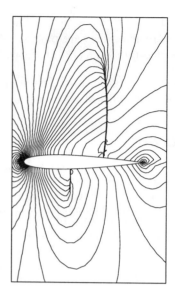

Figure 1: Refined mesh and Mach number isolines

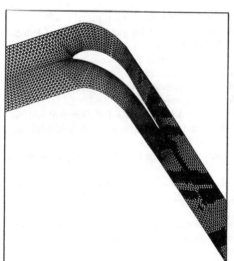 

Figure 2: Shock indicator and L^2-norm residual indicator

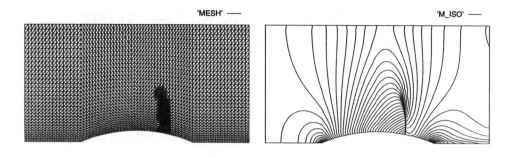

Figure 3: Refined mesh (shock indicator) and Mach number isolines

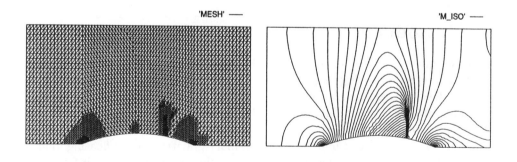

Figure 4: Refined mesh (H^{-1}-norm residual indicator) and Mach number isolines

Bifurcation, symmetry and parameter continuation in some problems about capillary-gravity waves

Loginov B.V., Ulyanovsk State Technical University
432027 Ulyanovsk, Severny Venets str.32, root@ulgpu.simbirsk.su

Karpova S.A., Ulyanovsk State Pedagogical University
432700 Ulyanovsk, root@ulgpu.simbirsk.su

Trenogin V.A., Moscow Institute of Steel and Alloys
117936,Moscow, Lenin prospekt 4, trenogin@swi.misa.ac.ru

Abstract

General results on the applications of group analysis methods to construction and investigation of branching equation in bifurcation theory are given. They are illustrated by capillart-gravity waves problems in magneto- and hydrodynamics having important industrial applications.

1 Introduction

Here we give the review of bifurcation theory under group symmetry conditions and its interaction with group analysis of differential equations methods (n.1,2). On the base of these general results we consider free boundary value problems of capillary-gravity waves in fluid layers (n.3). Briefly the problem about the relief of ferrofluid in magnetic field ,having the applications in cosmic technique,is described (n.3.A). Then one case of bifurcation of capillary-gravity waves in floating fluid is considered. By floating is named the fluid on the surface of which drifts a film of some substance i.e. free boundary is viewing as weighted with density. From physical point of view suuch situation arises in processes of mineral raw material refinement with the aid of the known flotation procedure.

Branching theory of solutions of nonlinear equations usually has the local treatment [1]. There are only single articles devoted to nonlocal statement of the general problem of branching theory [2-4]. Nonlocal approach is the most useful especially under group invariance conditions. In the present communication the nonlinear

equation
$$F(x,\lambda) = A(\lambda)x + R(x,\lambda), \qquad R(0,\lambda) \equiv 0$$
$$\|R(x,\lambda)\| = o(\|x\|), \qquad x \to 0 \text{ in } D \subset E_1 \qquad (1)$$

with values in E_2 (E_1 and E_2 are Banach spaces) is considered. Here $A(\lambda) \in L\{E_1 \to E_2\}$ and $R(x,\lambda)$ is nonlinear operator. It is supposed [5-9] that the equation (1) allows the group G, i.e. there exist the representations L_g in E_1 and K_g in E_2 such that

$$K_g F(x,\lambda) = F(L_g x, \lambda) \qquad (2)$$

Before the study of problems about capillary-gravity waves we give here some of our theoretical results concerning with group analysis methods [10,11] in branching theory [12-20] and existence theorems for solutions invariant relative to subgroups $H \in G$ [20].

Terminology and notations of the general branching theory follow to the works [1,9].

These investigations are supported by Novosibirsk University Grant Centrum by the first author (grant 96-15) and Russian Fundamental Science Foundation by the first and third author (grant 96-01-00512).

2 Group analysis methods in branching theory

It is known [7,9] that the branching equation (BEq) inherits the group symmetry of the corresponding nonlinear problem. Therefore the problem of construction of the general form of BEq by its group symmetry arises. For this problem the methods of group analysis of differential equations [10,11] are the most effective since they allow to construct the full explicit form of BEq, not only its main part. The BEq for stationary branching $0 = f(\xi, \varepsilon) = \{f_j(\xi, \varepsilon)\}_1^m : \Xi^n \to \Xi^m$ allows the group G (or invariant with respect to G) if for its certain representations $\mathcal{A}_g$ in Ξ^n and $\mathcal{B}_g$ in Ξ^m we have

$$f(\mathcal{A}_g \xi, \varepsilon) = \mathcal{B}_g f(\xi, \varepsilon). \qquad (3)$$

The equality (3) means that for the transformation

$$\tilde{\xi} = \mathcal{A}_g \xi, \qquad \tilde{f} = \mathcal{B}_g f \qquad (4)$$

the manifold $\mathbf{F} : f - f(\xi) = 0$ in the vector space Ξ^{n+m} is an invariant manifold. Considering the l-parametric transformation group (4) we shall suppose that F is its nonsingular manifold, i.e. if $\{(X_\nu; F_\nu)\}_{\nu=1}^l$ is the basis of the Lie algebra of the infinitesimal operators of the group (4), then the $rank\{(X_\nu; F_\nu)\}\big|_{\mathbf{F}}$ of the matrix $M\{(X_\nu; F_\nu)\} = [\eta_\nu^j; \xi_\nu^k]$, $\nu = 1 \ldots l; j = 1 \ldots n; k = n+1 \ldots n+m$; (ν is the number of rows and j, k are the numbers of columns) on the manifold $\mathbf{F}$ coincides with its general rank r_*. If now

$$I_1(\xi, f), \ldots, I_{n+m-r_*}(\xi, f) \qquad (5)$$

is a complete system of functionally independent invariants of the group (4), then according to [10,11] the manifold **F** may be presented in the form

$$\Psi^s\bigl(I_1(\xi,f),\ldots,I_{n+m-r_*}(\xi,f)\bigr) = 0, \quad s = 1,\ldots,m, \tag{6}$$

and the necessary condition for the possibility of construction of the general form of BEq is $rank\bigl[\frac{\partial I_k}{\partial f_j}\bigr] = m$, which means the independence of the system (5) with respect to the variables f_j. This condition can be replaced by the requirement $r_*(X,F) = r_*(X)$ ([11],p.250) and makes it possible to express from (4) the variable $f_1,\ldots,f_m$ through the variables $\xi_1,\ldots,\xi_n$. Consequently we obtain the scheme of the construction of the general form of BEq with the aid of the complete system of functionally independent invariants [12-18].

It should be noted that for the construction of analytic BEq's the use only functionally independent invariants in the form of monoms of a minimal degree in ξ reduces generally to omition of certain monomial summands in the expansion of BEq with respect to ξ. In order to take all possible summands into account it is necessary to use additional invariants, which leads to the repetition of monomial summands. This repetition can be removed by factorization on relations between the invariants, which is denoted by the symbol $[\ldots]^{out}$. In applications it is usually considered the real BEq, i.e.

$$f_{2k}(\xi,\varepsilon) = \overline{f_{2k-1}(\xi,\varepsilon)}. \tag{7}$$

In the work [19] we suggest the development of this scheme for nonstationary branching (Andronov-Hopf bifurcation).

Group analysis methods allow to develop the technique for the construction of solutions (1) which are invariant relative to subgroups of the group G and to prove the existence theorems for such solutions [20],[9,ch.3.2] (particular cases was considered in [21]). These results permit the global treatment [9] on the base of parameter continuation methods.

3   Applications to some problems of capillary-gravity waves

On the base of stated above results we had investigated the problem of crystallization of fluid phase state in statistical crystal theory [9,12], the free surface problem about spatial periodic capillary-gravity waves over the flat bottom [9,13,18] (and on the interface of two fluids [22]), problem about periodical relief of ferrofluid layer at the action of magnetic field [23] (this problem has applications in cosmonautics [24]). For all indicated problems we have obtained both asymptotics of small solutions and on the base of the theorem 1 the asymptotics of small solutions invariant relative to normal divisors of the allowing finite group symmetry. All indicated problems may be considered as symmetry breaking ones,

A. Here we give a short description concerning with the last problem. The layer of ferrofluid bounded from below by flat bottom and from above by vacuum is considered. When on the layer acts the vertically directed magnetic field of sufficient strength H, on the upper boundary of ferrofluid arises two-periodical relief, subjected to definition together with magnetic potentials of media. Ferrofluid is supposed to be incompressible, having the finite depth h and free from exterior currents. In dimensionless variables we have the system

$$-\Delta\Phi = 0, \ f(x,y) < z < 1; \quad \frac{\partial \Phi}{\partial z}|_{z=1} = 0;$$

$$-\Delta\varphi = 0, \ -1 < z < f(x,y); \quad \frac{\partial \varphi}{\partial z}|_{z=-1} = 0;$$

$$-\gamma f - \frac{1}{2}[|\nabla(\Phi + \mu Hz)|^2 - \mu|\nabla(\varphi + Hz)|^2] + (\frac{\partial \Phi}{\partial z} + \mu H)\nabla(\Phi + \mu Hx) \cdot \vec{n} -$$
$$\mu(\frac{\partial \varphi}{\partial z} + H)\nabla(\varphi + Hz) \cdot \vec{n} - \frac{1}{2}\mu(\mu - 1)H^2 + (\nabla, \frac{\nabla f}{\sqrt{1+|\nabla f|^2}}) = 0, \ z = f(x,y);$$

$$\Phi - \varphi + (\mu - 1)Hz = 0, \ z = f(x,y); \quad \nabla(\Phi - \mu\varphi) \cdot \vec{n} = 0, \ z = f(x,y).$$

Here Φ and φ are magnetic potentials of upper and lower media, $f(x,y)$ is a free boundary close to horizontal plane $z = 0$, $\vec{n}$ is the normal to it, μ_o is the magnetic constant, μ is the magnetic permeability of the ferrofluid, $\gamma = (\delta\rho)gh^2/\sigma$, $\delta\rho$ is the difference of the densities of media, g is the acceleration of gravity, σ is the surface tension coefficient. It is posed the problem of the construction of periodical solutions with the periods $\frac{2\pi}{a} = a_1$ and $\frac{2\pi}{b} = b_1$ along the coordinate axes, and Π_0 is the rectangle of periodicity.

After straightening of the free boundary $\zeta = \frac{z-f}{1-zf}$ and setting $H = H_o + \varepsilon$ we obtain nonlinear system linear part of which represents Fredholm [25] operator B : $C^{2+\alpha}([1,0] \times \Pi_o) \dotplus C^{2+\alpha}([-1,0] \times \Pi_o) \dotplus C^{2+\alpha}(\Pi_o) \to C^{\alpha}([1,0] \times \Pi_o) \dotplus C^{\alpha}([-1,0] \times \Pi_o) \dotplus C^{\alpha}(\Pi_o)$. The bifurcation point H_o is defined by the dispersive relation

$$\frac{\mu(\mu-1)^2}{\mu+1}H_o^2 \frac{shs_{mn}}{chs_{mn}} = \gamma + s_{mn}^2, \ s_{mn}^2 = m^2n^2 + n^2b^2 \tag{8}$$

We are showing that they are possible: two-multiple degeneration (rolls), four-multiple one (the interaction of two degenerate lattices, or one rectangular lattice), 6-multiple one (hexagonal lattice), 8-multiple (the interaction of rectangular and hexagonal lattices) and 12-multiple (double hexagon–four lattices of periodicity, two of which are degenerate). Note, that the solutions of the last type were not observed until now in synergetics. The asymptotics of periodical solutions are constructed and their stability is investigated.

B. More detailed we consider here one case of the problem about periodical capillary-gravity waves on the surface of spatial layer of floating fluid. Potential flows of incompressible heavy capillary floating fluid in a spatial layer with free upper boundary are studied. By floating is named the fluid on the surface of which drifts a film of some substance, i.e. free boundary is viewing as weighted with distribution of density $\rho_0 > 0$. Bifurcating from basic motion with constant velocity V along Ox-axis direction flows are described by the following system in dimensionless parameters

$$\Delta\Phi = 0, \quad -1 < z < f(x,y); \quad \Phi_z|_{z=-1} = 0;$$

$$\Phi_z - f_x - \nabla f \cdot \nabla_{xy}\Phi = 0, \quad z = f(x,y);$$

$$\Phi_x + \frac{1}{2}|\nabla\Phi|^2 + F^2 f + \frac{k}{\sqrt{1+|\nabla f|^2}}\left[F^2 - (\nabla f \cdot \nabla_{xy} - \frac{\partial}{\partial z})(\Phi_x + \frac{1}{2}|\nabla\Phi|^2)\right] - \quad (9)$$

$$-\gamma F^2 \mathrm{div}\left(\frac{\nabla f}{\sqrt{1+|\nabla f|^2}}\right) = \mathrm{const}, \quad z = f(x,y)$$

Here $\varphi = Vx + f$ is potential of fluid motion, $f(x,y)$ is free weighted boundary close to the horizontal plane $z = 0$, ρ_0 is the surface density of floating substance, ρ is the flind density, σ is the surface tension coefficient, h is the thickness of the layer, g is the acceleration of gravity and $k = \frac{\rho_0}{\rho h}$, $\gamma = \frac{\sigma}{\rho g h^2}$, $F = \frac{\sqrt{hg}}{V}$ is the inverse to Froud number magnitude.

Applying the standard reception of straightening of free boundary we make the following change of variables in (9)

$$\zeta = \frac{z - f(x,y)}{1 + f(x,y)}, \quad f(x,y,f(x,y) + \zeta(1+f(x,y))) = u(x,y,\zeta)$$

Setting $F^2 = F_0^2 + \xi$, where F_0^2 is critical value of Froud number we obtain the problem of finding of periodic solutions of the system

$$\Delta u = w_0(u,f), \quad -1 < \zeta < 0;$$

$$\left.\frac{\partial u}{\partial \zeta}\right|_{\zeta=-1} = 0; \quad \frac{\partial u}{\partial \zeta} - \frac{\partial f}{\partial x} = w_1(u,f), \quad \zeta = 0;$$

$$\frac{\partial u}{\partial x} + k\frac{\partial^2 u}{\partial x \partial \zeta} + F_0^2 f - \gamma F_0^2 \Delta f = w_2(u,f), \quad \zeta_1 = 0$$

Here $w_i(u,f), i = 1,2,3$, are nonlinearities. The corresponding nonlinear system defines Fredholm [25] operator B:

$$C^{2+\alpha}(\Pi_0 \times [0,1]) \dotplus C^{2+\alpha}(\Pi_0) \to C^{\alpha}(\Pi_0 \times [0,1]) \dotplus C^{\alpha}(\Pi_0) \dotplus C^{\alpha}(\Pi_0), \quad 0 < \alpha < 1.$$

where Π_0 is the rectangle of periodicity with the sides $\frac{2\pi}{a} = a_1$ and $\frac{2\pi}{b} = b_1$ along Ox and Oy axes Representing $f(x,y)$ by its Fourier series

$$f(x,y) = \sum_{m,n}(a_{mn}\cos max \cos nby + b_{mn}\cos max \sin nby +$$

$$c_{mn}\sin max \cos nby + d_{mn}\sin max \sin nby)$$

and solving the first three equations of homogeneous linearized system by Fourier separating variables method from the last equation we obtain the following dispersing relation

$$F_0^2(1 + \gamma s_{mn}^2) = \frac{m^2 a^2}{s_{mn}}\left(\frac{\operatorname{ch} s_{mn}}{\operatorname{sh} s_{mn}} + k s_{mn}\right), \quad s_{mn}^2 = m^2 a^2 + n^2 b^2, \qquad (10)$$

m and n are nonnegative itegers. Investigation of (10) shows that they are possible: two-multiple degeneration (rolls), four-multiple one (the interaction of two degenerate lattices, or one rectangular lattice of periodicity), 6-multiple one (irregular hexagon), 8-, 10-, and 12-multiple degenerations of Fredholm operator B. It is shown that regular hexagonal lattices of periodicity are impossible.

For breavity we consider here the construction of bifurcating solutions asymptotics only for the simplest case of 4-dimensional degeneracy.

The original nonlinear system (9) allows the two-parametric group of shifts $L(\beta_1, \beta_2) f(x,y) = f(x + \beta_1, y + \beta_2)$ by x and y and the diescrete group of reflections

$$S_x : x \to -x, \quad f(x,y,z) \to -f(-x,y,z), \quad f(x,y) \to f(-x,y);$$

$$S_y : y \to -y, \quad f(x,y,z) \to -f(x,-y,z), \quad f(x,y) \to f(x,-y)$$

Theorem 1 *The problem (9) in a neighbourhood of defined by (10) 4-multiple bifurcation point F_0^2 has to within the transformation $y \to -y$ two 2-parametric families of periodical solutions*

$$\{u^{(1)}, f^{(1)}\} = \left[-\frac{A}{B+C}(F^2 - F_0^2)\right]^{\frac{1}{2}}\{w(\zeta)\sin[ma(x+\alpha) - nb(y+\beta)],$$

$$\frac{\sqrt{ab}}{\pi}\cos[ma(x+\alpha) - nb(y+\beta)]\} + o(|F^2 - F_0^2|^{\frac{1}{2}}), \quad \operatorname{sign}(F^2 - F_0^2) = \operatorname{sign} A(B+C),$$

$$\{u^{(2)}, f^{(2)}\} = \left[-\frac{A}{B}(F^2 - F_0^2)\right]^{\frac{1}{2}}\{w(\zeta)\cos ma(x+\alpha)\sin nb(y+\beta),$$

$$\frac{\sqrt{ab}}{\pi}\sin ma(x+\alpha)\sin nb(y+\beta)\} + o(|F^2 - F_0^2|^{\frac{1}{2}}), \quad \operatorname{sign}(F^2 - F_0^2) = \operatorname{sign} AB,$$

We do not indicate here the values of the coefficients B and C ($A = -(1+\gamma s_{mn}^2) < 0$) in view of their inconvenience.

References

[1] Vainberg M.M., Trenogin V.A. Branching Theory of Solutions of Nonlinear Equations. Moscow. Nauka. 1969; English transl., Noordhoff Int. Publ., Leyden,1974.

[2] Rabinowitz P. Some aspects of nonlinear eigenvalue problems// Rocky Mountain J.Math.,3,(1973),pp.161-202.

[3] Fonarev A.A. On parameter continuation of solutions of nonlinear equations. Matematitsheskie Zametki,26,n.3(1979),pp.691-698.

[4] Alexander J.C.,Jorke J.A. The homotopy of certain spaces of nonlinear operators,and its relation to global bifurcation of the fixed points of parameterized condensing operators, //J.Funct.Anal.,34(1979),pp.87-106.

[5] Loginov B.V., Trenogin V.A. On application of continuous groups in branching theory. Doklady Akad.Nauk SSSR, 197,n.1(1971),pp.36-39; English transl. in Soviet Math.Dokl.12(1971)

[6] Loginov B.V., Trenogin V.A. On the usage of group properties for the determination of multiparameter families of solutions of nonlinear equations. Mat.Sbornik 85,n.3(1971),pp.440- 454; English transl. in Math. USSR Sbornik 14,n.3,(1971),pp.438-452.

[7] Loginov B.V., Trenogin V.A. On the usage of group invariance in branching theory. Different. Uravneniya 11,n.8(1975),pp.1518-1521; English transl. in Differential Equations 11(1975).

[8] Loginov B.V., Trenogin V.A. Ideas of group invariance in branching theory, Fifth Kazakhstan. Interuniv. Conf. Math.Mech., Abstracts of Reports, part 1, Alma-Ata,pp. 206-208 (1974) (Russian).

[9] Loginov B.V. Branching Theory of Solutions of Nonlinear Equations under Group Invariance Conditions. Tashkent,"Fan",1985,184p. (Russian).

[10] Ovsyannikov L.V. Lectures on the Theory of Group Properties of Differential Equations. Novosibirsk University, 1966, 131p. (Russian)

[11] Ovsyannikov L.V. Group Analysis of Differential Equations. Moskwa, Nauka, 1978, 400p.; English transl., Academic Press, New York,1982.

[12] Loginov B.V.,Rakhmatova Kh.R.,Yuldashev N.N. On the construction of branching equation by its group symmetry (crystallographic groups). in " Mixed Type Equations and free Boundary Value Problems" (M.S.Salakhitdinov and T.D.Dzuraev, eds) Tashkent, "Fan", (1987), pp.183-195 (Russian).

[13] Loginov B.V.,Kuznetsov A.O. On the construction of periodic solutions of 3-dimensional problem about capillary-gravity waves over a flat bottom in the cases of high degeneracy. Proceedings of 11-th Internat.Conf.Nonlinear Oscillations (Budapest,1987; M.Farkas et al., editors), Janos Bolyai Math. Soc.,Budapest,1987, pp.668-671.

[14] Loginov B.V. On the construction of the general form of branching equation by its group symmetry. EQUADIFF -VII. Enlarged Abstracts. Praha (1989),pp.48-50.

[15] Loginov B.V., Sabirova S.G. On the construction of periodic solutions of nonlinearly perturbed elliptic equations. Izvestiya Akad. Nauk Uzbek SSR, Ser. Fiz.-Mat.,n.4,(1988),pp.2-37 (Russian).

[16] Loginov B.V. Group analysis methods for construction and investigation of the bifurcation equation. Applications of Mathematics 37,n.4(1992),pp.241-248.

[17] Loginov B.V., Juldashev N.N. On the construction of branching equation allowing the group SO(3). Uzbek Math.J.,n.5,(1992),pp.52-58 (Russian).

[18] Loginov B.V.,Kuznetsov A.O. Capillary-gravity waves over the flat bottom.European Journal of Mechanics. B/Fluids,15,2(1996),pp.259-280.

[19] Loginov B.V., On determination of the bifurcation equation by its group symmetry - Andronov-Hopf bifurcation. Nonlinear Analysis. TMA,23(1997).

[20] Loginov B.V. On invariant solutions in branching theory.Doklady Acad. Nauk SSSR 246,n.5(1979),pp.1048-1051; English transl.in Sovied. Acad.Sci. Dokl.Math. 20,n.3(1979),pp.586-590.

[21] Cicogna G. Symmetry breakdown from bifurcation. Lett.Nuovo Cimento,31(1981),pp.600-602.

[22] Loginov B.V., Trofimov E.V. The calculation of capillary-gravity vawes asymptotics on the interface of two fluids of finite depth. In "Differential Equations of Math. Phys. and their Appl." Tashkent, Fan (1989),pp. 57-66.

[23] Abdullayeva A.D. Branching and Stability of Solution of the System of DEq for the Definition of Magnetic Fluid Free Surface. Kandidate Dissertation. Tashkent 1993.

[24] Rosensweig R.E. Ferrohydrodynamics. Cambridge Univ. Press 1987

[25] Agranovich M.S. Elliptic Operators on Closed Varieties. Modern Problems of Math. Fundamental Directions. 63(1990) Moscow, VINITI.

Wacker prize

The price was established by ECMI in memory of its founding member Hansjörg Wacker (1939-1991), to be awarded every second year for the best mathematical thesis on an industrial project written by a student from an ECMI institution.

In 1996 the prize was awarded to *Alberto Mancini*, University of Firenze.

Evolution of Sedimentation Profiles in the Transport of Coal Water Slurries through a Pipeline

A. Mancini

Dipartimento di Matematica "Ulisse Dini", University of Firenze, Italy

Abstract

A coal-water-slurry (CWS) is a highly concentrated mixture of finely ground coal particles and water, fluidized by the presence of a small amount of a chemical additive. The technology for the preparation and transport of this material through pipelines has been fully developed by Snamprogetti (Fano, Italy). The rheology of CWS has been the object of an extensive research (both experimental and theoretical) performed in collaboration between the Department of Mathematics "U. Dini" (Firenze) and Snamprogetti (see e.g. [3] [4] [8] [9]). The aim of this paper is to present a mathematical model for the growth and the evolution of sediments in CWS pipelining, still a research carried out in collaboration with Snamprogetti.

1 Introduction

The difficulties in investigating sedimentation of impurities in CWS are twofold: the fluid carrying imputities is Non-Newtonian and optical observations are impossible. We refer to [2] for a comprehensive exposition of technical details. The mathematical model presented here improves the one proposed in [7] and is the basis for a numerical code, property of Snamprogetti, which can be used to determine the strategy of periodical sediments removal, a rather complex and costly operation.

Despite its complex rheological behaviour, for our purposes a pipelined CWS can be assimilated to a Bingham fluid with constant rheological parameters. In the simple geometry we are considering a Bingham fluid is characterized by the following relations between the *shear rate* $\dot{\gamma}$ and the *stress* τ

$$(\tau - \tau_0)_+ = \eta_B \dot{\gamma},$$

where τ_0 is the so called *yield stress*, η_B is the *Bingham viscosity* and $(\bullet)_+$ denotes the *positive part* of the argument. Even in 1-D geometries nonsteady Bingham

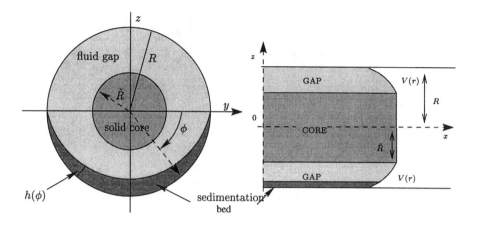

Figure 1: Qualitative behaviour.

flows give rise to difficult free boundary problems (see [1] for an extensive study of the planar case). Here we suppose that the flow is stationary and that the thickness of the sedimentation bed is always sufficiently small (as required in practice), so that we can neglect its influence on the main flow. Therefore the CWS velocity field in the pipeline produced by a pressure gradient G (with the coordinate system as given in *Figure* 1) is

$$v(r) = \begin{cases} -\dfrac{R^2}{4\eta_B}G(1-\dfrac{r^2}{R^2}) - \dfrac{\tau_0 R}{\eta_B}(1-\dfrac{r}{R}) & \text{per } r \geq \hat{R} \\ -\dfrac{1}{4\eta_B}R(R-\hat{R}) & \text{per } r \leq \hat{R} \end{cases} \quad (1)$$

where $\hat{R} = \dfrac{2\tau_0}{G}$, is the radius of the inner rigid core and R is the radius of the pipeline.

The evolution of the sediment profile is resulting from two concurrent phenomena: *settling trough the flow* and *transport of the sediment* which we will analize separately.

2 Settling through the flow

Considering a particle $P = P(\delta, \rho_s, [0, y_0, z_0])$ of radius δ and density ρ_s, entering the pipeline at point $[0, y_0, z_0]$ we can write its trajectory as:

$$\begin{cases} \dot{x} = v(r) & \text{main flow} \\ \dot{z} = -v_s(\delta, r) & \text{settling} \\ r = \sqrt{z^2 + y_0^2} \end{cases}$$

where $v(r)$ is given by (1). According to the theory developed in [2], the settling velocity is $v_s(\delta, r) = \alpha(\dot{\gamma}(r))\delta^2$. The function $\alpha(\dot{\gamma})$ has to be determined on the

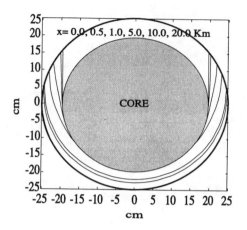

Figure 2: Some $\Gamma(x^*, \delta)$ graphs for Q=250.0 m^3/h, $\delta = 0.0067$ cm. The slopes of the U-shaped curves increase with x^*. The steepest U-shaped curve corresponds to $x^* = +\infty$.

basis of laboratory tests (see [2] for details). We can thus assume hereafter that the function α is a given function of the shear rate $\dot\gamma$.

Now we definine $\Gamma(x, \delta)$ as the set on the plane $x = 0$ of the entering points of the particles with radius δ which will end their trajectory on the pipe wall at the distance x. By means of a geometrical argument we prove that the settling rate per unit length at distance x from the initial cross-section due to particles with radii between δ and $\delta + d\delta$ is

$$S(x;\delta)d\delta = \frac{4}{3}\pi\delta^3 \rho_s N(\delta)d\delta \int_{-y_M}^{+y_M} v_s(\sqrt{y_0^2 + (\mathcal{Z})^2}, \delta)dy_0, \qquad (2)$$

where $\mathcal{Z} = \mathcal{Z}(y_0; x, \delta)$ is a parametrization of $\Gamma(x^*, \delta)$, $N(\delta)d\delta$ is the number of settling particles with radii between δ and $\delta + d\delta$ per unit volume of CWS and $\pm y_M$ are the endpoints of $\Gamma(x, \delta)$. Integrating over the particle distribution we obtain finally

$$S_T(x) = \int_{\delta_m}^{\delta_M} S(x;\delta)d\delta. \qquad (3)$$

- Numerical Simulations Part I: The Settling Rate Function.

Based on this results a code has been developed for the computation of Γ curves (see *Figure* 2) and the settling rate function S_T (see *Figure* 3). From the numerical results, and of course from equations (2) (3), we note that S_T is a rapidly decreasing function of x and is practically zero few kilometers far from the origin: in our simulations with a known population of sand particles ($\delta_{min} = 0.0035$cm, $\delta_{max} = 0.0113$cm), $\rho_s = 2.67$ g/cm^3 and with a CWS with known rheological characteristic parameters ($\tau_0 = 8.89$P, $\eta_B = 0.16$ Ps) we found $S_T \simeq 0$ at $x = 10$ km for a flow rate Q of $\simeq 100$m$^3 h$ and at $x = 60$km for a flow rate Q of $\simeq 450$ m$^3 h$ (pipe radius $R = 25$ cm). From *Figure* 3 we observe also that $S_T(x; Q)$ is an increasing function of the flow-rate Q. Now, because experimental observations prove that

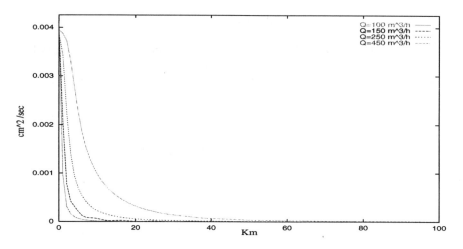

Figure 3: The function S_T for some values of the flow rate.

the thickness of the sedimentation bed is not an increasing function of Q, we can conclude that we cannot neglect the phenomenon of transport of sediment in the pipeline due to its interaction with the main flow. This mechanism is absolutely critical for getting the correct prediction of the evolution of the sediment profile.

3 Sediment Transport

First of all we have to remark that accurate dynamical measurements are not actually available even by using a sophisticated instrument based on a γ-rays scanning procedure. Therefore we chose to formulate a model describing the behaviour of some average quantities playing a basic role in this extremely complex phenomenon.

This choice is translated in the following assumptions:

A. The maximum thickness of the bed remains always small enough so that the flow geometry is not significantly modified. The perturbation of the velocity field and of the core profile due to a perturbation of the pipe outer profile has been studied in [11].

B. The radial thickness of the bed, at a given cross section, is a smooth symmetric monotone increasing function of the azimuthal coordinate ϕ chosen as in *Figure* 1 which attains its maximum at $\phi = \dfrac{\pi}{2}$.

C. The above quantity, denoted by $h(x,t,\phi)$, is proportional to the cross sectional area $a(x,t)$ of the sediment via

$$h(x,t,\phi) = C(\phi)a(x,t), \qquad (4)$$

where $C(\phi)$ is to be suitably chosen.

D. The bed consists of a *static* component and a *dynamic* one [1], separated by a sharp interface. The thickness of the moving layer can never exceed a limit value Δ. In other words, in the first stage, the whole bed flows. But once the critical level Δ is reached the build up of the static layer begins underneath the dynamic one.

From the physical point of view we expect that $C(\phi)$ has to be such that:

a. $C(\phi)$ is an increasing and continously differentiable function for $\phi \in [0, \frac{\pi}{2})$

b. $C(0)=C(\pi)=0$

c. $C(\phi)$ is such that the sediment profile is symmetric with the z-axis

d. $\dot{C}(\phi) \leq \{\frac{R}{a} - C(\phi)\}tg(\phi)$ (guaranteeing the monotonicity of the sediment profile in each cross-section).

By defining $C(\phi) = y_0 g(\frac{2}{\pi}\phi)$ we are able to find a family of functions C_k that satisfy conditions (a) trough (d) for $a(x,t)$ sufficiently small.

Thus we can write the volumetric flow rate in the x direction per unit area *due to the moving layer* as:

$$q(x,t) = 2\int_0^{\frac{\pi}{2}} \tilde{q}(x,t,\phi)d\phi, \qquad (5)$$

$$\tilde{q}(x,t,\phi) = \begin{cases} \lambda_1 Rh(x,t,\phi) & h(x,t,\phi) < \Delta \\ \lambda_2(a(x,t))R\Delta & h(x,t,\phi) \geq \Delta. \end{cases} \qquad (6)$$

The quantity $\tilde{q}(x,t,\phi)d\phi$ is just the volume of sediment passing through a section of width $d\phi$ per unit time, consistently with the hypotheses $h \ll R$ and $\Delta \ll R$.

The parameter λ_1 and the function $\lambda_2(a)$ have to be chosen so that $\tilde{q}(x,t,\phi)$ is continuous. The balance equation for the cross-sectional area of the bed is thus

$$\frac{\partial a}{\partial t} + \frac{\partial q(x,t)}{\partial x} = \tilde{S}_T(x), \quad \tilde{S}_T(x) = \frac{1}{[\rho_s(1-\varepsilon)]}S_T(x), \qquad (7)$$

ρ_s and ε being the density of the settled material and the porosity of the bed respectively (see [6]).

---

[1] This is actually the fundamental assumption and is fully supported by experimental data: the presence of the static (and solid) layer is exactly what we observe, at rest, when, after a test on the pipeline, we remove all the fluid. Essentially the sediment removal process is to be done when this layer reaches some prescribed thickness.

Equation (7) is naturally coupled with the following initial-boundary conditions

$$a(x,0) = a(0,t) = 0. \tag{8}$$

By means of some computations we can now prove that $q(x,t)$ is an explicit function of $a(x,t)$:

$$q(a) = \begin{cases} \lambda_1 a & a < a_0 \\ \pi R \Delta \{\lambda_1 \frac{a}{a_0} G(\frac{2}{\pi}\widehat{\phi}(a)) + \lambda_2(a)(1 - \frac{2}{\pi}\widehat{\phi}(a))\} & a \geq a_0 \end{cases} \tag{9}$$

$$a = a(x,t), \quad a_0 = \Delta/C(\pi/2)$$

$$G(s) = \int_0^s g(u)du, \quad \widehat{\phi}(a) = (\pi/2)g^{(-1)}(a_0/a).$$

Equation (7) takes the form

$$\frac{\partial a}{\partial t} + \dot{q}(a)\frac{\partial a}{\partial x} = \tilde{S}_T(x) \quad x > 0, \ t > 0. \tag{10}$$

- **Existence and Uniqueness theorem**

By using the method of characteristics we are able to prove that there exists a unique clasical solutution for (10) (8) locally in time and that there are, as shown in *Figure* 4, four different regions of the (x,t) plane characterized by different behaviours of the function $a(x,t)$. We refar to [8] for the mathematical details.

In regions I and II (10) is linear and $a(x,t)$ is an increasing function of x in region I, whereas in II it is decreasiing.

In regions III and IV (10) is non linear and we proved that in III the characteristics are divergeing, whereas in IV, if $\ddot{q}(a) > 0$, they can be converging. In this case we prove that the solution exists until

$$-\frac{1}{\lambda_1} + \tilde{S}_T(\xi) \int_{x_\gamma(\xi)}^{x} \frac{\ddot{q}(q^{-1}(\int_\xi^\eta \tilde{S}_T(\nu)d\nu))}{[\dot{q}(q^{-1}(\int_\xi^\eta \tilde{S}_T(\nu)d\nu))]^3} d\eta < 0$$

where $\xi = \xi(x,t)$ is implicitly defined via:

$$t - t_\gamma(\xi) = \int_{x_\gamma(\xi)}^{x} \frac{d\eta}{\dot{q}(q^{-1}(\int_\xi^\eta \tilde{S}_T(\nu)d\nu))},$$

$$\begin{cases} a_0 = \frac{1}{\lambda_1}\int_\xi^{x_\gamma(\xi)} \tilde{S}_T(\nu)d\nu \\ t_\gamma(\xi) = \frac{1}{\lambda_1}(x_\gamma(\xi) - \xi). \end{cases}$$

In region III a(x,t) is increasing respect to x and in IV it is decreasing.

- **Numerical Simulations Part II: The Sediment Profile**

Based on the formulas of the previous section we solved numerically our model for low, medium, and high values of the flow rate ($Q = 100, 150, 250, 450 \ m^3/h$).

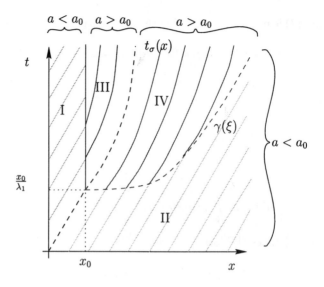

Figure 4: Qualitative behaviour of the characteristics.

The results obtained are shown in the next table where t_{cr} and x_{cr} are respectively the time when the maximum thickness of the (static) sedimentation bad reaches 1cm and the distance where this maximum is reached, respectively. They result in good agreement with experimental[2] data.

Numerical Results

Q	Δ	t_{cr}	x_{cr}
m^3/h	cm	days	Km
100	0.7	0.4	0.48
150	1.0	0.5	1.24
250	1.8	—	—
450	3.2	—	—

In particular we observe that the low flow rates, despite a lower value of the sedimentation rate, produce a large growt of sediment in the first part of the pipeline in a very short time. This fact is due to the strong dependence of the sediment transport mechanism on the flow rate.

---
[2] All data we used come from a 170 m pilot plant run by Snamprogetti in their Laboratories in Fano (Italy) and from a 250 Km long Industrial plant from Belovo to Novosibirsk in Siberia.

References

[1] E. Comparini, "*A one-dimensional Bingham flow*", J. Math. Anal. Appl. 169 (1992), pp. 127-139.

[2] E. De Angelis "*Modelli Stazionari e non per Fluidi di Bingham in Viscosimetro e in Condotta*"(PhD Thesis), Dipartimento di Matematica "U.Dini", Università degli studi di Firenze, 1994-95.

[3] E. De Angelis, A. Fasano, M. Primicerio, F. Rosso, E. Carniani, D. Ercolani, "*Modelling sedimentation in CWS*"in *Hydrotransport 12* Proc. of the 12th international conf. on slurries handling and pipeline transport, edited by C. A. Shook, MEP Pubbl. (1993), pp. 399-414.

[4] E. De Angelis, A. Fasano, M. Primicerio, F. Rosso, E. Carniani, D. Ercolani, "*Sedimentation bed dynamics for fluids with yield stress in a pipe*", Proc. of the 4th international conf. "*Fluidodinamica multifase nell'impiantistica industriale*", (1994), pp. 85-93.

[5] E. De Angelis, A. Mancini, "*A Model for the Dynamics of Sediments in a Pipe* ", to appear.

[6] E. De Angelis, F. Rosso, "*A functional approach to the problem of evaluating the velocity for a population of particles settling in a liquid*", Proc. of the 9th ECMI 1993, A. Fasano, M. Primicerio eds., B. G. Teubner (1994), pp. 191-198.

[7] A. Fasano, "*A mathematical model for the dynamics of sediments in a pipeline*", Proc. of the 11th ECMI 1994, H. Neunzert eds., to appear.

[8] A. Fasano, E. Manni, M. Primicerio, "*Modelling the dynamics of fluidizing agents in coal-water slurries*", Proc. of the *International Symposium on Non-linear Problems in Engineering and Science*, edited by S. Xiao and X. Hu, Science Press, Beijing, China, pp. 64-71.

[9] A. Fasano, M. Primicerio, "*Modelling the rheology of a coal-water slurry*", Proc. of the 4th Europ. conf. on *Mathematics in Industry*, edited by H. J. Wacker and W. Zulhener, B. G. Teubner and Kluwer Academic Pubbl. (1991), pp. 269-274.

[10] A. Mancini "*Evoluzione di profili di sedimentazione nel trasporto di sospensioni concentrate in condotta*"(Degree Thesis), Dipartimento di Matematica "U.Dini", Università degli studi di Firenze, 1994-95.

[11] S. Parrini, "*Moto stazionario di un fluido di Bingham in una condotta con sezione trasversale non circolare,*"(Degree Thesis), Dipartimento di Matematica "U.Dini", Università degli studi di Firenze, 1994-95.

TEUBNER-TASCHENBUCH der Mathematik

Das vorliegende »TEUBNER-TASCHENBUCH der Mathematik« ersetzt den bisherigen Band – Bronstein/Semendjajew, Taschenbuch der Mathematik –, der mit 25 Auflagen und mehr als 800.000 verkauften Exemplaren bei B. G. Teubner erschien.

In den letzten Jahren hat sich die Mathematik außerordentlich stürmisch entwickelt. Eine wesentliche Rolle spielt dabei der Einsatz immer leistungsfähigerer Computer. Ferner stellen die komplizierten Probleme der modernen Hochtechnologie an Ingenieure und Naturwissenschaftler sehr hohe mathematische Anforderungen.

Diesen aktuellen Entwicklungen trägt das »TEUBNER-TASCHENBUCH der Mathematik« umfassend Rechnung. Es vermittelt ein lebendiges und modernes Bild der heutigen Mathematik und erfüllt aktuell, umfassend und kompakt die Erwartungen, die an ein Nachschlagewerk für Ingenieure, Naturwissenschaftler, Informatiker und Mathematiker gestellt werden. Im Studium ist das »TEUBNER-TASCHENBUCH der Mathematik« ein Handbuch, das Studierende vom ersten Semester an begleitet; im Berufsleben wird es dem Praktiker ein unentbehrliches Nachschlagewerk sein.

Begründet von
I. N. Bronstein und
K. A. Semendjajew

Weitergeführt von
G. Grosche, V. Ziegler
und **D. Ziegler**

Herausgegeben von
Prof. Dr. **Eberhard Zeidler**
Leipzig

1996. XXVI, 1298 Seiten.
14,5 x 20 cm.
Geb. DM 59,–
ÖS 431,– / SFr 53,–
ISBN 3-8154-2001-6

Preisänderungen vorbehalten.

Aus dem Inhalt
Wichtige Formeln, graphische Darstellungen und Tabellen – Analysis – Algebra – Geometrie – Grundlagen der Mathematik – Variationsrechnung und Optimierung – Stochastik – Numerik

B. G. Teubner Stuttgart · Leipzig

TEUBNER-TASCHENBUCH der Mathematik Teil II

Mit dem »TEUBNER-TASCHENBUCH der Mathematik, Teil II« liegt eine vollständig überarbeitete und wesentlich erweiterte Neufassung der bisherigen »Ergänzenden Kapitel zum Taschenbuch der Mathematik von I. N. Bronstein und K. A. Semendjajew« vor, die 1990 in 6. Auflage im Verlag B. G. Teubner in Leipzig erschienen sind. Dieses Buch vermittelt dem Leser ein lebendiges, modernes Bild von den vielfältigen Anwendungen der Mathematik in Informatik, Operations Research und mathematischer Physik.

Aus dem Inhalt

Mathematik und Informatik – Operations Research – Höhere Analysis – Lineare Funktionalanalysis und ihre Anwendungen – Nichtlineare Funktionalanalysis und ihre Anwendungen – Dynamische Systeme, Mathematik der Zeit – Nichtlineare partielle Differentialgleichungen in den Naturwissenschaften – Mannigfaltigkeiten – Riemannsche Geometrie und allgemeine Relativitätstheorie – Liegruppen, Liealgebren und Elementarteilchen, Mathematik der Symmetrie – Topologie – Krümmung, Topologie und Analysis

Herausgegeben von
Doz. Dr.
Günter Grosche
Leipzig
Dr. **Viktor Ziegler**
Dorothea Ziegler
Frauwalde
und Prof. Dr.
Eberhard Zeidler
Leipzig

7. Auflage. 1995. Vollständig überarbeitete und wesentlich erweiterte Neufassung der 6. Auflage der »Ergänzenden Kapitel zum Taschenbuch der Mathematik von
I. N. Bronstein und
K. A. Semendjajew«.
XVI, 830 Seiten mit 259 Bildern.
14,5 x 20 cm.
Geb. DM 58,–
ÖS 423,– / SFr 52,–
ISBN 3-8154-2100-4

Preisänderungen vorbehalten.

B. G. Teubner Stuttgart · Leipzig

Neunzert
Progress in Industrial Mathematics at ECMI '94

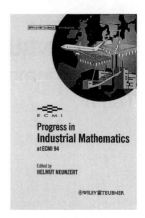

This book aims to bridge the gap between academic and industrial mathematics. A wide range of precise and up-to-date articles cover mathematics for the aerospace, automotive and chemical industries, as well as chip production, metallurgic processes, geodesy and applied control.

The articles were presented at the 1994 ECMI (European Consortium for Mathematics in Industry) conference in Kaiserslautern (Germany). More than 60 contributions were selected to give a complete and current overview of mathematics in industry. An introduction to each chapter summarizes important facts and makes special contributions more accessible.

Edited by Prof. Dr.
Helmut Neunzert
Universität Kaiserslautern

1996. X, 563 pages.
16,2 x 23,5 cm.
Bound DM 124,–
ÖS 905,– / SFr 112,–
ISBN 3-519-02603-1

(European Consortium for Mathematics in Industry)

Preisänderungen vorbehalten.

B. G. Teubner Stuttgart · Leipzig